de Gruyter Expositions in Mathematics 10

de Gruyter Expositions in Mathematics

1 The Analytical and Topological Theory of Semigroups, *K. H. Hofmann, J. D. Lawson, J. S. Pym* (Eds.)

2 Combinatorial Homotopy and 4-Dimensional Complexes, *H. J. Baues*

3 The Stefan Problem, *A. M. Meirmanov*

4 Finite Soluble Groups, *K. Doerk, T. O. Hawkes*

5 The Riemann Zeta-Function, *A. A. Karatsuba, S. M. Voronin*

6 Contact Geometry and Linear Differential Equations, *V. R. Nazaikinskii, V. E. Shatalov, B. Yu. Sternin*

7 Infinite Dimensional Lie Superalgebras, *Yu. A. Bahturin, A. A. Mikhalev, V. M. Petrogradsky, M. V. Zaicev*

8 Nilpotent Groups and their Automorphisms, *E. I. Khukhro*

9 Invariant Distances and Metrics in Complex Analysis, *M. Jarnicki, P. Pflug*

The Link Invariants of the Chern-Simons Field Theory

New Developments in Topological Quantum Field Theory

by

Enore Guadagnini

Walter de Gruyter · Berlin · New York 1993

Author

Enore Guadagnini

Dipartimento di Fisica
Università di Pisa
Piazza Torricelli, 2
56100 Pisa, Italy
e-mail: Guadagni@ipifidpt.difi.unipi.it

1991 Mathematics Subject Classification: Primary: 57-02; 81-02.
Secondary: 57M25, 57R65; 81Q20, 81T10
PACS numbers: 11.15.−q, 02.40.+m, 11.17.+y, 11.15.Tk
Keywords: Topological field theory, Wilson lines, knot theory, 3-manifold invariants

♾ Printed on acid-free paper which falls within the guidelines of the ANSI to ensure permanence and durability.

Library of Congress Cataloging-in-Publication Data

Guadagnini, Enore, 1953–
The link invariants of the Chern-Simons field theory / by Enore Guadagnini.
p. cm. – (De Gruyter expositions in mathematics ; 10)
Includes bibliographical references and index.
ISBN 3-11-014028-4
1. Quantum field theory. 2. Mathematical physics. I. Title. II. Series.
QC174.45.G88 1993
530.1′435–dc20 93-4729

Die Deutsche Bibliothek – Cataloging-in-Publication Data

Guadagnini, Enore:
The link invariants of the Chern-Simons field theory : new developments in topological quantum field theory / by Enore Guadagnini. – Berlin ; New York : de Gruyter, 1993
(De Gruyter expositions in mathematics ; 10)
ISBN 3-11-014028-4
NE: GT

Printed in Germany.
Disk Conversion: D. L. Lewis, Berlin. Printing: Ratzlow Druck, Berlin.
Binding: Dieter Mikolai, Berlin. Cover design: Thomas Bonnie, Hamburg.

**to my parents
Elena and Eugenio**

Preface

For almost two decades, several topological arguments have been used systematically in quantum field theory. These methods have been applied mainly in the study of the semiclassical approximation of the theory and, also, in the description of physical phenomena at low energy. Apart from a few exceptions, like the discovery of the general structure of anomalies, the applicability of these arguments was controlled by the validity of certain limits in some given parameters. So, even if the use of topological arguments has produced several important results, the fundamental structure of the field theory models has never been interpreted in purely topological terms.

In this context, the revolution of the last few years is represented by the quantum Chern-Simons field theory. In the Chern-Simons model, the applicability of topological arguments is not restricted to a region in which the semiclassical approximation is valid. In fact, no kind of approximation is at all necessary. We need not consider either low-energy limit or any other limit. The quantum Chern-Simons theory is a "true" topological field theory; any observable and any result obtained in this model has exclusively a topological origin and a topological meaning. The Chern-Simons theory is exactly soluble at the full quantum level not only in $\mathbb{R}^3$ and in S^3 but also in any closed, connected and orientable three-manifold $\mathcal{M}$.

This series of lecture notes is devoted to the discussion of several aspects of the quantum Chern-Simons field theory. In the first part of these notes, the relevant properties of the model in $\mathbb{R}^3$ are studied by means of the standard methods used in quantum field theory.

In the second part, it is shown how the observables of the Chern-Simons theory, associated with knots and links in $\mathbb{R}^3$, can equivalently be described in more abstract and purely topological terms. The exact solution of the model is obtained by combining the general properties of the observables, which are consequences of the symmetries of the system, with the numerical information derived in the canonical approach. The expectation values of the observables have the form of polynomials and represent the values of an ambient isotopy invariant of framed links in $\mathbb{R}^3$ (or S^3). A constructive method for the computation of the observables is presented and the reconstruction theorems for the non-Abelian $SU(N)$ Chern-Simons theory are proved. The link polynomial defined by the Chern-Simons field theory is characterized by the existing relations between satellites and their companions. The structure of these relations is universal and is described by the representation rings of the Lie algebras associated with compact simple Lie groups. For this reason, the invariant obtained in the Chern-Simons theory is called the universal link polynomial.

In the last part of these notes, the solution of the theory in any closed, connected and orientable three-manifold $\mathcal{M}$ is constructed by means of the operator surgery method. The case in which the gauge group is $SU(2)$ is studied in detail. A certain set of models based on the $U(1)$ gauge group is also considered. The three-manifold invariant, naturally associated with the Chern-Simons theory, is defined and its values are computed for several examples of three-manifolds. The present notes are essentially self-contained and include some introductory reviews of knot theory and surgery on three-manifolds.

The topological and algebraic structures described in these lectures are known in mathematics. The purpose of the present notes is to show how these structures emerge from the physical point of view. Intuition and working hypotheses play a fundamental role in physics; for this reason, the solution of the quantum Chern-Simons field theory is presented on the basis of inductive and constructive methods. For example, new rules for the computation of the universal link polynomial are introduced; these rules are derived from the properties of the field theory and are illustrated by means of several examples. Quite often, physical arguments will be used in the place of more rigorous but technical derivations. Consequently, some of the proofs have been simplified and their completion is left to the reader.

The quantum Chern-Simons field theory provides an intrinsic three-dimensional description of the link invariants. Thus, several features of the link polynomials admit a simple physical interpretation which is a consequence of the symmetry properties of the field theory. The interplay between the symmetries of the field theory and the properties of the link polynomials plays a crucial role in our discussion. Symmetry arguments will be used to derive satellite relations and to construct three-manifold invariants. The rigorous and formal proofs on the subject, which can be found in mathematical literature, are essentially based on the algebraic properties of certain modular Hopf algebras. Our approach, instead, gives prominence to the symmetry principles. On the one hand, this shows that the different constructions of three-manifold invariants, which have been produced in literature, represent slightly different versions of the same general structure. On the other hand, we will see how symmetry arguments can conveniently be used to simplify the explicit computation of these invariants.

The quantum Chern-Simons theory is the first example of a non-trivial gauge field theory whose exact solution can be explicitly produced in any three-manifold $\mathcal{M}$. The algebraic operations, involved in the computation of the observables, are surprisingly simple and elementary. The resulting theory is remarkable; its structure presents several new quantum field theory aspects and has deep connections with different fields of physics and mathematics. It is not clear how useful, in the description of physical phenomena, the possible applications of the Chern-Simons theory will be. Certainly, the Chern-Simons model represents an important starting point for new developments in quantum field theory.

The present book is an extended version of the lectures delivered at LAPP, Laboratoire d'Annecy-le-Vieux de Physique des Particules (France), in April 1991. I wish to thank Paul Sorba for having kindly invited me to give those lectures and I should also like to thank all the members of the Theory Division for their warm hospitality.

Most of the material presented here appears in published articles and deep gratitude is due to R. Cappuccio, P. Cotta-Ramusino, N. Maggiore, M. Martellini, S. Panicucci, S.P. Sorella and F. Volpi for their valuable collaboration. I thank beyond measure Mihail Mintchev; with Mihail I have shared the efforts of several computations and the joys for the first surprising results obtained in perturbative Chern-Simons theory. I also wish to thank Riccardo Benedetti and Corrado De Concini for several illuminating discussions on various topological aspects of three-manifolds and on the algebraic structure of quantum groups.

It is a pleasure to acknowledge the admirable work of Lucy Amasanti who improved the exposition.

These lecture notes would never have been written without Raymond Stora's constant interest in the subject. My debt to Raymond is enormous; many of the achievements obtained in the topological Chern-Simons theory are the results of innumerable discussions characterized by his insistent and tireless desire to get a deeper understanding of quantum field theory.

Pisa 1992 Enore Guadagnini

Table of Contents

Chapter 6
Properties of the expectation values

Chapter 7
Ordering fermions and knot observables

Chapter 8
Braid group

Chapter 9
R–matrix and braids

Chapter 10
Chern-Simons monodromies

Chapter 11
Defining relations

Chapter 12
The extended Jones polynomial

Chapter 13
General properties

Chapter 14
Unitary groups

Chapter 15
Reduced tensor algebra

Chapter 16
Surgery on three-manifolds

Chapter 1

Introduction

Quantum field theories provide a very accurate description of physical phenomena. Local gauge invariance turns out to be one of the basic principles underlying the construction of realistic particle models. On the other hand, the inclusion of quantum gravitational effects within some unified theory of the fundamental processes should be based on the principle of general covariance. In these lectures, we shall consider the case in which both gauge invariance and general covariance are realized at the full quantum level in a particular class of non-trivial field theories in three dimensions.

The models we shall consider and which are defined by a pure Chern-Simons action are solvable and possess quite remarkable properties. The expectation values of the basic observables of these systems can conveniently be expressed in terms of polynomials of a certain variable which is a function of the coupling constant of the theory. These polynomials, which are associated with knots and links, encode some universal features of physics in two and three dimensions and, at the same time, provide a useful set of invariants of three-manifolds. This introductory chapter contains a preliminary discussion of some of the essential concepts we shall encounter.

1.1 Quantum physics and classical electromagnetism

We shall begin our discussion with an example of quantum field theory computation in which some notions of classical electromagnetism are involved. Before introducing the action of the model in which we are interested, it is useful to recall a simple method of evaluating path-integrals.

There is a particular situation in which one can easily perform path-integral computations; namely, when the system is characterized by a "quadratic" action of the type

$$S_0[\phi] = \int \tfrac{1}{2}\,\phi\,\Delta\,\phi\,. \tag{1.1}$$

In equation (1.1), ϕ denotes the field (or the set of fields) of the model and Δ is some differential operator. For the moment, we do not need to specify the explicit form of Δ nor the number of dimensions of the space-time manifold in which the theory is defined. Let us introduce some external source coupled to the fields; we shall be interested in computing the vacuum expectation value

$$\langle e^{i\int J\phi} \rangle \equiv Z^{-1} \int d\phi \; e^{iS_0[\phi]} \; e^{i\int J\phi} , \tag{1.2}$$

where

$$Z = \int d\phi \; e^{iS_0[\phi]} . \tag{1.3}$$

The simplest way of computing the expression (1.2) is to perform a linear change in the integration variables in the numerator so that the free path-integral factorizes and then cancels out with the denominator. The whole idea is to put

$$\phi = \widetilde{\phi} + \psi , \tag{1.4}$$

where $\widetilde{\phi}$ is a fixed classical configuration. Then, in the path-integral, it is assumed that $d\phi = d\psi$ because this is precisely the fundamental property which should hold in any reasonable definition of integral. Now, the clever choice of $\widetilde{\phi}$ is the one for which ψ and $\widetilde{\phi}$ decouple; this means

$$\frac{\delta}{\delta\phi}\left(S_0[\phi] + \int J\phi \right)\Bigg|_{\phi=\widetilde{\phi}} = \Delta\widetilde{\phi} + J = 0 . \tag{1.5}$$

Finally, in terms of the solution $\widetilde{\phi}$ of equation (1.5), the expression (1.2) takes the form

$$\langle e^{i\int J\phi} \rangle = \exp\left(\frac{i}{2} \int J \widetilde{\phi} \right) . \tag{1.6}$$

Of course, when no solution of equation (1.5) exists or when several inequivalent solutions (with the same fixed boundary conditions) exist, this simple method of computing the expectation value (1.2) may need some improvements. For the purposes of the present introductory section, however, it is not necessary to discuss in detail these more complicated situations.

Let us now consider the Abelian Chern-Simons action [1,2]

$$S_0 = \frac{k}{8\pi} \int d^3x \, \epsilon^{\mu\nu\rho} A_\mu \partial_\nu A_\rho . \tag{1.7}$$

In this example, the vector field A_μ is the analogue of ϕ , the differential operator $(k/4\pi)\, \epsilon^{\mu\nu\rho}\, \partial_\nu$ is the analogue of Δ and the source term is chosen to be

$$\int J\,\phi \;\to\; \int d^3x\, J^\mu\, A_\mu \;=\; e_1 \oint_{C_1} A_\mu dx^\mu + e_2 \oint_{C_2} A_\mu dx^\mu\,. \tag{1.8}$$

In the expression (1.8), the line integrals are performed along the two oriented non-intersecting closed paths C_1 and C_2 in $\mathbb{R}^3$ shown in Fig. 1.1.

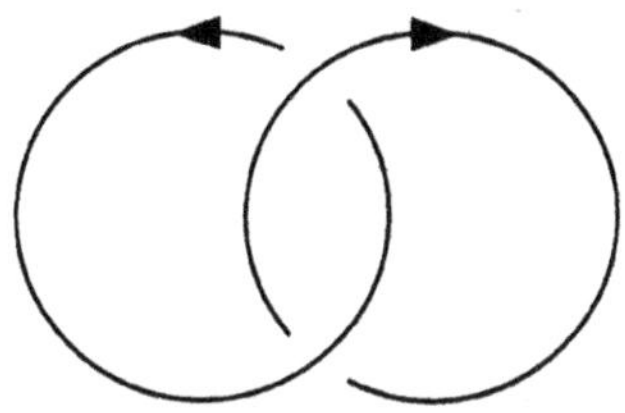

Figure 1.1.

Let us introduce a parametrization $y^\mu(s)$ $(0 \le s \le 1)$ for C_1 and $z^\mu(t)$ $(0 \le t \le 1)$ for C_2. The source J^μ appearing in equation (1.8) can be written as

$$J^\mu(x) \;=\; e_1 \int_0^1 ds\, \dot{y}^\mu(s)\, \delta^3(x - y(s)) + e_2 \int_0^1 dt\, \dot{z}^\mu(t)\, \delta^3(x - z(t))\,. \tag{1.9}$$

In order to compute the expectation value $\langle e^{i \int d^3x\, J^\mu A_\mu} \rangle$, we shall use the same method as before. The vector field is decomposed into two parts

$$A_\mu(x) \;=\; B_\mu(x) + C_\mu(x)\,, \tag{1.10}$$

where the classical configuration $B_\mu(x)$ should satisfy the analogue of equation (1.5), that is

$$\epsilon^{\mu\nu\rho}\, \partial_\nu\, B_\rho(x) \;=\; -\left(\frac{4\pi}{k}\right) J^\mu(x)\,. \tag{1.11}$$

This equation is well known in classical electromagnetism; it shows the connection between the magnetic field B_μ and the stationary current density which originates the field itself. Finding the solution of equation (1.11) means, in our case, finding the expression of the magnetic field in the presence of two filamentary wires (shown in Fig. 1.1) which carry the currents

$$I_1 \;=\; -e_1 \frac{c}{k} \qquad I_2 \;=\; -e_2 \frac{c}{k}\,. \tag{1.12}$$

The solution of this problem is expressed by the so-called Ampère Law: the total magnetic field is the linear combination

$$B_\mu(x) = B_\mu^{(1)}(x) + B_\mu^{(2)}(x) , \tag{1.13}$$

of the two components

$$B_\mu^{(1)}(x) = -\frac{e_1}{k} \int_0^1 ds\, \epsilon_{\mu\nu\rho}\, \dot{y}^\nu(s)\, \frac{(x-y(s))^\rho}{|x-y(s)|^3} , \tag{1.14}$$

and

$$B_\mu^{(2)}(x) = -\frac{e_2}{k} \int_0^1 dt\, \epsilon_{\mu\nu\rho}\, \dot{z}^\nu(t)\, \frac{(x-z(t))^\rho}{|x-z(t)|^3} , \tag{1.15}$$

generated by the wires C_1 and C_2 respectively.

In terms of the resulting magnetic field (1.13), the Chern-Simons vacuum expectation value of the source term takes the form

$$\langle e^{i\int d^3x\, J^\mu A_\mu} \rangle = \exp\left(i\frac{e_1}{2} \oint_{C_1} B_\mu\, dx^\mu + i\frac{e_2}{2} \oint_{C_2} B_\mu\, dx^\mu \right) . \tag{1.16}$$

The expression (1.16), which is the analogue of equation (1.6), presents a problem: each line-integral of the magnetic field along a wire contains a self-interaction part which has certain ambiguities. The solution to this problem will be discussed in detail in Chapter 3. For the moment, one can conclude that, *neglecting self-interactions*, the desired expectation value is given by

$$\langle e^{i\int d^3x\, J^\mu A_\mu} \rangle = \exp\left[-2ie_1e_2 \left(\frac{2\pi}{k}\right) \chi(C_1,C_2) \right] , \tag{1.17}$$

where

$$\chi(C_1,C_2) = \frac{1}{4\pi} \oint_{C_1} dx^\mu \oint_{C_2} dy^\nu\, \epsilon_{\mu\nu\rho}\, \frac{(x-y)^\rho}{|x-y|^3} . \tag{1.18}$$

Note that, in the Abelian Chern-Simons theory, there are several field configurations satisfying equation (1.11). Indeed, the longitudinal part of B_μ is not constrained by equation (1.11). This means that any B_μ configuration, which differs from the classical expression (1.13) by the gradient of an arbitrary function, still satisfies equation (1.11). Apparently, one has a situation in which the general method leading to equation (1.6) presents some ambiguities. In the Chern-Simons model, however, these ambiguities are completely harmless because of the particular choice of the source, equation (1.9). Since we are considering closed paths, the longitudinal part of B_μ is totally irrelevant (see equation (1.16)) and the result shown in equation (1.17) is unique. In fact, as we shall see, equation (1.17) represents the exact result (neglecting self-interactions, of course).

With the usual (right-handed) convention in the definition of the antisymmetric tensor $\epsilon^{\mu\nu\rho}$ ($\epsilon^{123} = 1$), the value of $\chi(C_1, C_2)$ for the paths shown in Fig. 1.1 is

$$\chi(C_1, C_2) = 1 . \tag{1.19}$$

The result (1.17) has a simple physical interpretation. The exponent in the expression (1.17) contains the circuitation along one closed path, say C_1, of the magnetic field generated by the second wire. This quantity is precisely the energy gain $\mathcal{E}$ of an imaginary magnetic monopole moving along C_1 in the presence of the magnetic field generated by C_2. For each "winding" of the magnetic monopole around C_2, the energy increases by a definite amount which, in our units, is given by $\mathcal{E} = -2e_1e_2(2\pi/k)$. For arbitrary non-intersecting closed paths C_1 and C_2, the value of the expression (1.18) (called the ***Gauss integral***) is an integer representing exactly how many times C_1 "winds" C_2.

One of the most remarkable features of the Abelian Chern-Simons theory is that the expectation value (1.17) is invariant under smooth deformations of the paths C_1 and C_2. We would now like to understand the reasons for this behaviour and construct new models possessing the same property.

1.2 Abelian Chern-Simons action

All the information on the physics of the different systems is encoded in their action; so, let us reconsider the Abelian Chern-Simons (CS) action [1,2]

$$S_0 = \frac{k}{8\pi} \int d^3x \, \epsilon^{\mu\nu\rho} \, A_\mu \partial_\nu A_\rho \, . \tag{1.20}$$

The functional (1.20) is invariant under (Abelian) gauge transformations acting on the vector field $A_\mu(x)$. The quantization of the system defined by the action (1.20) will be discussed in Chapter 3.

The action S_0 is also invariant under general coordinate transformations with $A_\mu(x)$ transforming as a covariant vector. This last property is called ***general covariance***. Note that, in our example, general covariance is realized in quite a peculiar way. In ordinary field theories, the action is not invariant under general coordinate transformations (acting on the fields of the theory) unless the metric also is a dynamical variable. In the Abelian CS model, the metric that one can introduce on the three-manifold is not a variable (or a "field") of the theory. Nevertheless, the action is invariant under general coordinate transformations for the simple reason that S_0 does not depend on the metric at all. In fact, S_0 can be understood as the integral of a three-form on a three-manifold.

Gauge invariance and general covariance are the real reasons for the properties of the expectation value (1.17) that we have observed. Gauge invariance forced us to choose the external source to be expressed in terms of closed paths (conserved external currents), since only gauge-invariant quantities have an intrinsic meaning in gauge theories. Because of general covariance, the final result (1.17) only depends on the topological structure of the closed contours. This is why there is invariance under smooth deformations of the paths in $\mathbb{R}^3$.

In the previous section, the source term was represented by the simple two-component link shown in Fig. 1.1. But one can consider more complicated links, of course; an example is shown in Fig. 1.2.

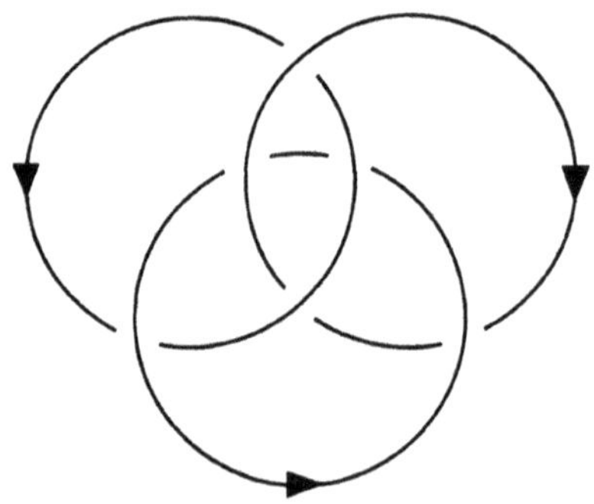

Figure 1.2.

Exercise. Consider the Abelian CS theory with a source term corresponding to the link shown in Fig. 1.2. In this case, what is the expression (neglecting self-interactions) of the vacuum expectation value $\langle e^{i\int d^3x\, J^\mu A_\mu} \rangle$?

1.3 Non-Abelian Chern-Simons action

The action (1.20) can be generalized [1,3,4] to the case in which the gauge group G is a non-Abelian. The corresponding CS action reads

$$S_{CS} = \frac{k}{4\pi} \int d^3x\, \epsilon^{\mu\nu\rho}\ \mathrm{Tr}\left(A_\mu \partial_\nu A_\rho + i\,\tfrac{2}{3}\, A_\mu A_\nu A_\rho\right), \tag{1.21}$$

where $A_\mu = A^a_\mu T^a$, $\{T^a\}$ are the Hermitian generators of a compact simple Lie group G in its defining representation and the real parameter k is the coupling constant of the model. The integral appearing in equation (1.21) has to be performed in $\mathbb{R}^3$.

The expectation values $\{ \langle W(L) \rangle \}$ of Wilson line operators associated with oriented links $\{L\}$ represent a useful set of gauge-invariant observables for the CS theory. Given an oriented knot C and an irreducible representation ρ of G, the associated Wilson line operator is

$$W_\rho(C) = W(C;\rho) = \mathrm{Tr}\ \mathrm{P}\exp\left(i \oint_C A^a_\mu(x)\, T^a_{(\rho)}\, dx^\mu \right) , \tag{1.22}$$

where the path ordering is performed along C and $\{ T^a_{(\rho)} \}$ are the generators of G in the ρ representation. Consider now an oriented link L with m components $\{C_1, \ldots, C_m\}$ and let ρ_i be the irreducible representation of G associated to the i-th component C_i of L. The vacuum expectation values

$$\langle W(L) \rangle \equiv \frac{\langle 0 |\, W_{\rho_1}(C_1) \cdots W_{\rho_m}(C_m)\, | 0 \rangle}{\langle 0 | 0 \rangle} , \tag{1.23}$$

defined for generic links and arbitrary representations $\{\rho_i\}$, are the gauge-invariant observables in which we are interested.

We will show that the expectation values (1.23) are well defined. The proof is divided into two parts: first, one has to verify that the quantum CS theory is renormalizable (the renormalized correlation functions satisfy an action principle based on the functional (1.21)) and, second, the precise meaning of the composite operator $W(C;\rho)$ at the quantum level must be specified. We will see that, in order to preserve general covariance, the Wilson line operators must be defined on framed knots.

As a consequence of general covariance, $\langle W(L) \rangle$ is invariant under smooth deformations of the framed link L in $\mathbb{R}^3$. Therefore, the set $\{ \langle W(L) \rangle \}$ defines a link invariant of the same type considered in knot theory and the main problem is how to compute its values on different framed links in closed form. We will derive the rules which permit the computation of $\langle W(L) \rangle$ with a finite number of operations. It turns out that $\langle W(L) \rangle$ takes the form of a polynomial in a certain complex variable which is a function of the coupling constant k of the theory.

After $\{ \langle W(L) \rangle \}$ have been found, the main issue is to identify these polynomials. It turns out that the link invariants obtained in the CS theory are those associated with the braid group representations described by the quasi-tensor category of quasi-triangular quasi-Hopf (QTQH) algebras associated to the quantum deformations of ordinary Lie algebras. In this sense, the obtained link polynomials are not completely unknown. In quite general terms, the link invariants described by the quasi-tensor category of QTQH algebras are called by Drinfeld the universal link (or knot) invariants [5]. For this reason, the polynomial $E(L)$, associated to the expectation value $\langle W(L) \rangle$, is called the ***universal link polynomial***.

The link invariant $E(L)$ is of particular interest for knot theory because it is the natural generalization of the Jones polynomial [6]. The most important aspect of this generalization is based on the existence of a Lie algebra structure underlying the construction of the universal link polynomial. The expectation values $\{ \langle W(L) \rangle \}$, computed in the case in which all the components of the links are associated with the fundamental representation of the gauge group $G = SU(2)$, essentially give the Jones polynomial. The first generalization consists of associating arbitrary representations of $G = SU(2)$ to the different components of the links. In this case, one obtains a new polynomial, which can be called the extended Jones polynomial because the $SU(2)$ Lie algebra remains the same. A further step consists of taking a generic real simple Lie algebra as the Lie algebra of the gauge group G and computing $\{ \langle W(L) \rangle \}$ when the different components of the links are associated with arbitrary representations of G. This is the case considered in these lectures. The general properties of the universal link polynomial associated with an arbitrary simple Lie algebra are obtained and the proof of the complete reconstruction of $E(L)$ in the case of $\{A_n\}$ algebras is presented.

The physics of the model described by the CS action is quite peculiar. In a (2+1) decomposition of $\mathbb{R}^3$, any fixed-time plane may intersect a given link at a certain number of points. These punctures can be interpreted as point-like particles whose world-lines represent the components of the link. As time goes by, these particles may move on the space plane; furthermore, two of them may either annihilate or be produced. A possible world-line configuration is shown in Fig. 1.3.

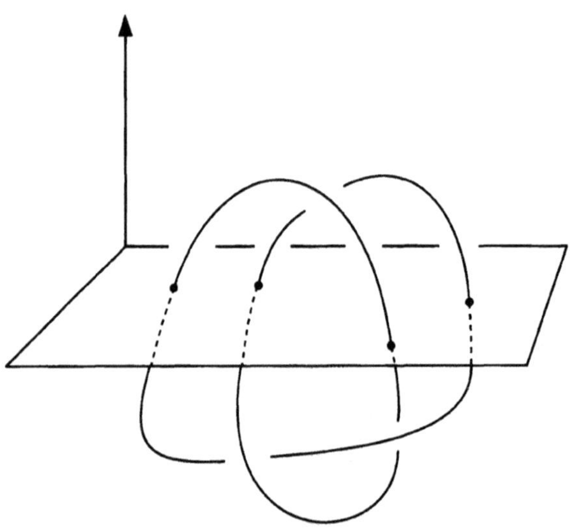

Figure 1.3.

The expectation value $\langle W(L) \rangle$ can be interpreted as the quantum mechanical amplitude associated with the whole process described by the link L. For generic values of k, each particle (puncture) is characterized by a quantum number which labels the inequivalent irreducible representations of the gauge group G. For integer values of k, something special happens: the number of the different kinds of particles is finite. For example, when $G = SU(2)$ and $k > 2$, only $|k| - 2$ different kinds of particles exist. For $|k| = 2$, the state space is one-dimensional and any state vector is proportional to the vacuum vector. When $|k| = 1$, the state space is two-dimensional; in addition to the vacuum state, there is only one nontrivial particle state.

A further generalization of the CS theory consists of considering the model, described by the action (1.21), in a generic three-manifold $\mathcal{M}$ which is closed, connected and orientable. In this case, the computation of the expectation values, denoted by $\{ \langle W(L) \rangle |_{\mathcal{M}} \}$, can be performed by means of an operator realization of surgery and the expectation values in $\mathcal{M}$ can be expressed as appropriate linear combinations of the expectation values in $\mathbb{R}^3$ which must be evaluated for integer values of the coupling constant k. We shall give the details of the surgery construction in the case in which $G = SU(2)$. The three-manifold invariant defined by the CS action is also constructed and the values taken by this invariant in some examples of manifolds are given.

The importance of the action (1.21) for knot theory and the study of three-manifolds has been pointed out by Schwarz [1] and Atiyah [3]. Further developments on the subject, together with a large number of possible connections of the CS theory with different areas of mathematics and physics, have been discussed by Witten in Ref.[4], in which several conjectures have been formulated. The first rigorous mathematical construction of the three-manifold invariants, whose existence was conjectured in [4], has been produced by Reshetikhin and Turaev [7,8]. These authors have given a precise definition of these invariants by means of certain modular Hopf algebras. We shall use a different approach based on the Feynman path-integral. But our final results perfectly agree with the Reshetikhin-Turaev invariants. Only recently has the exact solution of the quantum CS theory been produced [9,10,11]. Its possible realistic applications in physics have not been completely explored.

Chapter 2

Basic notions of knot theory

The full program of classifying and studying the properties of knots and links that one can construct in $\mathbb{R}^3$ was formulated on the basis of physical motivations in the second half of the nineteenth century. The interest in this subject mainly originated from the vortex-atoms model proposed by J.C. Maxwell, P.G. Tait and W. Thomson around 1867. Several excellent results have been obtained in knot theory and important developments, connecting different fields of mathematics, have taken place. In this chapter, a few definitions and results are briefly recalled; a more detailed and complete exposition can be found, for instance, in [12,13] and in the references quoted there. We shall begin by considering links in $\mathbb{R}^3$ or, equivalently, in the three-sphere S^3. The more general case of an arbitrary closed, connected and orientable three-manifold $\mathcal{M}$ will be discussed in Chapter 16.

2.1 Ambient and regular isotopy

A smooth non-intersecting closed path C in $\mathbb{R}^3$ is called a ***knot***. Since the definition of the holonomy requires an orientation for the path, we will always consider oriented knots. An oriented ***link*** L with m components is the union of m oriented non-intersecting closed paths. The m components of L will be denoted by $\{C_1, C_2, \ldots, C_m\}$. Smooth deformations in the ambient space do not modify the "topological" properties of links. Two links L_1 and L_2 in $\mathbb{R}^3$ are called ***ambient isotopic*** if L_1 can be smoothly connected with L_2 in $\mathbb{R}^3$. If one is interested in the topological properties of links, only the equivalence classes of ambient isotopic links are relevant, of course.

A convenient description of links is given in terms of ***diagrams*** obtained by projecting the links on a plane. In order to avoid all ambiguities, one usually considers link diagrams containing only simple crossing points; at each crossing point the choice of over/under crossing is specified, as shown for example in Fig. 1.1 and Fig. 1.2. Given two link diagrams D_1 and D_2, the associated links L_1 and L_2 are ambient isotopic if and only if a finite sequence of ***Reidemeister moves*** (shown in Fig. 2.1) which transforms D_1 into D_2 exists.

Reidemeister moves (RM) are very important in knot theory because they encode the symmetry structure which is relevant for the link classification problem. Indeed,

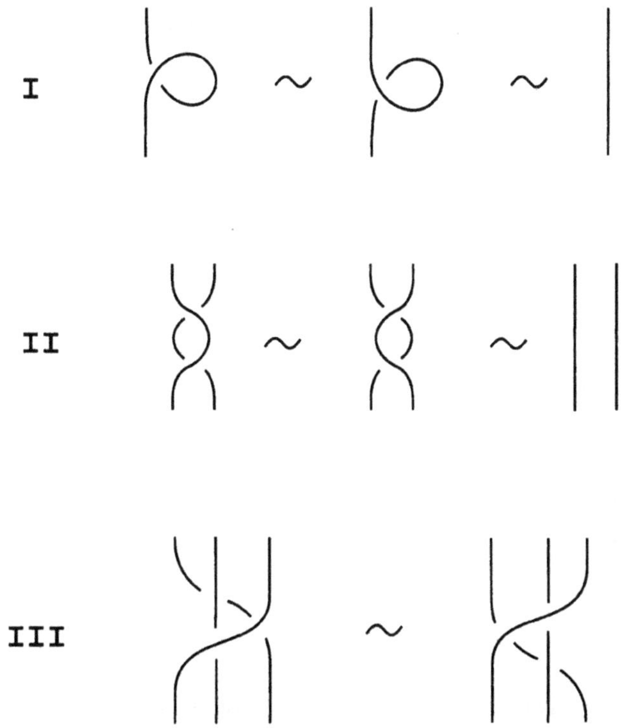

Figure 2.1.

constructing link invariants of ambient isotopy means finding invariants of the symmetry group generated by the RM. Several different methods of constructing link invariants have been discovered and some of them will be mentioned here. Before describing some explicit examples, let us analyse the RM a little more carefully. Reidemeister moves of type I are very special. In fact, one can eliminate them from the list of admissible moves. In this way, one can define an interesting structure which plays an important role in the construction of the universal link polynomial.

Two link diagrams D_1 and D_2 related by RM of types II and III only are called ***regular isotopic***. Consider now the equivalence classes of regular isotopic link diagrams. A useful invariant of regular isotopy is the ***writhe*** number $w(D_L)$ which is defined for any link diagram D_L by

$$w(D_L) = \sum_p \epsilon(p) . \tag{2.1}$$

The sum in equation (2.1) is performed over all the crossing points of the link diagram D_L and

$$\epsilon\,(L_\pm) = \pm 1\,, \tag{2.2}$$

where $L_\pm$ are shown in Fig. 2.2. The configurations L_+ and L_- will be called overcrossing and undercrossing respectively.

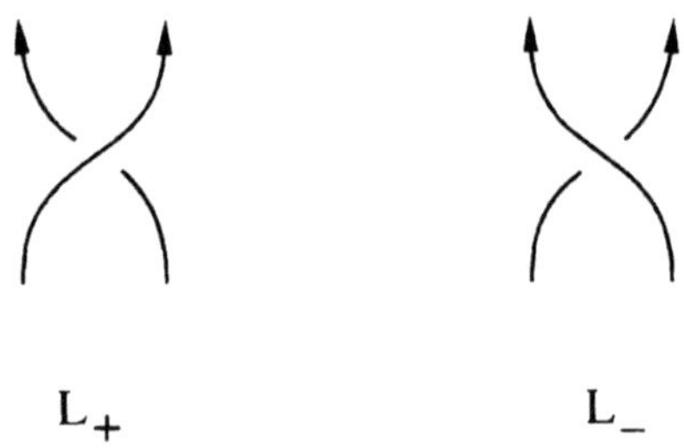

Figure 2.2.

The proof that $w(D_L)$ is a regular isotopy invariant is very simple; as a matter of fact, it is immediately verified that $w(D_L)$ is invariant under RM of types *II* and *III*.

The concept of regular isotopy is useful because, by eliminating the RM of type *I*, one does not lose any information concerning the topology of links; on the contrary, one gains a free variable for each component of the link. In fact, each equivalence class of ambient isotopy contains, by definition, all the equivalence classes of regular isotopy which are connected by RM of type *I*. Each RM of type *I* acts only on a single line of the diagrams; therefore, it can modify only the writhe number of a single component of the link diagrams. This being the case, each equivalence class of ambient isotopy corresponding to a link in $\mathbb{R}^3$ contains an infinite number of equivalence classes of regular isotopy, which are labelled by the writhe numbers $\{w(C_i)\}$ of the different components $\{C_i\}$ of the link. The crucial point is that one can give [13] the following interpretation of the above conclusion. The equivalence classes of regular isotopic link diagrams describe ambient isotopy classes of links in $\mathbb{R}^3$ in which each component is characterized by an integer number which, in turn, is an ambient isotopy invariant.

Now, suppose that one replaces links made of, say, strings with links made of oriented ***bands***. The topology of the links is not modified; the only change is that for each component C_i of the link we now have an extra variable $T(C_i)$ telling us how many times the oriented band is twisted. The ***twist*** T of the band is an ambient isotopy invariant and therefore we are precisely in the same conditions as above. In conclusion, one can represent the equivalence classes of ambient isotopic links made of bands with the equivalence classes of regular isotopic link diagrams. The only thing which remains to be fixed is the connection between the writhe number $w(C_i)$ and the the value of the twist $T(C_i)$ of each component C_i of the links. The simplest choice is

$$T(C_i) = w(C_i). \tag{2.3}$$

With the convention (2.3), the band shown in Fig. 2.3(a), for example, is represented by the link diagram shown in Fig. 2.3(b).

The importance of regular isotopy for the CS theory is due to the fact that, in studying the properties of the expectation values $\{\langle W(L)\rangle\}$, one has to consider ***framed links***. This means that for each component C of the links one has to introduce another closed and oriented path C_f called the framing of C. We will discuss this point in detail in Chapter 3. For the moment, imagine that C_f lies within an infinitesimal neighbourhood of C, with the condition that C and C_f never intersect. Moreover, C_f is always oriented in the same direction as C (i.e., C and C_f coincide in the limit in which the thickness of the neighbourhood vanishes). An example of a framed knot is shown in Fig. 2.4.

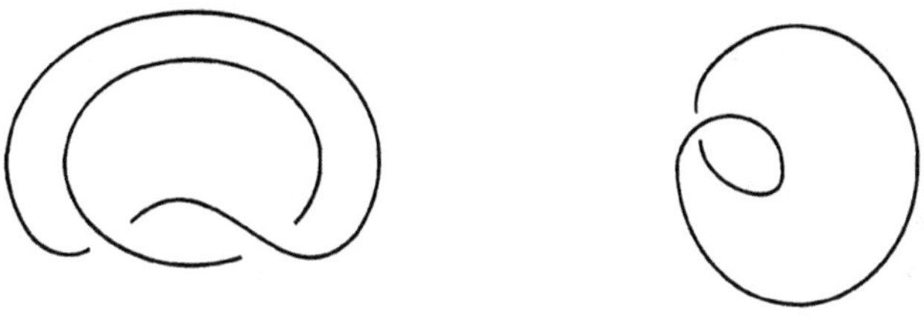

(a) (b)

Figure 2.3.

Since C and C_f can be considered to be the two components of the boundary of an oriented band, framed links can be interpreted as links made of bands. Consequently, we will represent oriented framed links in $\mathbb{R}^3$ with the regular isotopy classes of link diagrams with the identification (2.3); this representation will be called ***vertical framing***. In order to simplify the notations, framed links in $\mathbb{R}^3$ and their corresponding link diagrams in vertical framing will be indicated by the same symbol.

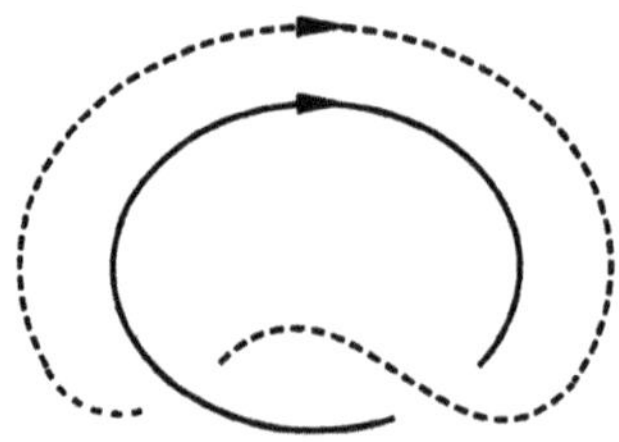

Figure 2.4. The unknot (solid line) with a particular framing (dashed line).

RM of type I are very peculiar for another reason too; as we shall see, they are the only moves in which the partial closure of braids is involved. As far as RM of types II and III are concerned, one should note that they essentially determine the algebraic structure of the Artin braid group B_n. For open braids, the invariance under RM of type II is quite trivial because this invariance is automatically satisfied in terms of the B_n generators. However, in considering the closure of braids, RM of type II have the important effect of associating the conjugacy classes of B_n to the links. Finally, let us consider the RM of type III; they represent the main feature of the braid group and enter directly the defining relations satisfied by the B_n generators. Finding a complete classification of the inequivalent realizations of the RM of type III is still an open problem. In considering matrix representations of B_n and with an appropriate choice of the form of the generators, RM of type III give origin to the famous quantum Yang-Baxter equation. Some relevant properties of the braid group will be considered in Chapter 8.

2.2 Link invariants

In this section some examples of link invariants are reported. Consider a two-component oriented link L in $\mathbb{R}^3$ with components C_1 and C_2 and let D_L, D_1 and D_2 be the associated diagrams. As we have stated above, $w(D_L)$ is a regular isotopy invariant; it is easy to see that also $w(D_1)$ and $w(D_2)$ are separately regular isotopy invariants. Let us now try to construct an invariant of ambient isotopy by combining these three writhe numbers. Under a RM of type I, $w(D_L)$ and $w(D_1) + w(D_2)$ transform as

$$\Delta w(D_L) = \Delta [w(D_1) + w(D_2)] = \pm 1 . \tag{2.4}$$

Therefore, the combination

$$\chi(C_1, C_2) = \frac{1}{2}[\, w(D_L) - w(D_1) - w(D_2)\,] \tag{2.5}$$

is an ambient isotopy invariant. The invariant $\chi(C_1, C_2)$ is called the ***linking number*** of C_1 and C_2. Roughly speaking, the value of $\chi(C_1, C_2)$ tells us how many times C_2 winds around C_1. This quantity can also be expressed in terms of the Gauss integral

$$\chi(C_1, C_2) = \frac{1}{4\pi}\oint_{C_1} dx^\mu \oint_{C_2} dy^\nu\, \epsilon_{\mu\nu\rho}\, \frac{(x-y)^\rho}{|x-y|^3}\,, \tag{2.6}$$

where the distance $|x - y|$ is computed by means of the ordinary flat metric in $\mathbb{R}^3$.

This example shows that the same link invariant can be constructed by using very different methods. In equation (2.5), χ is obtained through computations performed by looking at the link diagrams, whereas equation (2.6) provides a more direct geometrical interpretation of χ. The expression (2.5) has to be computed on a specific link diagram but the result, being ambient isotopic, does not depend on the particular choice of this diagram. Similarly, in the expression (2.6) the integrand depends on the metric $\delta_{\mu\nu}$ of $\mathbb{R}^3$, but the result of the integral is metric-independent: it depends exclusively on the topology. It is obvious that, for explicit computations, equation (2.5) is more practical to use than the expression (2.6). Usually, the invariants constructed by operating on the link diagrams are easier to compute than those constructed by means of geometrically more intrinsic methods.

Equations (2.5) and (2.6) are useful for illustrating the strategy pursued in solving the CS theory. The contributions to $\langle\, W(L)\,\rangle$ obtained, for instance, by computing the Feynman diagrams are the analogue of the expression (2.6) and are in general quite difficult to evaluate. The idea is to find the analogue of equation (2.5); that is, all the contributions of the Feynman diagrams will be expressed in terms of simple algebraic operations based on the structure of the link diagrams.

As we have already said, a framed oriented knot C with framing C_f can be interpreted as a knot made of a band; the twist T of this band is simply given by

$$T = \chi(C, C_f)\,. \tag{2.7}$$

Among the several link invariants that have been discovered, there are the so-called ***link polynomials***. Some of them are strictly connected with the universal link polynomial described by the CS theory and their construction is simply formulated in terms of the concept of the ***skein relation***. Three oriented links L_+, L_- and L_0 are skein related if they have diagrams which are identical except for a small part

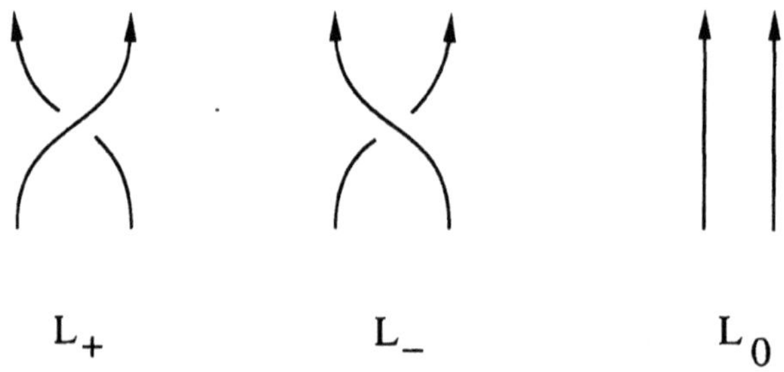

Figure 2.5.

contained inside a fixed open disc of the plane of the pictures. Moreover, inside this disc the three diagrams look as shown in Fig. 2.5.

The Alexander-Conway polynomial $\nabla(L; z)$, associated to the link L, is defined by [14]

(i) ambient isotopy invariance,
(ii) $\nabla(U; z) = 1$,
(iii) $\nabla(L_+; z) - \nabla(L_-; z) = z\,\nabla(L_0; z)$, (2.8)

where U is the unknotted knot and condition (ii) fixes the normalization. With our notations, $\nabla(L; z)$ is also called the Conway potential function and it is a (finite) Laurent polynomial in the variable z with integer coefficients. Of course, if $\nabla(L_1; z) \neq \nabla(L_2; z)$, then the two links L_1 and L_2 are not ambient isotopic. However, it is easy to find examples of links which are not ambient isotopic but have the same Alexander-Conway polynomial.

The Jones polynomial $V(L; q^{1/2}) \in Z[q^{\pm 1/2}]$ is defined by [6]

(i) ambient isotopy invariance,
(ii) $V(U; q^{1/2}) = 1$,
(iii) $qV(L_+) - q^{-1}V(L_-) = (q^{1/2} - q^{-1/2})\,V(L_0)$. (2.9)

The $V(L)$ polynomial is more selective than $\nabla(L)$; however, there are still ambient non-isotopic links with the same Jones polynomial. Our convention on the form of the exchange relation (2.9) should be noted; the dependence of equation (2.9) on the variable q is not standard. In these notations, the field theory results take a simple form. In Chapter 12, we shall compare our definition of the Jones polynomial with the standard convention used in mathematics.

The two-variable HOMFLY polynomial $P(L; t, z)$ is defined by [15]

$$\begin{aligned}
&\text{(i)} \quad \text{ambient isotopy invariance,} \\
&\text{(ii)} \quad P(U; t, z) = 1 \quad , \\
&\text{(iii)} \ t\, P(L_+) - t^{-1} P(L_-) = z\, P(L_0)\,,
\end{aligned} \tag{2.10}$$

and it represents essentially the most general polynomial [16] constructed by means of the skein relation involving the configurations shown in Fig. 2.5. In fact, $P(L; t, z)$ reduces to the Alexander-Conway and Jones polynomials with the obvious choices for the values of the variables t and z. The HOMFLY polynomial also does not provide a complete classification of knots or links.

A common feature of all these polynomials is that, by means of the skein relation, they can be easily constructed by analysing the link diagrams. By using the conditions (i) and (iii) recursively, the polynomial of whatever link can be written in terms of the polynomial of the unknotted knot U, which is conventionally taken to be the identity. At this stage, it is not completely obvious that the construction based on the recursive use of the skein relation is well defined. Several different proofs of the internal consistency of the defining conditions (i), (ii) and (iii) have been produced in literature, see for instance [6,15,17].

By looking at the defining conditions (2.8)-(2.10), one notes that the progress in the construction of the "classical" link polynomials has been made by modifying the skein (or exchange) relation. However, along this line it is hard to imagine how to improve the HOMFLY polynomial significantly. In fact, the strategy leading to the universal link polynomial is to come back to the Jones polynomial and provide it with a Lie algebra interpretation.

We conclude this section by considering a link polynomial of regular isotopy which is of particular interest for the CS theory. The $S(L; \alpha, \beta, z)$ polynomial [18] is defined by

$$\begin{aligned}
&\text{(i)} \quad \text{regular isotopy invariance,} \\
&\text{(ii)} \quad S(U_0) = 1\,, \\
&\text{(iii)} \ S(L^{(+)}) = \alpha\, S(L^{(0)}), \qquad S(L^{(-)}) = \alpha^{-1}\, S(L^{(0)})\,, \\
&\text{(iv)} \ \beta\, S(L_+) - \beta^{-1} S(L_-) = z\, S(L_0)\,,
\end{aligned} \tag{2.11}$$

where $L^{(\pm)}$ and $L^{(0)}$ are shown in Fig. 2.6 and U_0 is the unknotted knot with zero writhe. The use of a simplified notation should cause no problems here. From the definition (2.11) it is clear that $S(L)$ is defined on the equivalence classes of regular isotopic link diagrams or, equivalently, on the set of framed links.

The polynomial $S(L)$ was introduced in Ref.[18] to describe the behaviour of the Wilson line operators when the link components are associated with certain representations of the gauge group. $S(L)$ also can be constructed by using the skein relation recursively and is related [18] to the HOMFLY polynomial by

$$P(L; t = \alpha\beta, z) \;=\; \alpha^{-w(L)}\, S(L; \alpha, \beta, z)\,. \tag{2.12}$$

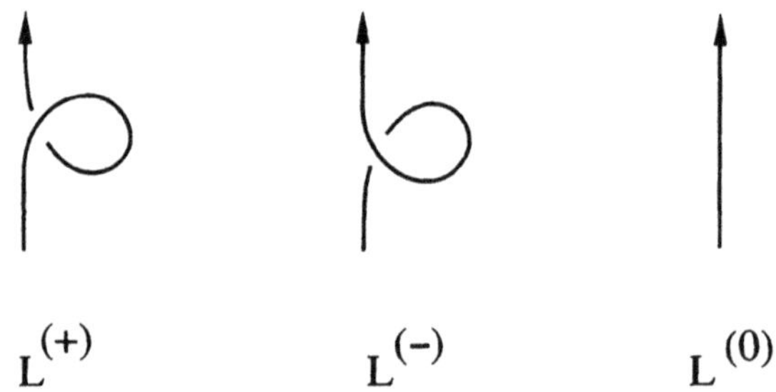

Figure 2.6.

The factor multiplying $S(L)$ in equation (2.12) just compensates the covariant variation of $S(L)$ under RM of type I. Because of the identity (2.12), the information concerning the link classification problem contained in $S(L)$ and $P(L)$ is essentially the same. However, the $S(L)$ polynomial is particularly significant; as we shall see in Chapter 8, $S(L)$ is in a way the ancestor of all the polynomials which are obtained in Hecke algebra representations of the braid group. Moreover, the structure of the relations (2.11) naturally extends to the general case described by $E(L)$.

Clearly, all the link polynomials defined by a skein relation represent invariants of links in $\mathbb{R}^3$ and of links in S^3 as well. Indeed, in the construction of the invariant what is important is the fact that any link is contained in the interior of a three-ball where the ordinary skein relation holds. We shall return to this important point later.

2.3 Framing and satellites

This section completes our rapid review of knot theory and introduces several definitions and concepts which are useful for our discussion. Consider a knot $C \subset \mathbb{R}^3$ or in S^3 and the two-dimensional disc D^2 (i.e. the unit ball in $\mathbb{R}^2$ centered at the origin). If we represent D^2 in the complex plane, its points have coordinates $(r\, e^{i\theta})$ with $0 \leq r \leq 1$. An embedding $f : C \times D^2 \to \mathbb{R}^3$, such that $f(x,0) = x$ for $x \in C$, defines a ***tubular neighbourhood*** of C. In other words, a tubular neighbourhood of C is just a solid torus whose core is C, as shown in Fig. 2.7.

A ***solid torus*** V is a three-dimensional space homeomorphic with $S^1 \times D^2$; a specified homeomorphism $h : S^1 \times D^2 \to V$ is called a ***framing*** of V. In Chapter 16, we shall analyse in detail the main properties of solid tori; for the moment, only some useful definitions are recalled. A ***meridian*** of V is a simple closed curve on the boundary ∂V of the solid torus which is essential

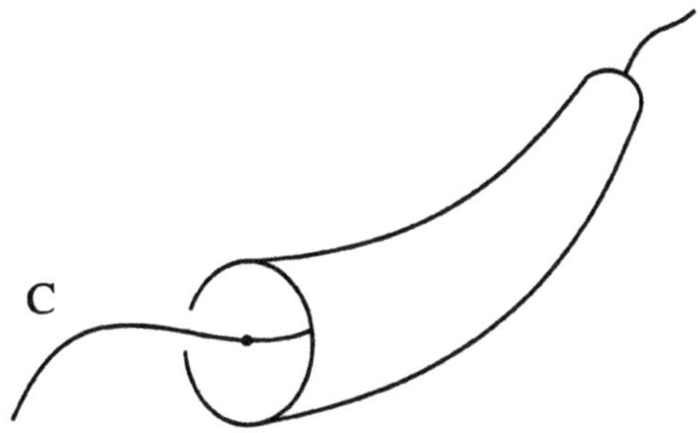

Figure 2.7. Tubular neighbourhood of the knot C

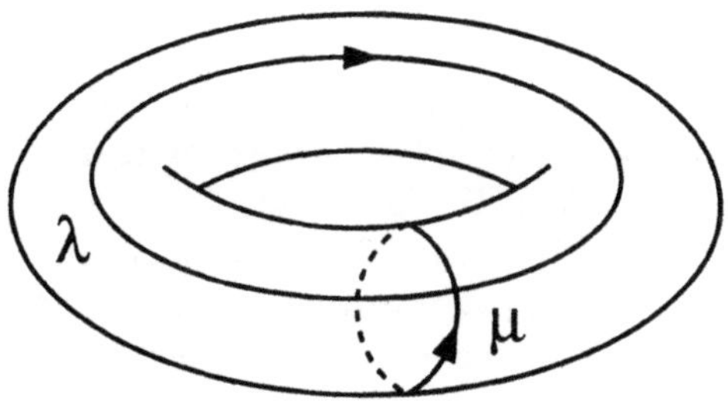

Figure 2.8.

in ∂V and is homotopically trivial in V. The curve μ, shown in Fig. 2.8, is a meridian. A ***longitude*** of V is a closed curve in ∂V intersecting some meridian of V (transversally) in a single point. The curve λ, shown in Fig. 2.8, is a longitude. For any given framing h of V, $h(S^1 \times 1)$ is a longitude. It should be noted that any two meridians are ambient isotopic in V; in this sense, the meridian is an intrinsic part of a solid torus. On the other hand, there is an infinite number of ambient isotopy classes of longitudes. These classes can be classified by the linking number of the longitude and the core of the solid torus.

Given a tubular neighbourhood N of the knot C and a specific framing h of N, the longitude $h(S^1 \times 1)$ defines a framing C_f of C. One can always assume that the image of the core of $S^1 \times D^2$, namely $h(S^1 \times 0)$, coincides with the knot C. Therefore, a given orientation for the knot C defines an orientation for S^1 and, consequently, $C_f = h(S^1 \times 1)$ also is oriented. In conclusion, a framing h of a tubular neighbourhood of a knot C defines a framing C_f of C. Viceversa, a given framing C_f of C determines, up to ambient isotopy, a unique framing h for a tubular neighbourhood of C. If $\chi(C, C_f) = 0$, C_f is called a ***preferred framing*** for C.

A two-dimensional disc D in a solid torus V is called ***meridinal*** if its boundary is a meridian of V and a closed subset X of V is ***essential*** in V if X intersects every meridinal disc in V.

Let C be a component of a framed link L and let N be a tubular neighbourhood of C. We denote by V a solid torus standardly embedded in $\mathbb{R}^3$ as shown in Fig. 2.8. Let $h^\diamond$ be a homeomorphism $h^\diamond : V \to N$ such that the image under $h^\diamond$ of the core of V is the component C and $h^\diamond(\lambda) = C_f$, where C_f is the framing of C. Up to ambient isotopy, this homeomorphism $h^\diamond$ is unique. Suppose now that the component C is removed from L and is replaced by $h^\diamond(P)$,

$$C \Longrightarrow h^\diamond(P), \tag{2.13}$$

where P is a (framed) link which is essential in V. The new link L', obtained according to this instruction, is called a ***satellite*** of L; L is a companion of L' and P is called the ***pattern link***. For convenience, we will represent the pattern link P in the complement solid torus of the unknot U in S^3. For example, the diagram of Fig. 2.9 represents (in vertical framing) a satellite of the trefoil, shown in Fig. 2.10(a), obtained with the pattern link P shown in Fig. 2.10(b).

Figure 2.9. Satellite of the trefoil

Note that the above definition of satellites is based on a specific choice of the homeomorphism $h^\diamond : V \to N$. As we shall see, with this convention the satellite rules will take a simple form. Consider the framed pattern link B, with two components, described by the diagram of Fig. 2.11.

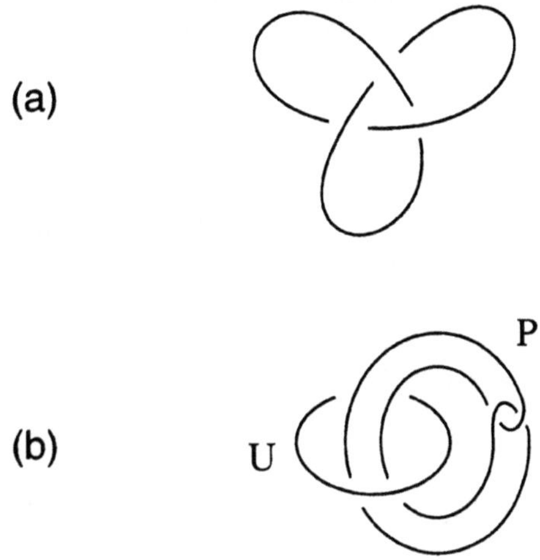

Figure 2.10.

Figure 2.11. Pattern link B

The satellite L' of L, constructed with the pattern link B, is obtained from L simply by "doubling" one of its components C:

$$C \Longrightarrow h^{\diamond}(B) \equiv C^{(2)}\,. \qquad (2.14)$$

The linking number of the two components of $h^{\diamond}(B) = C^{(2)}$ is equal to the linking number of C and its framing C_f. Moreover, the linking number of each component of $C^{(2)}$ and its framing is equal to the linking number of C and C_f. This means that the satellites constructed with the pattern link B are described (in vertical framing) by link diagrams in which the component C is doubled in such a way that the two new components are parallel on the plane of the diagram. Examples of satellites constructed with the pattern link B are shown in Fig. 2.12; their companions are described by: (a) the unknot with zero writhe, (b) the unknot with writhe $+1$ and (c) the right-handed trefoil with writhe $+3$.

Some final remarks are in order. It is clear that, for the construction of a satellite of some link L, one must specify (in addition to the component C which has to be substituted) the homeomorphism $h^\diamond$ and the pattern link P in V. In our conventions, $h^\diamond$ is specified by the framing of C. The satellites constructed with the pattern link B, shown in Fig. 2.11, will play a crucial role in our discussion.

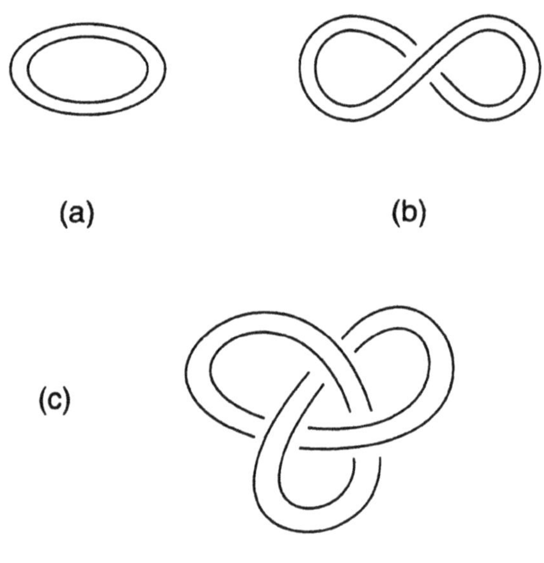

Figure 2.12.

Instead of representing the pattern link directly inside the solid torus V of Fig. 2.8, we have shown the pattern links (see Fig. 2.10(b) and Fig. 2.11) with respect to the circle U in S^3. The point is that the complement of a tubular neighbourhood of U in S^3 is again a solid torus and, for graphical reasons only, it is convenient to represent the pattern link in it. Finally, in the construction of satellites of oriented links, the pattern link P also must be oriented; moreover, the relative orientation of P and of the original component C must be specified.

Chapter 3

Framing in field theory

The problem of the computation of the Wilson line expectation values is meaningful provided $\{\langle W(L)\rangle\}$ are well defined. Usually in field theories, the expectation values of composite operators present in general several problems related, for instance, to the occurrence of divergences or of ambiguities in their definition. We must specify then what the precise meaning of the expression (1.22) is. In these lectures, we shall consider the case in which the expression (1.22) is supplemented by two further requirements: first, that in the quantum CS theory gauge invariance is preserved and, second, that general covariance is also preserved. In the quantization schemes used to perform the explicit computations reported here, both these conditions are always satisfied. Many of the properties of the expectation values $\{\langle W(L)\rangle\}$ follow from general covariance; in particular, the necessity of introducing a framing prescription for the definition of the Wilson line operators is a consequence of general covariance. The precise definition of framing in the CS model is certainly a somewhat technical issue; and yet, this is also one of the crucial aspects of the theory. The importance of framing cannot be overestimated. Indeed, as we shall see, finding the solution to the problem of framing essentially means finding the solution to the CS theory.

3.1 Abelian Chern-Simons theory

In order to understand what framing means in the field theory context, it is useful to consider first of all the simple case of the Abelian CS theory in $\mathbb{R}^3$ defined by the action

$$S_0 = \frac{k}{8\pi} \int d^3x \, \epsilon^{\mu\nu\rho} \, A_\mu \partial_\nu A_\rho \, . \tag{3.1}$$

The action (3.1) is invariant under gauge transformations

$$\Delta A_\mu(x) = \partial_\mu \Lambda(x) \, ; \tag{3.2}$$

the way to deal with gauge invariance at the quantum level has been amply discussed in literature.

General covariance follows from the fact that the expression (3.1), which is invariant under general coordinate transformations performed on the vector field $A_\mu(x)$, is metric-independent. This means that in the physical system described by the action (3.1) there is no notion of the distance between two points because there is no particular metric for defining a distance. This being the case, general covariance would suggest that $\langle W(L) \rangle$ is an ambient isotopy invariant for the links $\{L\}$. As it stands, this classical argument fails at the quantum level. Still, it is possible to recover general covariance in a slightly modified version.

In order to define the quantum theory, the well-known BRS formalism [19] can be used. In the Landau gauge, the total action reads

$$S = \frac{k}{4\pi} \int d^3x \left[\frac{1}{2} \epsilon^{\mu\nu\rho} A_\mu \partial_\nu A_\rho - B \partial^\mu A_\mu + \overline{c}\, \partial^\mu \partial_\mu c \right], \tag{3.3}$$

where B is a bosonic auxiliary field and c and $\overline{c}$ are the ghost fields. In the gauge-fixing and ghost terms, the ordinary flat metric of $\mathbb{R}^3$ has been used. The introduction of a metric in the Lagrangian has to be confronted with the requirement of general covariance. In our case, since the metric enters in the gauge-fixing procedure only, for gauge-invariant and metric-independent observables general covariance is preserved. The propagators obtained from (3.3) are [20]

$$\langle A_\mu(x) A_\nu(y) \rangle = \frac{i}{k} \epsilon_{\mu\nu\rho} \frac{(x-y)^\rho}{|x-y|^3}, \tag{3.4}$$

$$\langle A_\mu(x) B(y) \rangle = -\frac{i}{k} \frac{(x-y)_\mu}{|x-y|^3}, \tag{3.5}$$

$$\langle c(x)\, \overline{c}(y) \rangle = -\frac{i}{k} \frac{1}{|x-y|}. \tag{3.6}$$

The quadratic action (3.3) defines a free theory and the correlation functions of the fields are well defined; they can obviously be expressed in terms of the expressions (3.4)-(3.6).

The integral of the one-form $A_\mu dx^\mu$ on a closed oriented path C in $\mathbb{R}^3$ is gauge-invariant and metric-independent; let us consider the associated expectation values. Given two non-intersecting oriented closed paths C_1 and C_2, one gets

$$\begin{aligned} \langle \oint_{C_1} A_\mu dx^\mu \oint_{C_2} A_\nu dy^\nu \rangle &= \frac{i}{k} \oint_{C_1} dx^\mu \oint_{C_2} dy^\nu \epsilon_{\mu\nu\rho} \frac{(x-y)^\rho}{|x-y|^3} \\ &= i \frac{4\pi}{k} \chi(C_1, C_2), \end{aligned} \tag{3.7}$$

where $\chi(C_1, C_2)$ is the linking number between C_1 and C_2. As expected, even if the $\mathbb{R}^3$ metric has been used in the intermediate steps of the computations, the final result (3.7) is metric-independent.

Since the action (3.3) is quadratic in the fields, it is clear that the expectation value of the product of an arbitrary number of $A_\mu dx^\mu$ integrals, performed on different paths, is a topological invariant. For instance, in the case of a four-component link one obtains

$$\begin{aligned} \langle \oint_{C_1} A_\mu dx^\mu \oint_{C_2} A_\nu dx^\nu \oint_{C_3} A_\rho dx^\rho \oint_{C_4} A_\tau dx^\tau \rangle = \\ = -\left(\frac{4\pi}{k}\right)^2 \Big[\chi(C_1, C_2)\chi(C_3, C_4) + \chi(C_1, C_3)\chi(C_2, C_4) \\ + \chi(C_1, C_4)\chi(C_2, C_3) \Big] . \end{aligned} \tag{3.8}$$

In conclusion, for any oriented link L in $\mathbb{R}^3$ the expectation value of the associated observable, obtained by integrating the one-form $A_\mu dx^\mu$ on the components of L, is well defined and is an ambient isotopy invariant.

3.2 Framed Wilson line operators

In considering Wilson line operators, one has to exponentiate the integral of the one-form $A_\mu dx^\mu$. Consequently, we must also analyse the case in which the product of two or more $A_\mu dx^\mu$ integrals, performed on the same path, occurs. Let us consider first the simplest case: namely, the expectation value of two $A_\mu dx^\mu$ integrals associated with the same oriented path C parametrized by $\{x^\mu(s)\,;\, 0 \leq s \leq 1\}$. The "naive" result is

$$\begin{aligned} \langle \oint_C A_\mu dx^\mu \oint_C A_\nu dy^\nu \rangle = \\ = \frac{i}{k} \int_0^1 ds \int_0^1 dt \; \epsilon_{\mu\nu\rho}\, \dot{x}^\mu(s)\, \dot{x}^\nu(t) \, \frac{(x(s) - x(t))^\rho}{|x(s) - x(t)|^3} . \end{aligned} \tag{3.9}$$

The integral appearing in equation (3.9) is well defined and finite [21], so there is no need to introduce any regularization. The problem is that, unlike the Gauss integral (3.7), the expression (3.9) is not an ambient isotopy invariant [21] and this is in contradiction with general covariance.

How has equation (3.9) to be interpreted ? At first sight, you could think that there is a breaking of general covariance in the quantization procedure that we have used. This is not the case, however. In reality, the problem connected with the expression (3.9) is related to a typical quantum effect which is well known in field theory and which shows that the classical arguments are quite often misleading in the quantum context. The whole point can be illustrated by an example; consider a renormalizable quantum field theory constructed in terms of a scalar field $\phi(x)$. At the classical level, the knowledge of the field $\phi(x)$ is sufficient for determining the values of $\{\phi^2(x), \phi^3(x), \ldots\}$. At the quantum level, the situation is quite different. The field operator $\phi(x)$ is well defined in the sense that the correlation functions of $\phi(x)$ are well defined. However, this is not enough for uniquely determining what the quantum operators $\{\phi^2(x), \phi^3(x), \ldots\}$ mean. This problem is not related to the choice of the quantization scheme which has been used to define the renormalized correlation functions of $\phi(x)$. In other words, even if the field operator $\phi(x)$ is well defined, one still has to define the precise meaning of the composite operators $\{\phi^2(x), \phi^3(x), \ldots\}$. In general, different definitions of the same power of $\phi(x)$ are possible and, of course, different definitions can lead to different composite operators.

Let us consider the CS theory now. The vector field $A_\mu(x)$ is the fundamental field of the model. In any consistent and general-covariance-preserving quantization scheme, the integral of the one-form $A_\mu dx^\mu$ on a closed path is a well-defined operator. This means that the expectation values of an arbitrary number of such operators, associated with non-intersecting different paths, are unique and well defined. This is precisely the result that we have obtained by using the covariant BRS formalism. However, the product of two $A_\mu dx^\mu$ integrals on the same path is a composite operator and we must define it. Different definitions of $\left(\oint_C A_\mu dx^\mu\right)^2$ are possible and, accordingly, different composite operators can be obtained. The first possibility is to define $\left(\oint_C A_\mu dx^\mu\right)^2$ precisely like its classical expression suggests, i.e.

$$\left(\oint_C A_\mu dx^\mu\right)^2 \Bigg|_{\text{naive}} = \int_0^1 ds \int_0^1 dt\, \dot{x}^\mu(s) A_\mu(x(s))\, \dot{x}^\nu(t) A_\nu(x(t)) \,. \tag{3.10}$$

Equation (3.10) is called the naive definition; the result (3.9) has been obtained by using equation (3.10). With the naive definition of $\left(\oint_C A_\mu dx^\mu\right)^2$, the vacuum expectation values are finite; unfortunately, general covariance is not maintained.

Another possibility is to define $\left(\oint_C A_\mu dx^\mu\right)^2$ by means of a framing procedure [22]. For each knot C parametrized by $\{x^\mu(s)\,;\ 0 \leq s \leq 1\}$, we introduce a framing contour C_f parametrized by

$$y^\mu(s) = x^\mu(s) + \varepsilon\, n^\mu(s) \quad , \quad (\varepsilon > 0 \quad , \quad |\vec{n}(s)| = 1)\,, \tag{3.11}$$

where n^μ is a vector field orthogonal to C. Then we define

$$\left(\oint_C A_\mu dx^\mu\right)^2 \Bigg|_{\text{framed}} = \lim_{\varepsilon \to 0} \oint_C A_\mu dx^\mu \oint_{C_f} A_\nu dy^\nu \,, \tag{3.12}$$

with the convention [22] that the $\varepsilon \to 0$ limit has to be taken after all the Wick-contractions and integrations have been performed. At the classical level, the definitions (3.10) and (3.12) coincide. At the quantum level, however, the two definitions lead to very different results. Indeed, according to (3.12) one obtains

$$\begin{aligned} \langle \left(\oint_C A_\mu dx^\mu \right)^2 \Big|_{\text{framed}} \rangle &= i\left(\frac{4\pi}{k}\right) \lim_{\varepsilon \to 0} \chi(C, C_f) \\ &= i\left(\frac{4\pi}{k}\right) \chi(C, C_f) . \end{aligned} \tag{3.13}$$

The result (3.13) is now an ambient isotopy invariant for the link defined by $\{C, C_f\}$ but its particular values depend on the choice of the framing C_f. So, by using the framing procedure (3.12), general covariance is maintained in the sense that the expression (3.13) is an ambient isotopy invariant of the ***framed*** knot C.

The problem of giving a topological meaning to the "linking number" of a single knot is known as the self-linking problem. The solution (3.13), obtained by means of framing, was already known in mathematics [23]. In the field theory context, the framing procedure is extended from the expectation value (3.13) to the composite operator $(\oint_C A_\mu dx^\mu)^2$ itself, equation (3.12). The improvement obtained by means of this definition is clear. Not only $\langle (\oint_C A_\mu dx^\mu)^2 \rangle$, but also any expectation value, involving an arbitrary number of $(\oint A_\mu dx^\mu)$ and $(\oint_C A_\mu dx^\mu)^2$ operators, now preserves general covariance.

At this stage, the definition of the composite Wilson line operator by means of the framing procedure is quite natural. For an oriented knot C characterized by a given value e of the "charge", the naive expression of the associated Wilson line operator would be

$$\begin{aligned} W(C; e) &= \exp\left(ie \oint_C A_\mu dx^\mu \right) \\ &= \sum_n \frac{(ie)^n}{n!} \oint_C A_{\mu_1}(x_1) dx_1^{\mu_1} \cdots \oint_C A_{\mu_n}(x_n) dx_n^{\mu_n} . \end{aligned} \tag{3.14}$$

In this equation, the products of the $A_\mu dx^\mu$ integrals performed on the same path C would lead to problems with general covariance. But we already know how to overcome this difficulty; it is sufficient to introduce a framing and take the limit of coincident paths at the end of the computations. To be more precise, in a generic product

$$\begin{aligned} &\oint_C A_{\mu_1}(x_1)\, dx_1^{\mu_1} \cdots \oint_C A_{\mu_n}(x_n)\, dx_n^{\mu_n} = \\ &= \int_0^1 ds_1 \cdots \int_0^1 ds_n\, \dot{x}^{\mu_1}(s_1)\, A_{\mu_1}(x(s_1)) \cdots \dot{x}^{\mu_n}(s_n)\, A_{\mu_n}(x(s_n)) \end{aligned}$$

appearing in equation (3.14), each term $\dot{x}^{\mu_i}(s_i)\, A_{\mu_i}(x(s_i))$ is replaced by

$$\dot{x}^{\mu_i}(s_i)\, A_{\mu_i}(x(s_i)) \rightarrow \left[\dot{x}^{\mu_i}(s_i)\, A_{\mu_i}(x(s_i))\right]_f , \tag{3.15}$$

$$\left[\dot{x}^{\mu_i}(s_i)\, A_{\mu_i}(x(s_i))\right]_f = (\dot{x}^{\mu_i}(s_i) + \varepsilon_i \dot{n}^{\mu_i}(s_i))\, A_{\mu_i}(x(s_i) + \varepsilon_i n(s_i)) , \tag{3.16}$$

where the vector field n^μ characterizes the choice of framing, as shown in equation (3.11), and the values

$$\{\varepsilon\} = \{\varepsilon_i \, , \, \varepsilon_i > 0\} \tag{3.17}$$

can be arbitrarily chosen provided that $\varepsilon_i \neq \varepsilon_j$ for $i \neq j$. Finally, the composite Wilson line operator associated with a framed knot C is defined as

$$\begin{aligned} W(C;e)\Big|_f = \lim_{\{\varepsilon\}\to 0} \sum_n \frac{(ie)^n}{n!} \int_0^1 ds_1 \cdots \int_0^1 ds_n \\ \left\{ \left[\dot{x}^{\mu_1}(s_1) A_{\mu_1}(x(s_1))\right]_f \cdots \left[\dot{x}^{\mu_n}(s_n) A_{\mu_n}(x(s_n))\right]_f \right\} . \end{aligned} \tag{3.18}$$

In the case of a non-Abelian gauge group, composite Wilson line operators associated with framed knots are similarly defined; in each term of the field expansion of the path-ordered exponential, the substitution (3.15) is performed and then the $\{\varepsilon\} \to 0$ limit is taken at the end of the computations. Of course, in the case of a generic link, the framing procedure described above must be applied to each component of the link. From now on, it is understood that Wilson line operators are defined by means of the framing procedure as shown in equation (3.18).

The effects produced by the framing procedure on the composite operator $W(C)$ are shown in Fig. 3.1. The original path C gets effectively transformed into a ***band***; the dashed lines represent the vector fields A_μ which must be integrated along the framed knot.

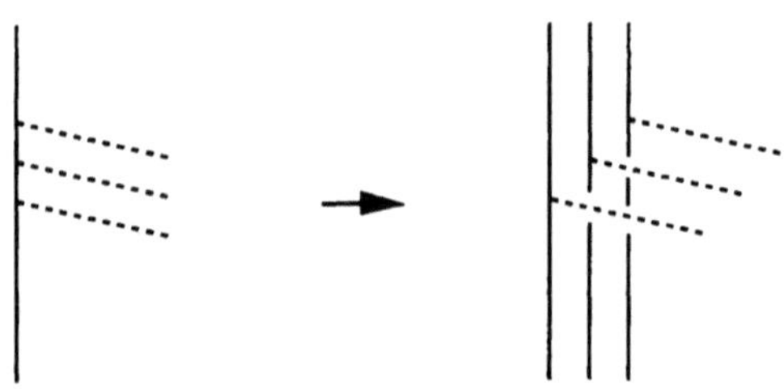

Figure 3.1.

We can now give the solution of the Abelian CS theory. Consider a generic oriented framed link L in $\mathbb{R}^3$ with m components $\{C_1, \ldots, C_m\}$ in which the i-th component

C_i has framing C_{if} and charge e_i. In each term of the expression (3.18), the Wick-contractions are defined by equation (3.4). As it is well known, the combinatorics of the resulting Feynman diagrams reproduce the expansion of an exponential term. The sum of all the contributions defined by the perturbative expansion gives the expectation value of the associated Wilson line operator $W(L)$

$$\langle W(L)\rangle = \exp\left\{-i\left(\frac{2\pi}{k}\right)\left[\sum_i e_i^2\,\chi(C_i, C_{if}) + 2\sum_{i<j} e_i\,e_j\,\chi(C_i, C_j)\right]\right\}. \tag{3.19}$$

Note that this result does not depend on the choice of the direction in the $\{\varepsilon\}$ space along which the $\{\varepsilon\} \to 0$ limit is performed.

Let the ***linking matrix*** L_{ij} be defined by

$$L_{ij} = \begin{cases} \chi(C_i, C_j) & \text{if } i \neq j; \\ \chi(C_i, C_{if}) & \text{if } i = j. \end{cases} \tag{3.20}$$

In terms of the linking matrix L_{ij}, the expression (3.19) can be written simply as

$$\langle W(L)\rangle = \exp\left[-i\left(\frac{2\pi}{k}\right)\sum_i\sum_j e_i\,L_{ij}\,e_j\right]. \tag{3.21}$$

As explained in Chapter 2, the framed link L can be represented by a particular link diagram in vertical framing that we indicate by the same symbol L. In vertical framing, the result (3.19) takes the form

$$\langle\, W(L)\,\rangle = \exp\left\{-i\left(\frac{2\pi}{k}\right)\left[\sum_i e_i^2\,w(C_i) + 2\sum_{i<j} e_i e_j\,\chi(C_i, C_j)\right]\right\}, \tag{3.22}$$

where $w(C_i)$ is the writhe of the component C_i.

The exact solution (3.19) of the Abelian CS theory shows that $\langle\, W(L)\,\rangle$ defines an ambient isotopy invariant for framed links in $\mathbb{R}^3$; equivalently, in vertical framing $\langle\, W(L)\,\rangle$ represents a regular isotopy invariant for link diagrams, equation (3.22). So, general covariance is manifestly preserved.

To sum up, framing is a way of defining the Wilson line operator at the quantum level. The introduction of framing is not motivated by the need to eliminate divergences but by the necessity of preserving general covariance. At the classical level and for each given vector field $A_\mu(x)$, the Wilson line (3.14) is a function on the space of knots in $\mathbb{R}^3$. At the quantum level, the Wilson line operator, obtained by means of the framing procedure, is defined on the space of knots and of their "first-order" neighbourhoods. In this sense, framing is similar to the blowing-up procedure. The only ambient isotopic characterization of these knot neighbourhoods is represented by the linking number $\chi(C, C_f)$ between C and its framing C_f. This is why the whole dependence of $\langle W(L) \rangle$ on the framing is expressed by $\chi(C, C_f)$ or, in vertical framing, by $w(C)$. Clearly, this general feature is common to both the Abelian and non-Abelian CS theories.

Chapter 4

Non-Abelian Chern-Simons theory

The action of the CS theory is constructed in terms of a vector field $A_\mu(x)$ which takes values on a real Lie algebra $\mathcal{A}$. Each Abelian $U(1)$ subfactor and each invariant simple subalgebra of $\mathcal{A}$ has its own coupling constant, just like in any gauge theory. The Abelian CS theory has already been analysed in Chapter 3. We shall now consider the case of a real simple Lie algebra $\mathcal{G}$. This algebra can be interpreted as the Lie algebra of a compact simple Lie group G. Following the standard language used in physics, we will often refer to the generic group G, but it is understood that most of the statements really concern the associated Lie algebra $\mathcal{G}$. In this chapter, we shall consider some general aspects of the non-Abelian CS theory in $\mathbb{R}^3$ and its renormalizability.

4.1 Covariant quantization

The vector field is written as $A_\mu(x) = A_\mu^a(x)\, T^a$, where $\{T^a\}$ are Hermitian matrices representing the generators of G in some defining representation. We use the convention

$$\mathrm{Tr}\left(T^a\, T^b\right) = \frac{1}{2}\,\delta^{ab}\,, \tag{4.1}$$

and the commutators

$$\left[T^a\,,\, T^b\right] = i\, f^{abc}\, T^c \tag{4.2}$$

define the structure constants $\{f^{abc}\}$. The CS action

$$S_{CS} = \frac{k}{4\pi}\int d^3x\, \epsilon^{\mu\nu\rho}\; \mathrm{Tr}\left(A_\mu \partial_\nu A_\rho + i\,\tfrac{2}{3}\, A_\mu A_\nu A_\rho\right) \tag{4.3}$$

can be rewritten as

$$S_{CS} = \frac{k}{4\pi}\int d^3x\, \epsilon^{\mu\nu\rho}\; \left\{\frac{1}{2}\, A_\mu^a \partial_\nu A_\rho^a - \frac{1}{6}\, f^{abc}\, A_\mu^a A_\nu^b A_\rho^c\right\}\,; \tag{4.4}$$

the real parameter k is called the coupling constant of the model. As in the case of the Abelian CS theory, gauge invariance and general covariance are the basic properties of the model. Under a gauge transformation

$$A_\mu(x) \rightarrow A_\mu^\Omega(x) = \Omega^{-1}(x)\, A_\mu(x)\, \Omega(x) - i\, \Omega^{-1}(x)\, \partial_\mu \Omega(x)\,, \tag{4.5}$$

the action (4.3) is not invariant. Indeed, it transforms as

$$S_{CS}[A^\Omega] = S_{CS}[A] + 2\pi\, k\, \Gamma_t[\Omega]\,, \tag{4.6}$$

where

$$\Gamma_t[\Omega] = \frac{1}{24\pi^2} \int_{\mathcal{M}} d^3x\, \epsilon^{\mu\nu\rho}\, \mathrm{Tr}\left[\Omega^{-1}\, \partial_\mu \Omega\, \Omega^{-1}\, \partial_\nu \Omega\, \Omega^{-1}\, \partial_\rho \Omega\right]. \tag{4.7}$$

It is easy to verify that the functional $\Gamma_t[\Omega]$ is invariant under smooth deformations of $\Omega(x)$. In fact, the value of Γ_t depends on the homotopy class only of the map Ω. When the value of $\Omega(x)$ is well defined also in the $|x| \rightarrow \infty$ limit, Ω defines a map $\Omega : S^3 \rightarrow G$, S^3 being represented by $\mathbb{R}^3 + \{\infty\}$. When the gauge group G is $SU(2)$, $\pi_3(SU(2)) \cong Z$ and the value of $\Gamma_t[\Omega]$ is an integer which coincides with the index of the class of Ω. Therefore, because of equation (4.6), gauge invariance of the CS theory in S^3 implies that k must be an integer. When the gauge group G is $SU(N)$, a standard embedding of $SU(2)$ into $SU(N)$ is sufficient for showing that, again, k must be an integer [4].

Note that, since $\Gamma_t[\Omega_1 \cdot \Omega_2] = \Gamma_t[\Omega_1] + \Gamma_t[\Omega_2]$, it is sufficient to compute the value of Γ_t on a generator $\Omega^{(1)}$. If we represent S^3 by $R^3 + \{\infty\}$, $\Omega^{(1)}$ can be taken to be (for $G = SU(2)$)

$$\Omega^{(1)}(\vec{x}\,) = \exp\left(i\,\pi\, \widehat{x} \cdot \vec{\sigma}\, \frac{|\vec{x}\,|}{|\vec{x}\,| + r}\right), \tag{4.8}$$

where $\widehat{x} = \vec{x}\,/|\vec{x}\,|$, r is a scale parameter with the dimension of a length, and σ_i are the Pauli matrices. A direct computation shows that, indeed, $\Gamma_t[\Omega^{(1)}] = 1$. The map $\Omega^{(1)}$, given in equation (4.8), also represents a homeomorphism of S^3 into $SU(2) \cong S^3$.

In ordinary perturbation theory in $\mathbb{R}^3$, the boundary conditions of the fields at infinity are fixed. Moreover, only the structure of the infinitesimal gauge transformations is relevant for defining the BRS [19] quantization procedure. In agreement with equation (4.6), the action S_{CS} is invariant under infinitesimal gauge transformations

$$\Delta\, A_\mu(x) = D_\mu \Lambda(x) = \partial_\mu \Lambda(x) + i\, [A_\mu(x), \Lambda(x)]\,. \tag{4.9}$$

General covariance simply follows from the metric independence of S_{CS}. The physical system defined by the action S_{CS} does not contain any kind of ordinary particles. Indeed, the classical energy-momentum tensor derived from the expression (4.4) vanishes identically. This fact reflects precisely the topological nature of the system.

The quantum CS theory formulated by means of the covariant BRS method is defined by the total action S which contains, in addition to S_{CS}, the gauge-fixing and ghost terms. In the Landau gauge, the total action is [20]

$$S = \frac{k}{4\pi} \int d^3x \left\{ \frac{1}{2} \epsilon^{\mu\nu\rho} A^a_\mu \partial_\nu A^a_\rho - \frac{1}{6} \epsilon^{\mu\nu\rho} f^{abc} A^a_\mu A^b_\nu A^c_\rho \right. \\ \left. - B^a \partial^\mu A^a_\mu + \overline{c}^a \partial^\mu \left(\partial_\mu c^a - f^{abc} A^b_\mu c^c \right) \right\}, \tag{4.10}$$

where, as usual, the $\delta_{\mu\nu}$ metric of $\mathbb{R}^3$ has been used to contract the vector indices in the gauge-fixing and ghost Lagrangian terms. The resulting non-vanishing propagators are [20]

$$\langle A^a_\mu(x) A^b_\nu(y) \rangle = \frac{i}{k} \delta^{ab} \epsilon_{\mu\nu\rho} \frac{(x-y)^\rho}{|x-y|^3}, \tag{4.11}$$

$$\langle A^a_\mu(x) B^b(y) \rangle = -\frac{i}{k} \delta^{ab} \frac{(x-y)_\mu}{|x-y|^3}, \tag{4.12}$$

$$\langle c^a(x) \overline{c}^b(y) \rangle = -\frac{i}{k} \delta^{ab} \frac{1}{|x-y|}. \tag{4.13}$$

The expression (4.10) is the action of a renormalizable quantum field theory, thus the correlation functions of the fields are well defined and can be constructed, for example, by means of ordinary perturbation theory.

In the covariant BRS formalism, the quantization condition on k does not play any relevant role in the definition of the quantum CS model in $\mathbb{R}^3$ by means of perturbation theory. Indeed, in perturbation theory $\lambda = \left(\frac{2\pi}{k}\right)$ can be considered a free parameter. Each term of the perturbative expansion in powers of λ respects all the symmetries of the action and, clearly, the quantization condition on k has no influence on it. Of course, in the renormalized theory k can be an integer; what has to be noted is that the construction of the quantum theory itself is not modified by this possibility. This does not mean that the quantization condition on k is not important. As we shall see, the fact that in S^3 k must be an integer is crucial for constructing the surgery operators. Moreover, for the internal consistency of the CS theory defined on a generic three-manifold, it is important that k is an integer.

The action (4.10) is invariant under the BRS transformations [20]

$$\delta A_\mu = \left(D_\mu c \right), \quad \delta B^a = 0, \tag{4.14}$$

$$\delta c^a = \tfrac{1}{2} f^{abc} c^b c^c , \quad \delta \bar{c}^a = B^a . \tag{4.15}$$

The associated BRS charge operator Q is off-shell nilpotent; this property can always be maintained at any order in the loop expansion because there are no gauge anomalies in three dimensions.

Exercise. Prove that there are no gauge anomalies in three dimensions. (Hint: write the most general expression of a potentially anomalous term; i.e., construct a local combination of the fields which has dimension three and is Lorentz invariant. Show then that a local counterterm whose variation cancels this term always exists.)

The definition of anomalous term, modulo the variation of local counterterms, is a good definition because, in this way, the anomaly does not depend on the particular choice of the regularization. In fact, an anomaly has physical consequences and, therefore, cannot depend on the arbitrariness in the choice of the regularization.

The symmetric energy-momentum tensor $\Theta_{\mu\nu}$, derived from the expression (4.10) is [20]

$$\begin{aligned} \Theta_{\mu\nu} = (k/4\pi) \big[& A_{\{\mu} \partial_{\nu\}} B^a - g_{\mu\nu} A^a_\rho \partial^\rho B^a \\ & - \partial_{\{\mu} \bar{c}^a (D_{\nu\}} c)^a + g_{\mu\nu} \partial_\rho \bar{c}^a (D^\rho c)^a \big] , \end{aligned} \tag{4.16}$$

and can be rewritten as

$$\Theta_{\mu\nu} = \big[Q , (k/4\pi) \big(\partial_{\{\mu} \bar{c}^a A^a_{\nu\}} - g_{\mu\nu} \partial^\rho \bar{c}^a A^a_\rho \big) \big] . \tag{4.17}$$

Since Q annihilates the physical states, the mean value of $\Theta_{\mu\nu}$ between physical states vanishes; this means that general covariance is preserved in the physical state space, as it should be.

4.2 One-loop effective action

The renormalizability of any field theory can easily be studied by constructing the effective action Γ. It is well known that Γ is BRS invariant and admits an expansion in powers of the fields. The terms of this expansion which are quadratic in the fields determine the inverse of the dressed propagators, whereas the terms of higher order in powers of the fields define the proper vertices. In this section, we shall consider the CS effective action at the one-loop level. Part of the material presented here is taken from some notes produced in collaboration with M. Mintchev. The divergent one-loop diagrams are shown in Fig. 4.1 in which the continuous and dashed lines represent the gauge and the ghost fields respectively. At this stage, a regularization must be introduced. The particular choice of the

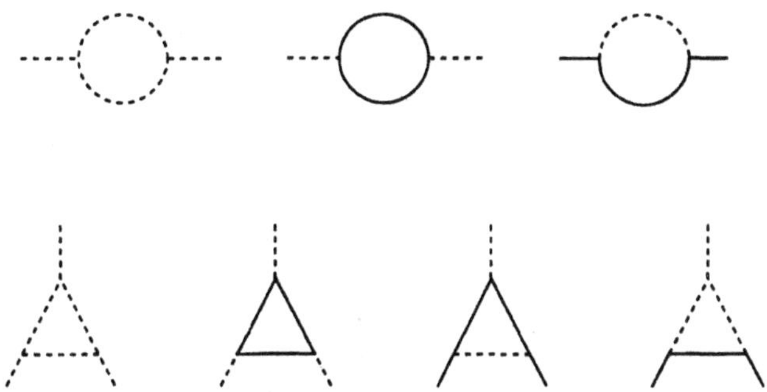

Figure 4.1.

regularization is totally irrelevant. In quantum field theory, physics never depends on the particular choice of the regularization because physics exclusively depends on (and is defined by) the renormalized theory. In what follows, we shall consider two different regularizations and we will show that both of them lead in fact to the same conclusions.

Let us introduce $\Delta(x)$ defined by

$$\Delta(x) = -\frac{1}{4\pi|x|} = \int \frac{d^3p}{(2\pi)^3}\, e^{ipx}\, \frac{1}{p^2} . \tag{4.18}$$

The propagators (4.11)-(4.13) can be rewritten as

$$\langle A^a_\mu(x) A^b_\nu(y) \rangle = i\,\frac{4\pi}{k}\,\delta^{ab}\,\epsilon_{\mu\nu\rho}\,\partial^\rho_x\,\Delta(x-y) , \tag{4.19}$$

$$\langle A^a_\mu(x) B^b(y) \rangle = -i\,\frac{4\pi}{k}\,\delta^{ab}\,\partial^x_\mu\,\Delta(x-y) , \tag{4.20}$$

$$\langle c^a(x)\,\bar{c}^b(y) \rangle = i\,\frac{4\pi}{k}\,\delta^{ab}\,\Delta(x-y) . \tag{4.21}$$

The first regularization procedure, called the method (A), consists of performing the substitution [20]

$$\Delta(x-y) \to \Delta_{reg}(x-y) , \tag{4.22}$$

where

$$\Delta_{reg}(x) = \int \frac{d^3p}{(2\pi)^3}\, e^{ipx} \left(\frac{1}{p^2} - \frac{1}{p^2+M^2} \right) , \tag{4.23}$$

in the expressions (4.19)-(4.21). The mass parameter M plays the role of a cut-off and, in the $M \to \infty$ limit, the original Feynman diagrams are recovered. This regularization prescription is particularly simple and has the great advantage of preserving, in the $M \to \infty$ limit, the local gauge invariance [24]. This fact just simplifies the construction of the effective action.

It should be clear that any other regularization, which eventually breaks (in the limit in which the cut-off is removed) gauge invariance, could also be used. In this case, one has simply to add the appropriate local counterterms which restore the gauge symmetry. One should remember that only anomalous symmetries cannot be restored and gauge invariance is not anomalous.

With the regularization (4.22), all the single diagrams shown in Fig. 4.1 are finite; moreover, the sum of these diagrams is finite [20] in the $M \to \infty$ limit. Consequently, there is no need to introduce divergent counterterms. Let us denote by k_0 the "bare" coupling constant and by k_1 the one-loop coupling constant; we shall denote by k the fully renormalized (or resummed to all orders) coupling constant. In the scheme (A), one has [20]

$$k_1 = k_0 \,. \tag{4.24}$$

As expected, the beta function and the anomalous dimensions of the fields are vanishing [20]. One of the most remarkable properties of the CS theory (in the Landau gauge) is that the two- and the three-point functions do not get radiative corrections [20] at one-loop. Actually, this property holds also at the two-loop level [20].

The second regularization prescription, called method (B), consists of using some kind of Pauli-Villars regularization [25]; in this case also, gauge invariance is preserved. By using method (B), gauge invariance is preserved even with finite values of the cut-offs. It is well known that this extra requirement is not necessary. Indeed, in order to have gauge invariance we only need to preserve gauge invariance in the limit in which the cut-offs are removed. Nevertheless, let us consider regularization (B). In addition to the ordinary Feynman diagrams obtained from the CS action, new one-loop diagrams are introduced with "massive" particles propagating inside the loop. We are interested in the value of the one-loop coupling constant. Because of gauge invariance, it is sufficient to consider the contributions to the two-point function; the corresponding Feynman diagrams (including those of the regulator fields) are shown in Fig. 4.2.

In this scheme, the regularization parameters, associated with the regulator fields, are [25] $\{b_j, M_j\}$ with $j = 1, 2, \ldots, J$ and $\{c_i, m_i\}$ with $i = 1, 2, \ldots, I$ and there are no constraints, coming from gauge invariance, on these parameters [25]. In the $M_j \to \infty$ and $m_i \to \infty$ limit, the sum of all the diagrams of Fig. 4.2 represents the quadratic part in the field expansion of a gauge-invariant quantity, the well known one-loop determinant. Gauge-invariance implies that the sum of all these contributions must be finite in the $M_j \to \infty$ and $m_i \to \infty$ limit. It has been noted in [25] that there is no need to subtract divergent counterterms. Indeed, from equations (3.34), (3.7)-(3.9) and (3.12) of Ref.[25], it follows that, by imposing

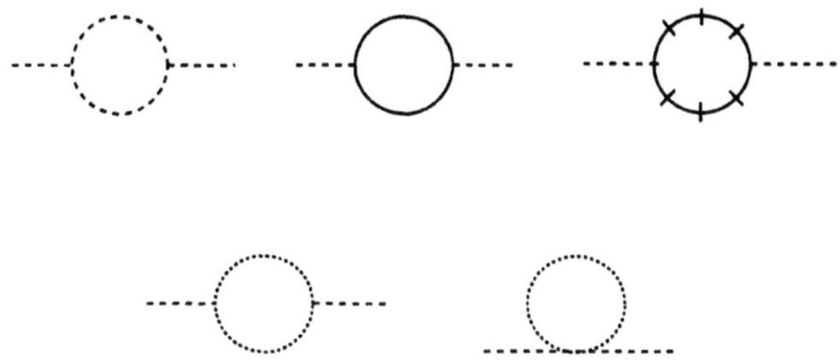

Figure 4.2.

the conditions

$$I = J\,, \quad m_j = M_j\,, \quad c_j = \tfrac{1}{2} b_j\,, \tag{4.25}$$

and

$$\sum_{j=1}^{J} \frac{b_j}{M_j} = 0\,, \qquad \sum_{j=1}^{J} \frac{b_j}{M_j^2} = 0\,, \tag{4.26}$$

the sum of all the contributions is finite in the $M_j \to \infty$ limit and reads

$$\Pi^{\mu\nu}_{ab}(p) = \left(\sum_{j=1}^{J} b_j \right) \left(\frac{4\pi}{k_0} \right)^2 \frac{c_v}{4\pi} \delta_{ab} \, \epsilon^{\mu\rho\nu} \, p_\rho\,, \tag{4.27}$$

where c_v is the value of the quadratic Casimir operator in the adjoint representation. A set of solutions to the conditions (4.25) and (4.26) (with $J = 3$), parametrized by a non-vanishing real constant ξ, is given by

$$\begin{aligned} b_1 &= 2\xi / c_v\,, \quad b_2 = \xi / c_v\,, \quad b_3 = -2\xi / c_v\,, \\ M_1 &= 3M\,, \quad M_2 = \frac{3}{4} M\,, \quad M_3 = M \end{aligned} \tag{4.28}$$

Consequently, the one-loop correction to the two-point function is given by

$$\Pi^{\mu\nu}_{ab}(p) = \xi \left(\frac{4\pi}{k_0} \right)^2 \frac{1}{4\pi} \delta_{ab} \, \epsilon^{\mu\rho\nu} \, p_\rho\,. \tag{4.29}$$

In conclusion, with the gauge invariant regularization (B), the one-loop coupling constant k_1 is given by

$$k_1 = k_0 + \xi\,, \tag{4.30}$$

where ξ is an ***arbitrary*** (non-vanishing) real parameter. The authors of Ref.[25] have chosen $\xi = c_v$; however, this choice is not motivated by gauge invariance nor by any other symmetry of the system.
In conclusion, we have proved that the statement

> *"If a gauge-invariant regularization is used, then the CS one-loop coupling constant k_1 depends on the bare coupling constant k_0 as $k_1 = k_0 + c_v$"*

is false. In fact, we have constructed an explicit counterexample in which, by using a gauge-invariant regularization and by putting for instance $\xi = 4.738/(1+\pi^2)$ in equation (4.30), one has $k_1 = k_0 + 4.738/(1+\pi^2)$. By varying the value of ξ in the real axis, we have also shown that there is in fact an infinite number of counterexamples of the above statement.

When all the cut-offs are removed, different regularizations may differ by finite local counterterms. In fact, the two methods (A) and (B) differ, at one-loop, by a local term which is proportional to the CS action itself. This term clearly preserves all the symmetries of the action (by definition). Therefore, no symmetry argument can be used for fixing the finite multiplicative coefficient in front of this possible finite local term which can be added at one-loop. Of course, the existence of this arbitrariness, which is intrinsic in any regularization, is well known in physics and is discussed in all the textbooks on the renormalization theory and the renormalization group. This apparent ambiguity is completely removed by performing the renormalization.

In both schemes (A) and (B), the normalization of the field A_μ is fixed by the gauge transformation law (4.9). By using the method (A), the three-point proper vertex at one loop turns out to be

$$-\frac{(k_0)}{24\pi}\int d^3x\, \epsilon^{\mu\nu\rho}\, f^{abc}\, A^a_\mu A^b_\nu A^c_\rho\,; \tag{4.31}$$

whereas, in the scheme (B), one has

$$-\frac{(k_0+\xi)}{24\pi}\int d^3x\, \epsilon^{\mu\nu\rho}\, f^{abc}\, A^a_\mu A^b_\nu A^c_\rho\,. \tag{4.32}$$

Let k_1 be the renormalized one-loop coupling constant; that is, the coefficient multiplying the three-point proper vertex. All the results (obtained in the two schemes (A) and (B)), expressed in terms of k_1, coincide. Therefore, the two regularization methods really describe the same physics.

The results obtained in the two schemes seem to be different when they are expressed in terms of the "bare" coupling constant k_0. This difference is completely illusory: the origin of this difference is that the same quantity (the one-loop coupling constant) has different names in the two schemes. If we agree to use the ***same symbol*** k_1 to denote the ***same quantity*** –the one-loop coupling constant– then the results obtained in the two schemes, expressed in terms of k_1, coincide. Whatever regularization method is adopted in the intermediate steps of the computations, the final results depend on the renormalized parameters in a unique and universal way.

It should be noted that, also in the scheme (B), the two- and three-point functions do not receive one-loop radiative corrections. Indeed, the one-loop local term (4.29) just represents a change of the name of the coupling constant. This is not a quantum mechanical induced effect; this is simply a (completely arbitrary) redefinition of the one-loop coupling constant with respect to an (arbitrarily fixed) "bare" coupling constant. The freedom of redefining, at each order of the loop expansion, the dependence of the renormalized coupling constant on the "bare" coupling constant is well known in field theory and has absolutely no physical consequences. The only relevant issue is to fix the value of the renormalized parameters. In our case, we must fix the value of the renormalized coupling constant k. In fact, in any given experiment, you cannot measure the value of the "bare" coupling constant (that is, the value of the coupling constant at the zero-loop level). Also, you cannot measure its value at the one-loop level, nor at the two-loop level and so on; what you see is always the renormalized (or resummed to all orders) value. In other words, physics only depends on the renormalized theory and, in the renormalized theory, all the ambiguities introduced in the intermediate steps of the computations (and originated by the choice of the regularization) are absent.

In conclusion, in any Lagrangian quantum field theory (topological or not) the meaningful parameters are never the "bare" parameters. The concept of "bare" parameters itself has not any intrinsic meaning. Only the physical or renormalized parameters have a definite meaning. The renormalization group equations are based precisely on this fact.

4.3 Higher order results

The main properties of the quantum field theory defined by the action (4.10) have been explored at second order of perturbation theory in [20] and have been analysed in [26,27]; the results obtained in [20] have been confirmed and extended to all orders of perturbation theory. It has been proved [26,27] that the theory is not only renormalizable but actually finite; both the β-function and the anomalous dimensions vanish. This means that there is an exact scale invariance in the correlation functions with A_μ transforming (at the full quantum level) exactly as a vector field. The vanishing of the β-function is in agreement with the general arguments presented in [4]. The vanishing of the anomalous dimension of A_μ is also in perfect agreement with the constraints imposed by the topological character of the theory. Indeed, $A_\mu dx^\mu$ should be a one-form in the quantum theory and the vanishing of the A_μ anomalous dimension guarantees that this is precisely the case.

The Landau gauge is particularly convenient because no external spurious scales (i.e. dimensioned parameters) are introduced in the action. Another interesting feature of the Landau gauge is that the total action (4.10) is invariant also under symmetry transformations whose generators carry a vector index. These generators connect commuting with anticommuting fields and, together with the BRS and anti-BRS operators, form an algebra which closes on-shell on the translations [28]. This allows for a supersymmetric interpretation of the model.

At first sight, the existence of this supersymmetry may appear to be a coincidence without any particular significance. On the contrary, this SUSY has really a deep physical meaning. Remember that the anticommuting variables appearing in (4.10) have been introduced to compensate for the unphysical pure-gauge degrees of freedom contained in the bosonic fields. The existence of a SUSY connecting bosonic and fermionic variables tells us that there are in fact no physical degrees of freedom. This phenomenon is a common feature of several different models [29]. Of course, the conclusion that there are no particles propagating in the CS model is not new; this is just a consequence of general covariance. The advantage of having a supersymmetric realization of general covariance at the quantum level is technically of great help in studying the properties of the theory [26].

Summarising, the non-Abelian CS quantum field theory in $\mathbb{R}^3$, defined by the action (4.10), exists, is finite and its correlation functions admit a well-defined expansion in powers of $\left(\frac{2\pi}{k}\right)$.

The paper [26] also contains the proof that all the symmetries of the action (4.10) are not anomalous and, therefore, can be maintained at the full quantum level. As expected, the only free parameter in the renormalized action turns out to be precisely the renormalized coupling constant k of the theory. The normalization condition on k can be fixed in several equivalent ways. Just as in any Lagrangian field theory, the proper vertices and the dressed propagators satisfy the renormalized Schwinger-Dyson equations following from an action principle.

The action principle is formulated in terms of the renormalized action which can be written in the form shown in equation (4.10) and this fixes the renormalized coupling constant k. Equivalently, consider the effective action Γ of the CS theory. Let Γ have an expansion of the form

$$\begin{aligned}\Gamma &= \int \frac{d^3p}{(2\pi)^3}\, A^a_\mu(p)\, M^{\mu\nu}_{ab}(p)\, A^b_\nu(-p) \\ &\quad + \int \frac{d^3p}{(2\pi)^3}\frac{d^3q}{(2\pi)^3}\, N^{\mu\nu\rho}_{abc}(p,q)\, A^a_\mu(p)\, A^b_\nu(q)\, A^c_\rho(-p-q) \;+\cdots\end{aligned} \tag{4.33}$$

As explained in Ref.[26], the exact form of $M^{\mu\nu}_{ab}(p)$, resummed to all orders of perturbation theory, is given by

$$M^{\mu\nu}_{ab}(p) = i\,\widetilde{\alpha}\,\delta^{ab}\,\epsilon^{\mu\nu\rho}\,p_\rho\,, \tag{4.34}$$

where $\widetilde{\alpha}$ is a real coefficient. Consider now the three-point proper vertex. Because of gauge invariance (the BRS symmetry is not anomalous), the normalization of the A_μ^a fields can always be chosen in such a way that the equality

$$p_\mu \, N_{abc}^{\mu\nu\rho} = -\frac{\widetilde{\alpha}}{3} \, f^{abc} \, \epsilon^{\mu\nu\rho} \, p_\mu \tag{4.35}$$

holds for the exact three-point proper vertex. With this convention on the vector field normalization, gauge transformations are given precisely by equation (4.9) and the coefficient $\widetilde{\alpha}$ appearing in equations (4.34) and (4.35) is related to the renormalized coupling constant k by

$$\widetilde{\alpha} = \frac{k}{8\pi} \, . \tag{4.36}$$

Equations (4.34)-(4.36) uniquely fix the ***normalization condition*** for k.

Equation (4.34) shows that there are no radiative corrections to the two-point function. The coupling constant k enters the canonical commutation relations between the fields as shown in equation (3.2) of [4] or equation (2.6) of [30]. Also, k appears in the Gauss Law constraint as shown in equation (3.4) of [4] or in equation (2.26) of [30]. Finally, the topological argument concerning the quantization of the CS coupling constant, presented in [4], refers precisely to the renormalized coupling constant k.

It is well known that any result obtained in quantum field theory has a definite and intrinsic meaning only when it is expressed in terms of the renormalized parameters. Consequently, all the formulae will be expressed in terms of the renormalized coupling constant k, as in [9,10,11,22,30].

Chapter 5

Observables and perturbation theory

Any expectation value can be expanded in powers of $\lambda = 2\pi/k$. Each term of this expansion can be written as a suitable combination of Feynman diagrams. In this chapter, we shall give the expression of the first three terms of this expansion, contributing to the vacuum expectation value of the Wilson line operators in $\mathbb{R}^3$.

5.1 Wilson line operators

Let us consider first of all the precise definition of a composite Wilson line operator in the non-Abelian case. The classical expression of the Wilson line $W_\rho(C)$, associated with an oriented knot C in $\mathbb{R}^3$ and an irreducible representation ρ of G, is shown in equation (1.22). Let C be parametrized by $\{x^\mu(s)\,;\ 0 \leq s \leq 1\}$; the classical expression (1.22) admits an expansion in powers of the vector field

$$\begin{aligned} W_\rho(C) = \mathrm{Tr}\Bigg\{ \mathbb{1} &+ i \int_0^1 ds\, A\,(s) - \int_0^1 ds \int_0^s dt\, A\,(t)\, A\,(s) \\ &- i \int_0^1 ds \int_0^s dt \int_0^t du\, A\,(u)\, A\,(t)\, A\,(s) + \cdots \Bigg\}, \end{aligned} \tag{5.1}$$

where

$$A\,(s) = \dot{x}^\mu(s)\, A^a_\mu(x(s))\, T^a_{(\rho)}\,, \tag{5.2}$$

and $T^a_{(\rho)}$ are the generators of the group G in the ρ representation. Equation (5.1) is the analogue of equation (3.14) obtained in the Abelian case. The only difference is that now the path-ordering is non-trivial, and therefore the order of the different $A(s)$ factors appearing in (5.1) cannot be modified (modulo cyclic permutations, of course). Just as in the Abelian case, at the quantum level, the product of $A_\mu dx^\mu$ integrals performed on the same path are composite operators and must be defined. In order to preserve general covariance, a framing procedure is adopted [22]. Let us introduce a framing contour C_f parametrized by

$$y^\mu(s) = x^\mu(s) + \varepsilon\, n^\mu(s)\,, \quad (\varepsilon > 0\,, \quad |\vec{n}(s)| = 1\,)\,. \tag{5.3}$$

Each $\dot{x}^\mu A_\mu$ term appearing in equation (5.1) is replaced by

$$A\,(s) \;\to\; [\,A\,(s)\,]_f\,, \tag{5.4}$$

where

$$[\,A\,(s)\,]_f \;=\; \left[\,\dot{x}^\mu(s)\,A^a_\mu(x(s))\,\right]_f\, T^a_{(\rho)} \tag{5.5}$$

has been defined in equation (3.16). The composite Wilson line operator $W_\rho(C)$ associated with the framed knot C is defined as

$$\begin{aligned} W_\rho(C)\Big|_f \;=\; \lim_{\{\varepsilon\}\to 0} \mathrm{Tr}\Bigg\{\; \mathbb{1} \;+ i\int_0^1 ds\;[\,A\,(s)\,]_f \\ -\int_0^1 ds\int_0^s dt\;[\,A\,(t)\,]_f\;[\,A\,(s)\,]_f + \cdots\Bigg\}\,, \end{aligned} \tag{5.6}$$

and the $\{\varepsilon\} \to 0$ limit has to be taken at the end of the computations.

At this point, we can give the precise definition of the vacuum expectation values $\{\,\langle\, W(L)\,\rangle\,\}$ for the non-Abelian CS theory. We shall consider framed links in $\mathbb{R}^3$; each framed component C_i $(i = 1, 2, \ldots, m)$ of L is oriented and is associated with an irreducible representation ρ_i of the gauge group G. The vacuum expectation value

$$\langle\, W(L)\,\rangle \;=\; \frac{\langle 0|\; W_{\rho_1}(C_1)\cdots W_{\rho_m}(C_m)\;|0\,\rangle}{\langle\, 0\,|\,0\,\rangle} \tag{5.7}$$

is computed by means of the action (4.10) and $W_{\rho_i}(C_i)$ are defined according to the framing prescription shown in equation (5.6). The expectation values $\{\,\langle\, W(L)\,\rangle\,\}$ are now well defined and describe an invariant of ambient isotopy for framed and oriented links [22]. In vertical framing, we will denote the expectation value by

$$E(L) \;=\; \langle\, W(L)\,\rangle\,. \tag{5.8}$$

By definition, $E(L)$ is an ***invariant of regular isotopy*** for oriented and coloured link diagrams. Equivalently, $E(L)$ represents an ***invariant of ambient isotopy for coloured, oriented and framed links*** in $\mathbb{R}^3$. At second non-trivial order of perturbation theory [22], it turns out that $\langle\, W(L)\,\rangle$ does not depend on the choice of the particular direction (in the $\{\varepsilon\}$ space) along which the limit (5.6) is performed. I will assume that this is indeed the case for all orders of perturbation theory.

5.2 Perturbative computations

Unlike the Abelian case considered in Chapter 3, the action (4.10) does not define a free theory. Consequently, by making use of the Feynman rules, the complete computation of $E(L)$ in closed form (to all orders in $\lambda = 2\pi/k$) is hard to perform. At each given order of perturbation theory, the contribution to $E(L)$ is well defined, of course. The difficult task consists in giving the numerical evaluation of the relevant Feynman diagrams for the different knots or links. For a given framed knot C in $\mathbb{R}^3$, let us consider the first three terms of the expansion of $E(C)$ in powers of $\lambda = (2\pi/k)$. Let ρ be the representation of G associated with the knot C. We shall denote by $Q(\rho)$ the value of the quadratic Casimir operator of G in the ρ representation

$$T^a_{(\rho)}\, T^a_{(\rho)} = Q(\rho) \cdot \mathbb{1} \,, \tag{5.9}$$

and by c_v the value of Q in the adjoint representation. By using the ordinary Feyman rules, one obtains [22]

$$\begin{aligned} E(C) = (\dim \rho)\Big[1 - i\lambda\, Q(\rho)\, w(C) - \frac{1}{2}\lambda^2\, Q^2(\rho)\, w^2(C) \\ + c_v\, \lambda^2\, Q(\rho)\, b(C)\Big] + O(\lambda^3)\,, \end{aligned} \tag{5.10}$$

where $w(C)$ is the writhe of C and

$$b(C) = b_1(C) + b_2(C)\,, \tag{5.11}$$

with

$$\begin{aligned} b_1(C) = -\frac{1}{32}\oint_C dx^\mu \int^x dy^\nu \int^y dz^\rho\, \epsilon^{\alpha\beta\gamma}\, \epsilon_{\mu\alpha\sigma}\, \epsilon_{\nu\beta\lambda} \\ \epsilon_{\rho\gamma\tau}\, I^{\sigma\lambda\tau}(x, y, z)\,, \end{aligned} \tag{5.12}$$

$$I^{\sigma\lambda\tau}(x, y, z) = \int d^3w\, \frac{(w-x)^\sigma}{|w-x|^3}\, \frac{(w-y)^\lambda}{|w-y|^3}\, \frac{(w-z)^\tau}{|w-z|^3}\,, \tag{5.13}$$

and

$$\begin{aligned} b_2 = \frac{1}{8\pi^2}\oint_C dx^\mu \int^x dy^\nu \int^y dz^\rho \int^z dw^\sigma\, \epsilon_{\sigma\nu\alpha}\, \epsilon_{\rho\mu\beta} \\ \frac{(w-y)^\alpha}{|w-y|^3}\, \frac{(z-x)^\beta}{|z-x|^3}\,. \end{aligned} \tag{5.14}$$

The integral $b(C)$, in spite of its complicated expression, is well defined and represents an ambient isotopy invariant. The value of $b(U)$ for the unknotted knot

U has been computed explicitly from expressions (5.11)-(5.14); the result is [22]

$$b(U) = -\frac{1}{12} . \tag{5.15}$$

For a generic knot C, the value of $b(C)$ is simply related to the second coefficient of the associated Alexander-Conway polynomial [22].

The first three terms of the perturbative expansion of $E(L)$ (associated with a generic link L) have been computed in Ref.[22]. The resulting expression of $E(L)$ has the same structure as $E(C)$ shown in equation (5.10).

Now, let $G = SU(N)$ and let all the components of the links be associated with the fundamental (N-dimensional) representation of $SU(N)$. In this case, the value of $E(L)$ (determined in perturbation theory) satisfies [22] (at order λ^2 included) the skein relation of the $S(L; \alpha, \beta, z)$ polynomial defined in Chapter 2. The particular values of the parameters α, β and z are consistent (at this order in λ) with the expressions

$$\alpha = q^{\frac{N^2-1}{2N}} , \tag{5.16}$$

$$\beta = q^{\frac{1}{2N}} , \tag{5.17}$$

$$z = q^{\frac{1}{2}} - q^{-\frac{1}{2}} , \tag{5.18}$$

where

$$q = \exp\left(-\frac{2\pi i}{k} \right) . \tag{5.19}$$

As we shall see, this is the precise solution to all orders in λ. Namely, for $G = SU(N)$ and for link components in the defining N-representation of $SU(N)$, $E(L)$ satisfies the skein relation of the $S(L)$ polynomial with parameters given in equations (5.16)-(5.19). In the definition (2.11) of $S(L)$, the normalization for the unknot was conventionally fixed to be the identity. The value $E(U_0; N)$ of the unknot U_0 with zero writhe, however, cannot be arbitrarily chosen because it is uniquely fixed by equations (5.7) and (5.8). As explained in Chapter 14, for the unknot U_0 (in the N-representation of $SU(N)$) one has

$$E(U_0; N) = \frac{q^{N/2} - q^{-N/2}}{q^{1/2} - q^{-1/2}} . \tag{5.20}$$

Note that the expression (5.20) admits a well-defined Taylor expansion in powers of $\lambda = (2\pi/k)$. The first three terms of this expansion coincide with the result (5.10) obtained in perturbation theory. Indeed, when the knot C in equation (5.10) is taken to be the unknot with zero writhe, the value of $b(C)$ is shown in equation (5.15); moreover, $c_v = N$ and, for the N-representation of $SU(N)$, the value of the quadratic Casimir operator is $(N^2 - 1)/2N$. In general, each term of the Taylor expansion of the expression (5.20) represents precisely the result that one would obtain in perturbation theory.

Unlike the case of ordinary gauge theories (and of ordinary field theories in general), the series defined by the perturbative expansion in the CS theory is summable. This also is a consequence of general covariance. In fact, there are no Dyson instabilities because, with a vanishing Hamiltonian, the CS theory is well defined and exists for both signs of k. But there are also no non-perturbative effects and there are no renormalons either. Quite remarkably, the perturbative approach to the CS model in $\mathbb{R}^3$ defines the theory exactly. I have no complete explanation for this except for the following plausible argument. In computing the expectation values $\{ \langle W(L) \rangle \}$, one considers a set of operators defined within a limited domain of $\mathbb{R}^3$. This means that, in the functional integral, only the values of the fields inside some finite region of $\mathbb{R}^3$ are important. On the other hand, there are no physical degrees of freedom in the CS theory; in fact, the equations of motion derived from the action S_{CS} are $F_{\mu\nu} = 0$. Essentially, the vector field A_μ is a pure gauge. Now, all the values that a pure-gauge configuration can take inside some finite region of $\mathbb{R}^3$ can be represented by gauge transformations connected to the identity in the whole $\mathbb{R}^3$. Consequently, only the group structure of the infinitesimal gauge transformations (4.9) is relevant. But this is precisely what is carefully taken into account by the covariant BRS formalism and this is why (in $\mathbb{R}^3$) the perturbative approach gives the correct and complete answer.

All the contributions to $E(L)$ found in perturbation theory have the structure shown in equation (5.10); e.g., the gauge group dependence appears in the form of Casimir operators of the group G multiplying rather complicated expressions which represent link invariants. In general, it is difficult to evaluate these expressions for generic links.

In principle, the expression (5.20) could be obtained by means of a direct re-summation of its perturbative expansion. In practice, this possibility is excluded because of obvious technical reasons. The problem then is how to produce the explicit expressions of $\{E(L)\}$. The idea is to combine the general properties of the CS field theory with the results obtained by means of some explicit computations. These computations are performed in a certain simple situation in which the complete answer (to all orders in λ) can easily be achieved. The whole point is that gauge invariance and general covariance impose very stringent constraints on the properties of the expectation values. These constraints, supplemented by some definite numerical results, uniquely fix $\{E(L)\}$.

Chapter 6

Properties of the expectation values

In Chapters 4 and 5, we have seen that $\{\langle W(L)\rangle\}$ have a precise meaning and define an invariant $E(L)$ of regular isotopy for link diagrams. In this chapter, we want to explore the main properties of $E(L)$ which are consequences of its particular field theory description. In doing this, we will not use perturbation theory. We shall concentrate instead on the general aspects which follow from the existence of the quantized CS theory. The results obtained in this section are consequences of the discrete symmetries of the field theory, of general covariance, of the properties of the ambient space, of gauge symmetry and of the particular structure of the Wilson line operators.

6.1 Holonomy matrix

For each classical value of the vector field $A^a_\mu(x)$ and for each irreducible unitary representation ρ of G, the holonomy matrix associated with an oriented open path γ in $\mathbb{R}^3$, connecting the points x and y as shown in Fig. 6.1(a), is given by

$$U_\rho(\gamma; x, y) = \mathrm{P}\exp\left(i \int_\gamma A_\mu(x)\, dx^\mu \right), \tag{6.1}$$

where $A_\mu(x) = A^a_\mu(x)\, T^a_{(\rho)}$ and $\{T^a_{(\rho)}\}$ are the Hermitian generators of G in the ρ representation. By definition, the holonomy satisfies

$$U_\rho(\gamma; x, y) = U_\rho(\gamma_1; x, z) \cdot U_\rho(\gamma_2; z, y), \tag{6.2}$$

where γ_1, γ_2 and z are shown in Fig. 6.1(b). Under gauge transformations

$$A_\mu \mapsto A^\Omega_\mu = \Omega^{-1} A_\mu \Omega - i\, \Omega^{-1} \partial_\mu \Omega , \tag{6.3}$$

$U_\rho(\gamma; x, y)$ transforms as

$$U^\Omega_\rho(\gamma; x, y) = \Omega^{-1}(x)\, U_\rho(\gamma; x, y)\, \Omega(y) . \tag{6.4}$$

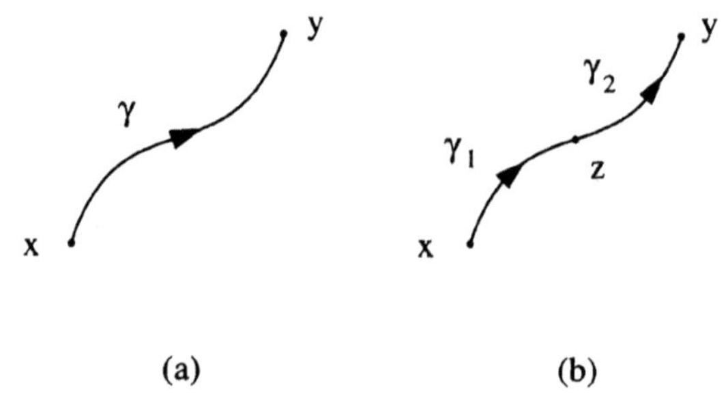

Figure 6.1.

Let γ be parametrized by $\{x^\mu(s)\,;\ 0 \le s \le 1\}$ with $x^\mu(0) = x^\mu$ and $x^\mu(1) = y^\mu$, then the field expansion of $U_\rho(\gamma; x, y)$ takes the form

$$U_\rho(\gamma; x, y) = \mathbb{1} + i \int_0^1 ds\, A(s) - \int_0^1 ds \int_0^s dt\, A(t)\, A(s) \\ - i \int_0^1 ds \int_0^s dt \int_0^t du\, A(u)\, A(t)\, A(s) + \cdots , \tag{6.5}$$

where

$$A(s) = \dot{x}^\mu(s)\, A_\mu(x(s)) . \tag{6.6}$$

Let us denote by γ^{-1} the path obtained by reversing the orientation of γ; from equation (6.1) it follows that

$$U_\rho(\gamma^{-1}; y, x) = U_\rho^{-1}(\gamma; x, y) , \tag{6.7}$$

and since ρ is unitary, one obtains

$$U_\rho^\dagger(\gamma; x, y) = U_\rho^{-1}(\gamma; x, y) . \tag{6.8}$$

When ρ is equivalent to its complex conjugate representation ρ^*,

$$T^a_{(\rho)} = - V\, T^{a\,*}_{(\rho)}\, V^{-1} , \tag{6.9}$$

one immediately finds

$$U_{\rho^*}(\gamma; x, y) = U_\rho^*(\gamma; x, y) = V^{-1}\, U_\rho(\gamma; x, y)\, V . \tag{6.10}$$

The classical Wilson line associated with an oriented closed path C is

$$W_\rho(C) = \mathrm{Tr}\big[\, U_\rho(C\,;x,x)\,\big]\,. \tag{6.11}$$

From equation (6.4), it follows that $W_\rho(C)$ is gauge invariant and, by means of equations (6.7) and (6.8), one gets

$$W^*_\rho(C) = W_\rho(C^{-1})\,. \tag{6.12}$$

In addition to equation (6.12), when ρ is equivalent to ρ^* one has

$$W_{\rho^*}(C) = W^*_\rho(C) = W_\rho(C)\,. \tag{6.13}$$

Let us now consider the Wilson line operators, defined by the framing prescription (5.6), at the quantum level. The first issue which has to be considered is gauge invariance. Unlike the Abelian case, the expression (5.6) is not gauge invariant before the $\{\varepsilon\} \to 0$ limit is performed. So the question is if gauge invariance is restored (in the $\{\varepsilon\} \to 0$ limit) in the expectation values. It is not difficult to see that this is indeed the case. The simplified argument is this: under gauge transformations, the expression (5.6) is not invariant (for finite $\{\varepsilon\}$) and its variation gives a non-vanishing breaking term Δ . The operator Δ contains products of line integrals performed on framed paths; the integrands are made of products of the one-form $A_\mu dx^\mu$ multiplied by numerical functions (expressed in terms of the gauge parameters) which vanish in the $\{\varepsilon\} \to 0$ limit like powers of $(\varepsilon_i - \varepsilon_j)$. On the other hand, in the $\{\varepsilon\} \to 0$ limit the expectation values of the product of $A_\mu dx^\mu$ integrals are in any case finite. Therefore, the insertion of the breaking term Δ inside the expectation values gives a contribution which necessarily vanishes in the $\{\varepsilon\} \to 0$ limit. Gauge invariance is in fact preserved by the framing procedure.

6.2 Discrete symmetries

The consequences of equations (6.12) and (6.13) for the expectation values $\{\langle W(L)\rangle\}$ can easily be derived. First of all it is clear that the algebraic properties of the generators of G are left unchanged by the introduction of the framing. Secondly, the "inversion" operation $C \to C^{-1}$ is well defined for framed knots, see equations (3.11), (3.16) and (5.6); moreover, $\chi(C, C_f) = \chi(C^{-1}, C_f^{-1})$, or $w(C) = w(C^{-1})$. Finally, the effect of complex conjugation in the functional integral is well known. Since the measure is real, the complex conjugation modifies the sign of the total action (4.10). In our case, this is equivalent to changing the sign of the renormalized coupling constant k. As a result, from equation (6.12) it follows that

$$E^*(L; -k) = E(L^{-1}; k). \tag{6.14}$$

Consider now the case in which a particular representation ρ_i, associated to the component C_i of L, is equivalent to its complex conjugate. Then, equations (6.12) and (6.13) imply that $E(L)$ is invariant under $C_i \to C_i^{-1}$. Therefore, when all the representations ρ_i associated to the components C_i of L are real, $E(L)$ is invariant under the "inversion" operation $C_i \to C_i^{-1}$ of any component of L. In this case, we will say that $E(L)$ represents a regular isotopy invariant for unoriented link diagrams.

The particular structure of the action (4.10) puts a further constraint on the dependence of $E(L)$ on k. As shown in [22], it is always possible to assign definite parity properties to the fields in such a way that a modification of the orientation of the three-manifold is equivalent to a change in the sign of the renormalized coupling constant k. Because of this, if $\widetilde{L}$ denotes the mirror image of the framed link L, one obtains [22]

$$E(\widetilde{L}; -k) = E(L; k). \tag{6.15}$$

From the definitions (5.7) and (5.8) it also follows that when $L = L_1 \cup L_2$, e.g. when the link L is the distant (disjoint) union of L_1 and L_2, one has

$$E(L_1 \cup L_2) = E(L_1)\, E(L_2). \tag{6.16}$$

Equation (6.16) is just a consequence of general covariance and of the uniqueness of the vacuum.

6.3 Satellite formulae

The particular structure of the Wilson line operator combined with the properties of the ambient space permit the derivation of very useful satellite formulae. The remaining part of this chapter is devoted to this subject. Let us start by considering an oriented framed link which is the distant union of two unknotted knots U_1 and U_2. Let ρ_1 and ρ_2 be the irreducible representations of G associated with U_1 and U_2 and let us consider the case in which the framings U_{1f} and U_{2f} of U_1 and U_2 are such that

$$\chi(U_1, U_{1f}) = 0 = \chi(U_2, U_{2f}). \tag{6.17}$$

From equation (6.16), we know that

$$\langle W_{\rho_1}(U_1)\, W_{\rho_2}(U_2) \rangle = \langle W_{\rho_1}(U_1) \rangle \langle W_{\rho_2}(U_2) \rangle. \tag{6.18}$$

One can imagine that equation (6.18) has been obtained in the following way. Because of general covariance, $\langle W_{\rho_1}(U_1)\, W_{\rho_2}(U_2) \rangle$ does not depend on how "far" U_2 is from U_1. In particular, one can send U_2 to infinity and then, by the unicity of the vacuum, equation (6.18) follows.

Now what happens in the opposite limit when U_2 gets closer and closer to U_1 and, at the end, coincides with U_1 ? Of course, this limit has to be carefully defined; in particular, one must take into account the two framings and the two path-orderings.

The limit in which U_2 coincides with U_1 is defined in the following way. First of all, U_1 and U_2 can always be represented by two planar circles which are parallel to each other, as shown in Fig. 6.2. Moreover, the circles U_1 and U_2 are placed in $\mathbb{R}^3$ in such a way that they define the two bases of a cylinder with eight δ ($\delta > 0$). We are interested in the $\delta \to 0$ limit, of course, and we already know that this limit exists because $\langle W_{\rho_1}(U_1)\, W_{\rho_2}(U_2) \rangle$ does not depend on δ.

Let U_1 be parametrized by $\{x^\mu(s)\,,\ 0 \leq s \leq 1\}$. Then one can parametrize U_2 by

$$y^\mu(s) \;=\; x^\mu(s) + \delta\, v^\mu\,, \qquad |\vec{v}\,| = 1\,, \tag{6.19}$$

where $\vec{d} = \delta\, \vec{v}$ is the constant vector which connects each point x of U_1 with its corresponding point y in U_2, as shown in Fig. 6.2.

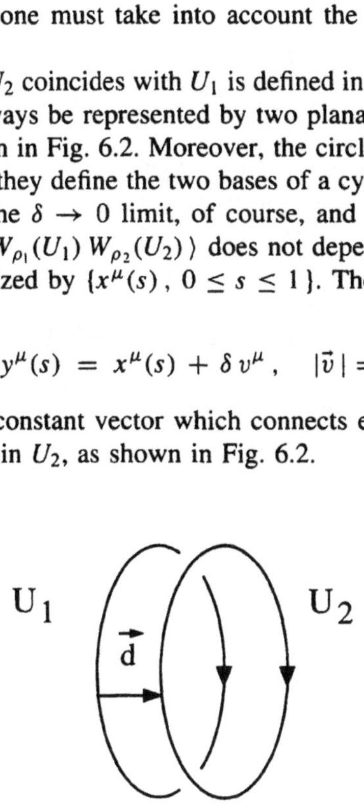

Figure 6.2.

With the parametrization (6.19) of U_2, the product of the two path-orderings on U_1 and U_2 can obviously be reduced to the single ordering defined by the s parameter. In fact, the path-ordering between the two one-forms

$$A^a_\mu(x)\, T^a_{(\rho_1)}\, dx^\mu \quad \text{and} \quad A^b_\nu(y)\, T^b_{(\rho_2)}\, dy^\nu\,,$$

which have to be integrated along U_1 and U_2 respectively, is totally ineffective because $\{T^a_{(\rho_1)}\}$ and $\{T^b_{(\rho_2)}\}$ act on different spaces and therefore commute.

Let us consider the framings U_{1f} and U_{2f}. Since we are in the conditions specified by equation (6.17), we can take the unit vector $\vec{v}$, appearing in equation (6.19), as the vector defining both the framings U_{1f} and U_{2f}. For instance, one can choose U_{1f} specified by

$$z^\mu(s) \;=\; x^\mu(s) + \delta_1 \, v^\mu \, , \quad \delta_1 > 0 \, , \tag{6.20}$$

and U_{2f} specified by

$$w^\mu(s) \;=\; y^\mu(s) + \delta_2 \, v^\mu \, , \quad \delta_2 > 0 \, . \tag{6.21}$$

According to the framing prescription (3.15)-(3.17) and (5.6), the Wilson operators $W_{\rho_1}(U_1)$ and $W_{\rho_2}(U_2)$ are characterized by the sets $\{\delta_1\}$ and $\{\delta_2\}$ and we are interested in the limit in which $\{\delta_1\} \to 0$, $\{\delta_2\} \to 0$ and then $\delta \to 0$. Here is the crucial point: this limit is none other than the limit (performed along a particular direction in the $\{\varepsilon\}$ space) entering the framing procedure which defines a single Wilson line operator W' in U_1 with framing U_{1f}, see Fig. 6.3. Therefore,

$$\lim_{U_2 \to U_1} \; W_{\rho_1}(U_1) \, W_{\rho_2}(U_2) \;=\; W'(U_1) \; . \tag{6.22}$$

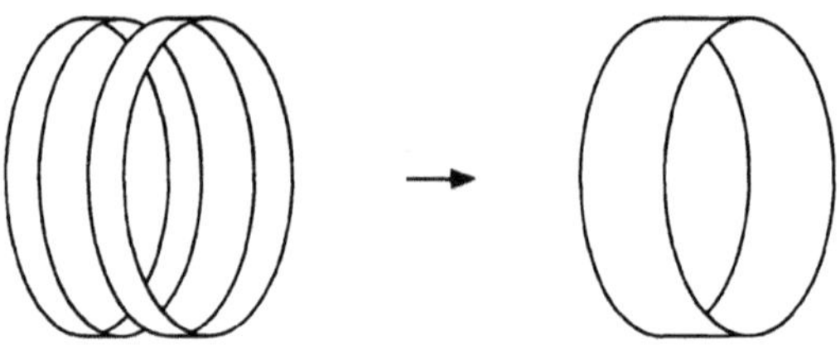

Figure 6.3.

The G representation associated to W' can immediately be obtained; since the generators are

$$T^a_{(\rho_1)} + T^a_{(\rho_2)} \;=\; T^a_{(\rho_1)} \otimes \mathbb{1} \,+\, \mathbb{1} \otimes T^a_{(\rho_2)} \, , \tag{6.23}$$

W' is defined in the $\rho = \rho_1 \otimes \rho_2$ representation of G. This is in agreement with the fact that

$$\mathrm{Tr}_{\rho_1} \, \mathrm{Tr}_{\rho_2} \;=\; \mathrm{Tr}_{\rho_1 \otimes \rho_2} \, , \tag{6.24}$$

In conclusion,

$$W' \;=\; W_{\rho_1 \otimes \rho_2}(U_1) \, , \tag{6.25}$$

with framing U_{1f} and then

$$\lim_{U_2 \to U_1} W_{\rho_1}(U_1)\, W_{\rho_2}(U_2) \;=\; W_{\rho_1 \otimes \rho_2}(U_1) \;. \tag{6.26}$$

The "fusion" property (6.26) of the Wilson line operators is illustrated in Fig. 6.4.

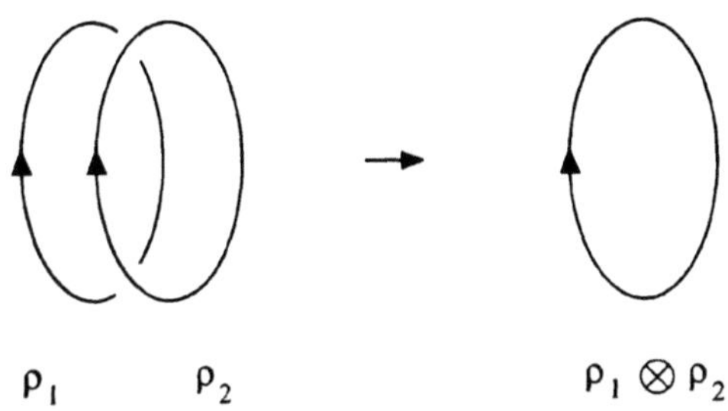

Figure 6.4.

Let the tensor product $\rho_1 \otimes \rho_2$ have the decomposition

$$\rho_1 \otimes \rho_2 \;=\; \oplus_t\, \rho(t) \tag{6.27}$$

in terms of the irreducible representations $\{\rho(t)\}$ of the group G. It follows then that

$$\langle\, W_{\rho_1}(U_1)\, W_{\rho_2}(U_2)\,\rangle \;=\; \langle\, W_{\rho_1 \otimes \rho_2}(U_1)\,\rangle \;=\; \sum_t \langle\, W_{\rho(t)}(U_1)\,\rangle \,. \tag{6.28}$$

Equation (6.28) combined with equation (6.18) finally give

$$\langle\, W_{\rho_1}(U_1)\,\rangle\, \langle\, W_{\rho_2}(U_2)\,\rangle \;=\; \sum_t \langle\, W_{\rho(t)}(U_1)\,\rangle \;. \tag{6.29}$$

Equation (6.29) gives origin to highly non-trivial constraints for the expectation values of the unknots. For the moment, of course, the only exact expression that we know for the unknots concerns the trivial ($\rho = 0$) representation of the group G: in this case $\langle\, W_{(\rho=0)}(U)\,\rangle = 1$. Clearly, for any knot C it is true that $\langle\, W_{(\rho=0)}(C)\,\rangle = 1$.

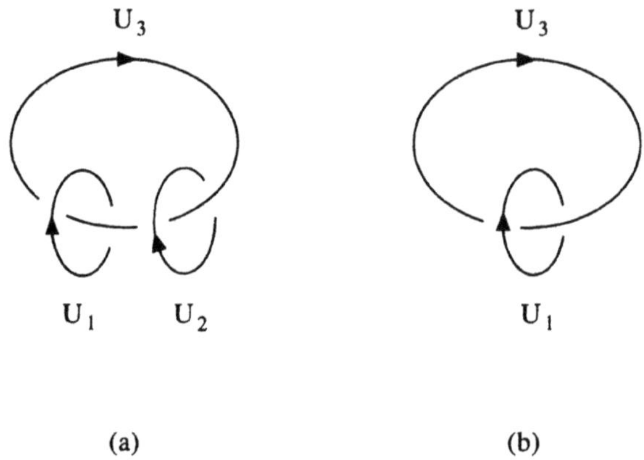

Figure 6.5.

Now equation (6.28) can easily be generalized. For example, suppose that U_1 and U_2 are two components of the link shown in Fig. 6.5(a) with framings specified by equation (6.17).
The whole argument used to derive equation (6.28) can be repeated without any modification and the result obtained in this case is

$$\langle\, W_{\rho_1}(U_1)\, W_{\rho_2}(U_2)\, W_{\rho_3}(U_3)\,\rangle \;=\; \sum_t \langle\, W_{\rho(t)}(U_1)\, W_{\rho_3}(U_3)\,\rangle\,, \tag{6.30}$$

where the two-component link appearing in the sum of the RHS of equation (6.30) is shown in Fig. 6.5(b); the irreducible representations $\{\rho(t)\}$ are those defined in equation (6.27).

Equations (6.28) and (6.30) are particular examples of satellite formulae. Let us briefly recall the definition of satellites given in Chapter 2. Consider a framed link L in $\mathbb{R}^3$ with m components $\{C_1, \ldots, C_m\}$. Starting from L one can construct another link L' called a satellite of L; L' is obtained by replacing one of the components, say C_i, of L with $h^\circ(P)$,

$$C_i \;\Longrightarrow\; h^\circ(P)\,, \tag{6.31}$$

where $h^\circ(P)$ is the image of the pattern link P under the action of the homeomorphism h° defined in Chapter 2. Remember that h° is uniquely determined (up to ambient isotopy) by the framing of C_i. We are interested now in the case in which the pattern link is the two component link B shown in Fig. 2.11. Since we are considering oriented links, the orientations of the two components of B are chosen to be collinear, as shown in Fig. 6.6. Moreover, the two components of

$$C^{(2)} = h^{\diamond}(B) \tag{6.32}$$

are oriented like C_i. The link B is framed; each component of B has preferred framing, as shown in Fig. 6.6 (vertical framing convention). Of course, in looking for satellite formulae for $\{\langle W(L)\rangle\}$, one has to take into account the different representations of G associated with the Wilson line operators and their framings too.

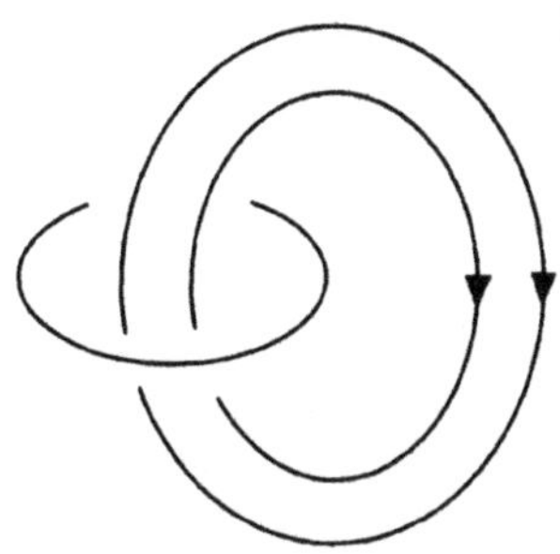

Figure 6.6. Oriented and framed pattern link B

The same construction, which has been used before for deriving equations (6.26) and (6.28), allows one to obtain a very important satellite formula based of the pattern link B shown in Fig. 6.6.

Consider a generic oriented framed link in $\mathbb{R}^3$ and let C be one of its components. We shall indicate by $\{W(L; C, C_f, \rho)\}$ the Wilson line operators associated with L when C_f is the framing of C and the irreducible representation ρ of G is associated with C; the framings and the representations of all the remaining components of L are kept fixed. Let us now concentrate on a particular satellite L' of L obtained by substituting the the two-component link $C^{(2)} = h^{\diamond}(B)$ for the component C. As explained in Chapter 2, with our choice of the homeomorphism $h^{\diamond}$ the linking number of the two components K_1 and K_2 of $C^{(2)}$ is

$$\chi(K_1, K_2) = \chi(C, C_f)\,. \tag{6.33}$$

Moreover, the framings K_{1f} and K_{2f} are given by

$$\chi(K_1, K_{1f}) = \chi(K_2, K_{2f}) = \chi(C, C_f)\,. \tag{6.34}$$

Let ρ_1 and ρ_2 be the irreducible representations associated with K_1 and K_2. We shall indicate by $W(L'; C^{(2)}, \rho_1, \rho_2)$ the Wilson line operator associated to the satellite L' of L obtained by substituting $C^{(2)} = h^{\diamond}(B)$ for C. As shown by equations (6.33) and (6.34), there is a perfect matching between the framing structure of the two components K_1 and K_2 and the linking structure of K_1 and K_2. This means that the analogue of the fusion mechanism shown in Fig. 6.3 takes place. As a result, a straightforward extension of the argument used to derive equation (6.26) shows that

$$\langle W(L'; C^{(2)}, \rho_1, \rho_2) \rangle = \sum_t \langle W(L; C, C_f, \rho(t)) \rangle , \tag{6.35}$$

where the sum is performed on the irreducible representations of G appearing in the decomposition (6.27).

Equation (6.35) represents one of the most important results in the CS theory. The framing structure of $W(L)$ is crucial for deriving this equation which, in turn, essentially permits us to solve the CS theory.

Equation (6.35) can also be understood in the following way. The fusion property of the Wilson line operators, shown in Fig. 6.4, can be applied directly to the pattern link B shown in Fig. 6.6. In this way, the two-component pattern link B can be replaced by the single oriented and framed knot K, shown in Fig. 6.7, and we have a sum over the irreducible representations (6.27) contained in the tensor product $\rho_1 \otimes \rho_2$.

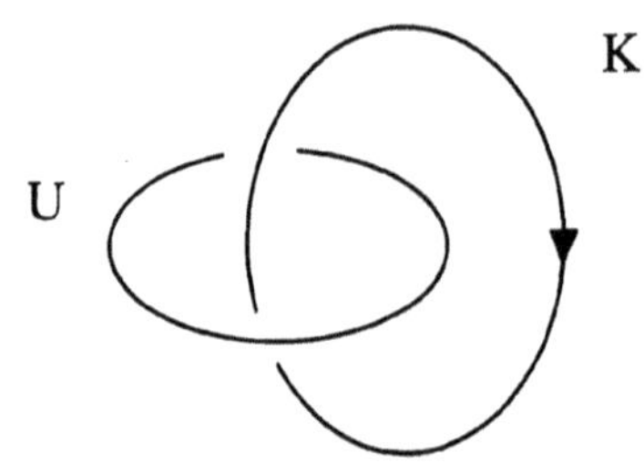

Figure 6.7.

The image of K under the homeomorphism $h^{\diamond}$ is precisely the knot C with framing C_f. Therefore, equation (6.35) is obtained.

The exact solution of the Abelian CS theory is shown in equations (3.19) and (3.22). In this particular case, one can check all the results obtained in this chapter. For an Abelian gauge group, the decomposition (6.27) takes the simple form

$$e_1 \otimes e_2 = (e_1 + e_2) , \tag{6.36}$$

where e_1 and e_2 are the charges of the Wilson lines. It is easy to see that equations (6.14)-(6.16) and equations (6.28), (6.30) and (6.35) are indeed satisfied.

In the non-Abelian case, the explicit expression of $\langle W(L) \rangle$ has been computed in [22] at second non-trivial order of perturbation theory. As expected, all the general properties of $\{ \langle W(L) \rangle \}$ discussed in this chapter are indeed satisfied (at order λ^2 included). Note that in the CS theory the naive use of a reflection positivity argument may lead to wrong conclusions (in particular for the unknot). Finding the correct version of reflection positivity in the quantum CS theory is an interesting open problem.

Summarizing, the link invariants obtained in the CS theory are of a very particular kind. First of all, the dependence of $E(L)$ on the renormalized coupling constant k is constrained by equations (6.14) and (6.15). Secondly, because of general covariance, the factorization property (6.16) must be satisfied. Finally, satellite formulae like equations (6.28) and (6.30) and, more generally, formulae of the form shown in equation (6.35) must hold. As we shall see, all these conditions impose such strong constraints on $E(L)$ that, by adding some extra numerical inputs to our data, the universal link polynomial $E(L)$ turns out to be uniquely determined.

The problem now is how to incorporate all this information in some definite scheme permitting the reconstruction of $E(L)$. One possibility is to operate directly on the link diagrams by using a set of simple rules derived from the three-dimensional quantum CS theory. For this purpose, it is convenient to introduce a more detailed and, in a certain sense, systematic description of link diagrams. This can be done of course by making use of the Artin braid group.

In Chapters 8 and 9, we shall consider the braid group and its connections with knot theory. We shall briefly review the known methods of constructing link invariants by means of certain braid group representations. These algebraic constructions are not based on the field theory approach; nevertheless, they are particularly interesting mainly for two reasons. First, we will learn what the basic ingredients for constructing link invariants are. In particular, the general properties of $E(L)$, derived in this chapter, do not identify $E(L)$ uniquely. We wish to understand what new information must be added in order to reconstruct $E(L)$. Secondly, the precise relation between the universal link polynomial $E(L)$ and the known polynomials will be produced.

Before considering the braid group approach to knot theory, let us make some general remarks about the knot observables constructed in the CS field theory.

Chapter 7

Ordering fermions and knot observables

The one-form $A_\mu dx^\mu$ represents the set of the fundamental fields of the CS theory, and the Wilson line operators $\{ W_\rho(C) \}$, appearing in equation (1.22), are the natural observables associated with knots. One can also construct new gauge-invariant observables associated with knots; some explicit examples are considered in this chapter. We shall analyse first their definition at the classical level and then we shall consider their extension at the quantum level. We will show that $\{ W_\rho(C) \}$ represents a complete set of knot observables.

7.1 Ordering fermions

In this section, some algebraic properties of the classical holonomy (6.1) are studied and we will show how to represent the holonomy by means of a path-integral over some auxiliary variables called the "ordering fermions".

Consider an oriented path γ in $\mathbb{R}^3$, as shown in Fig. 7.1, with some given parametrization $x^\mu(s)$ where s is a real variable. Given a classical $A^a_\mu(x)$ configuration, the holonomy associated with the open path described by $x^\mu(s)$, with $s_1 \leq s \leq s_2$, is

$$U(s_1, s_2) = \mathrm{P}\exp\left(i \int_{s_1}^{s_2} ds\, \dot{x}^\mu(s)\, A_\mu(x(s)) \right), \tag{7.1}$$

where $A_\mu = A^a_\mu T^a$ and $\{ T^a \}$ are the generators of the group G in some irreducible representation. Let $s_1 < s_2$, for fixed s_2 one has

$$\frac{\partial}{\partial s_1} U(s_1, s_2) = -i\, \dot{x}^\mu(s_1)\, A_\mu(x(s_1))\, U(s_1, s_2), \tag{7.2}$$

and, similarly, for fixed s_1 one gets

$$\frac{\partial}{\partial s_2} U(s_1, s_2) = i\, U(s_1, s_2)\, \dot{x}^\mu(s_2)\, A_\mu(x(s_2)). \tag{7.3}$$

Equation (7.2) can be written in the form (for $s_1 < s_2$)

$$\left[\frac{\partial}{\partial s_1} + i\,\dot{x}^\mu(s_1)\,A_\mu(x(s_1))\right] U(s_1, s_2) = 0\,, \tag{7.4}$$

and, together with the boundary condition

$$U(s_1, s_2)\Big|_{s_1=s_2} = \mathbb{1}\,, \tag{7.5}$$

determines the value of the holonomy $U(s_1, s_2)$. The operator appearing in equation (7.4) can be written as

$$\frac{\partial}{\partial s} + i\,\dot{x}^\mu(s)\,A_\mu(x(s)) = \dot{x}^\mu(s)\left[\partial_\mu + i\,A_\mu\right], \tag{7.6}$$

and represents the projection of the covariant derivative on the $\dot{x}^\mu(s)$ direction.

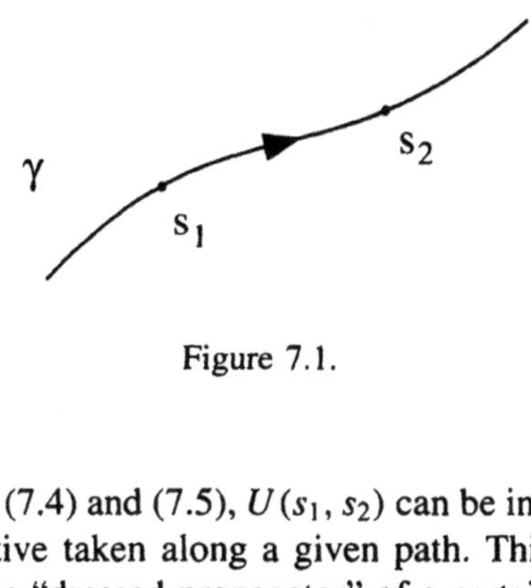

Figure 7.1.

Because of equations (7.4) and (7.5), $U(s_1, s_2)$ can be interpreted as the "inverse" of the covariant derivative taken along a given path. This suggests that $U(s_1, s_2)$ can be understood as the "dressed propagator" of a certain set of one-dimensional fields whose quadratic action is constructed with the operator (7.6).

In particular, consider the action [31]

$$S_F = \int ds\ \overline{\psi}(s)\left(-i\,\frac{d}{ds} + A(s)\right)\psi(s)\,, \tag{7.7}$$

where the field configuration

$$A(s) = \dot{x}^\mu(s)\,A_\mu(x(s))\,, \tag{7.8}$$

is fixed. The different components ψ_α of the anticommuting field ψ transform according to some unitary irreducible representation of G, with generators $\{\,T^a\,\}$, so that the action (7.7) is gauge invariant. The propagator

$$\langle \psi_\alpha(s)\,\overline{\psi}_\beta(s') \rangle_0 \;=\; i\,\Delta_{\alpha\beta}(s-s') \tag{7.9}$$

must satisfy

$$-i\,\frac{d}{ds}\,\Delta_{\alpha\beta}(s-s') \;=\; \delta_{\alpha\beta}\,\delta(s-s')\,. \tag{7.10}$$

Equation (7.10) admits several inequivalent solutions; if we want to recover the holonomy matrix, the good choice for $\Delta_{\alpha\beta}(s-s')$ is

$$i\,\Delta_{\alpha\beta}(s-s') \;=\; \delta_{\alpha\beta}\,\theta(s'-s)\,. \tag{7.11}$$

The fields ψ do not describe real particles but are just auxiliary fields in terms of which the holonomy can be constructed. Therefore, the exotic choice of the propagator (7.11) is perfectly legitimate; the fields ψ with propagator (7.11) are called ordering fermions. Consider now, for $s' > s$, the "dressed propagator"

$$\langle \psi_\alpha(s)\,\overline{\psi}_\beta(s') \rangle \;=\; \langle \psi_\alpha(s)\,\overline{\psi}_\beta(s')\,e^{i\int \overline{\psi}A\psi} \rangle_0\,. \tag{7.12}$$

A simple analysis of the perturbative expansion of the expression (7.12) shows that

$$\langle \psi_\alpha(s)\,\overline{\psi}_\beta(s') \rangle \;=\; \left[\,U(s,s')\,\right]_{\alpha\beta}\,. \tag{7.13}$$

The method of the ordering fermions can be used to represent the expectation values of Wilson line operators by means of sources (defined in terms of ψ and $\overline{\psi}$) coupled with the gauge fields. Another reason for introducing the functional (7.7) is that, by modifying the form of the propagator (7.11), one can easily construct new gauge-invariant observables associated with knots.

7.2 Antiperiodic boundary conditions

For any given oriented knot C in $\mathbb{R}^3$ with parametrization $x^\mu(s)$, where $0 \le s \le 1$ (and $x^\mu(1) = x^\mu(0) = x_0^\mu$), and for any irreducible representation ρ of the gauge group G, let us introduce the functional

$$S_C \;=\; \int_0^1 ds\;\overline{\psi}(s)\left(-i\,\frac{d}{ds} + A(s)\right)\psi(s)\,. \tag{7.14}$$

We shall consider S_C as the action for the anticommuting fields ψ and $\overline{\psi}$, whereas A_μ is a fixed classical configuration. The path-integral over ψ and $\overline{\psi}$, with the action (7.14), gives the partition function $Z[A]$ and the A_μ-dependent part of $Z[A]$ is

$$\mathcal{O}(C;A) = \frac{Z[A]}{Z[0]} = \frac{\mathrm{Det}\left(-i\,\frac{d}{ds} + A(s)\right)}{\mathrm{Det}\left(-i\,\frac{d}{ds}\right)} . \tag{7.15}$$

By definition, $\mathcal{O}(C;A)$ is gauge invariant because S_C also has this property; moreover, $\mathcal{O}(C;A)$ does not depend on the chosen parametrization of the knot C because S_C does not depend on it. In conclusion, $\mathcal{O}(C;A)$ represents a gauge invariant observable associated with the knot C.

When the propagator of ψ and $\overline{\psi}$ is taken to be equal to the expression (7.11), $\mathcal{O}(C;A)$ is trivial: it does not depend on A_μ at all. However, with a different choice of the propagator, the observable $\mathcal{O}(C;A)$ is non-trivial.

In order to give a precise meaning to the expression (7.15), one must first specify the boundary conditions of the anticommuting fields at $s = 0$ and $s = 1$. With periodic boundary conditions, the "free" determinant, appearing in the denominator of the expression (7.15), has a zero mode and, formally, the expression (7.15) is ill-defined. Therefore, some procedure for removing the zero mode must be introduced. In general, such a procedure gives origin to some problems with the reparametrization invariance and, for our purposes, it is not necessary to discuss this situation here. Therefore, we shall consider the more simple case of antiperiodic boundary conditions.

Since the system defined by the action (7.14) is one-dimensional, the explicit expression of $\mathcal{O}(C;A)$ can easily be produced. The determinants entering equation (7.15) can be computed by using, for instance, the method of the Floquet indices [32].

First, let us compute the ratio

$$R = \frac{\mathrm{Det}\left(-i\,\frac{d}{ds} + \sigma\right)}{\mathrm{Det}\left(-i\,\frac{d}{ds}\right)} , \tag{7.16}$$

where σ is a real constant. The eigenvalue equation

$$\left(-i\,\frac{d}{ds} + \sigma\right) f_n(s) = \lambda_n\, f_n(s) , \tag{7.17}$$

with antiperiodic boundary conditions

$$f_n(s+1) = -\, f_n(s) , \tag{7.18}$$

admits the solutions

$$\begin{aligned} f_n(s) &= \exp\left(i\,2\pi\,(n+\tfrac{1}{2})\,s\right) \\ \lambda_n &= 2\pi\,(n+\tfrac{1}{2}) + \sigma . \end{aligned} \tag{7.19}$$

Consequently,

$$\mathrm{Det}\left(-i\frac{d}{ds}+\sigma\right) = \prod_{n\in Z}\left(2\pi\,(n+\tfrac{1}{2})+\sigma\right), \tag{7.20}$$

and then [33]

$$\begin{aligned} R &= \prod_{n=0}^{\infty}\left[1-\frac{4(\sigma/2)^2}{\pi^2(2n+1)^2}\right] \\ &= \cos(\sigma/2)\,. \end{aligned} \tag{7.21}$$

The next step consists of computing

$$R' = \frac{\mathrm{Det}\left(-i\frac{d}{ds}+\sigma(s)\right)}{\mathrm{Det}\left(-i\frac{d}{ds}\right)}, \tag{7.22}$$

where $\sigma(s)$ is a non-trivial real function of $s\in[0,1]$ with $\sigma(1)=\sigma(0)$. Suppose that the solution $G_0(s)$ of the equation

$$\left(-i\frac{d}{ds}+\sigma(s)\right)\,G_0(s) = 0 \tag{7.23}$$

is known and satisfies

$$G_0(1) = e^{-i\alpha}\,G_0(0) \tag{7.24}$$

for some (index) α. Then, the solution of the eigenvalue equation

$$\left(-i\frac{d}{ds}+\sigma(s)\right)\,f_n(s) = \lambda_n\,f_n(s)\,, \tag{7.25}$$

with antiperiodic boundary conditions

$$f_n(s+1) = -\,f_n(s)\,, \tag{7.26}$$

is given by

$$\begin{aligned} f_n(s) &= G_0(s)\,\exp\left\{i\left[2\pi\,(n+\tfrac{1}{2})+\alpha\right]s\right\} \\ \lambda_n &= 2\pi\,(n+\tfrac{1}{2})+\alpha\,. \end{aligned} \tag{7.27}$$

At this point, the same procedure which has been used for deriving equation (7.21) gives

$$R' = \cos(\alpha/2)\,. \tag{7.28}$$

Let us now consider the observable $\mathcal{O}(C;A)$ defined in equation (7.15); in this case, the matrix $A(s)$ is the analogue of $\sigma(s)$ because $A(1)=A(0)$. In order to find the Floquet indices (there are several indices now because we have a multidimensional space) we must find the solution $\psi_0(s)$ to the equation

$$\left(-i\frac{d}{ds}+A(s)\right)\psi_0(s) = 0\,, \tag{7.29}$$

where $\psi_0(s)$ is a vector of the same dimension $D(\rho)$ as the considered representation ρ of G. From equation (7.4), it follows that

$$\psi_0(s) = U(s,1)\,\psi_0(1)\,, \tag{7.30}$$

where $U(s,1)$ is the holonomy (7.1) and $\psi_0(1)$ is a vector which does not depend on the variable s and belongs to a $D(\rho)$-dimensional space. Since $U(1,1) = \mathbb{1}$, the analogue of equation (7.24) is

$$\psi_0(1) = U^{-1}(0,1)\,\psi_0(0)\,, \tag{7.31}$$

where $U(0,1)$ is the holonomy associated with the oriented knot C and computed with respect to some reference point of C. By varying $\psi_0(1)$ in the complete set of the eigenvectors of $U(0,1)$, we find that the Floquet indices,

$$\{\alpha_1, \alpha_2, \ldots\}\,, \tag{7.32}$$

are determined by the eigenvalues,

$$\{e^{i\alpha_1}, e^{i\alpha_2}, \ldots\}\,, \tag{7.33}$$

of the holonomy matrix $U(0,1)$. Therefore,

$$\begin{aligned}\mathcal{O}(C;A) &= \prod_i \cos(\alpha_i/2)\\ &= (2)^{-D(\rho)}\,\mathrm{Det}^{-1/2}\,(U(0,1))\;\mathrm{Det}(\mathbb{1} + U(0,1))\,.\end{aligned} \tag{7.34}$$

The expression (7.34) is manifestly gauge-invariant (as it should be) and therefore $\mathcal{O}(C;A)$ does not depend on the particular choice of the reference point x_0 in C, used for defining $U(0,1)$.

Clearly, several gauge invariant observables associated with knots can be constructed at the classical level. Two main issues remain to be understood: the first is how to extend the definition of these observables at the quantum level and the second is how to find a complete set of these observables. The natural solution to these two problems is presented in the next section.

7.3 Knot observables

In the CS theory, the basic field functional associated with an oriented knot C is the holonomy matrix

$$U_\rho(C;0,1) = U_\rho(C;x_0,x_0)\,, \tag{7.35}$$

defined in terms of a given parametrization $x^\mu(s)$ of C, with $0 \le s \le 1$ and $x(1) = x(0) = x_0 \in C$, and in some representation ρ of the gauge group. At the classical level, $U_\rho(C;0,1)$ is reparametrization invariant but depends on the choice of the reference point $x_0 \in C$. Moreover, $U_\rho(C;0,1)$ is not gauge invariant and transforms as

$$U_\rho(C;x_0,x_0) \;\to\; \Omega_\rho^{-1}(x_0)\,U_\rho(C;x_0,x_0)\,\Omega_\rho(x_0)\,. \tag{7.36}$$

For any fixed vector field configuration $A_\mu(x)$, the holonomy matrix is an element of the gauge group G in some given representation. Since the representation can be arbitrarily chosen, the holonomy can be understood, in more abstract terms, as an element $g(C;x_0)$ of G associated with a given knot C and a reference point x_0 on it. Any function $F(g)$, which is invariant under conjugation of the element $g \in G$ (i.e., F is constant on the conjugacy classes of G), defines the gauge invariant observable $F(g(C,x_0))$. Of course, gauge invariance implies that $F(g(C,x_0))$ does not depend on the choice of the particular reference point $x_0 \in C$. For example, the A_μ-functionals

$$\mathcal{O}_{(n)}(C;\rho) = \mathrm{Tr}\left\{\left[\,U_\rho(C;x_0,x_0)\,\right]^n\right\}, \tag{7.37}$$

are gauge invariant knot observables for any integer n and any fixed representation ρ of G.

For any classical expression $F(g(C,x_0))$ of a knot observable, we would now like to give the precise definition of $F(g(C,x_0))$ as a quantum composite operator at the quantum level. Moreover, we would like to find a complete set of such observables. Both these problems have an intrinsic and simple solution. Let us recall that any function defined on a simple compact Lie group G admits a linear decomposition in terms of the matrix elements of the different inequivalent irreducible representations of G. In particular, any function $F(g(C,x_0))$, which is constant on the conjugacy classes of G, can be written as a linear combination of the characters of G

$$F(g(C,x_0)) = \sum_\rho f(\rho)\,\mathrm{Tr}\left[\,U_\rho(C;x_0,x_0)\,\right], \tag{7.38}$$

where the sum is performed on the inequivalent irreducible representations of G. At this point, one should note that $\mathrm{Tr}\left[\,U_\rho(C;x_0,x_0)\,\right]$ is precisely the Wilson line operator of equation (1.22)

$$\mathrm{Tr}\left[U_\rho(C; x_0, x_0) \right] = W(C; \rho)\,; \tag{7.39}$$

therefore, any gauge invariant knot observable can always be written as

$$F(g(C, x_0)) = \sum_\rho f(\rho)\, W(C; \rho)\,, \tag{7.40}$$

where $\{ f(\rho) \}$ are (in general complex) numerical coefficients which characterize the observable $F(g(C, x_0))$. In conclusion,

(a) equation (7.40) shows that the set $\{ W(C; \rho) \}$ represents a complete set of knot observables;
(b) since $\{ W(C; \rho) \}$ have been defined as quantum composite operators, see Chapters 3 and 5, equation (7.40) tells us how to define $F(g(C, x_0))$ at the quantum level.

These two results, combined with general covariance of the CS model, have far reaching consequences in the construction of the solution of the topological field theory. We shall use these results in Chapters 13, 15, 17 and 20.

Chapter 8

Braid group

As we have said in Chapter 2, the study of the topology of links in $\mathbb{R}^3$ and in S^3 can be reduced to the analysis of link diagrams; to be more exact, one has to consider the properties of the classes of link diagrams modulo the Reidemeister moves. A further step consists of formulating the problem in an algebraic context; the braid group enters precisely at this stage.

8.1 Artin braid group

The Artin braid group B_n has a presentation in terms of $(n-1)$ elements $\{g_1, g_2, \ldots, g_{n-1}\}$ which are called the generators and satisfy the relations [34]:

$$g_i\, g_j = g_j\, g_i\,, \qquad |i-j| > 1\,, \tag{8.1}$$

$$g_i\, g_{i+1}\, g_i = g_{i+1}\, g_i\, g_{i+1}\,, \qquad i < n-1\,. \tag{8.2}$$

Each generator or product of generators admits an interesting n-string graphical presentation. One can associate the picture shown in Fig. 8.1(a) to g_i, and the product of two generators, say $g_2^{-1}\, g_1$, can be represented as in Fig. 8.1(b). The graphs associated to the product of generators of B_n can be understood as diagrams corresponding to the projections on a plane of n strings in $\mathbb{R}^3$ oriented, for instance, in the upward direction.

Of course, the elements of B_n are in correspondence with the sets of the products of generators modulo the equivalence relation defined by equations (8.1) and (8.2). Because of the above-mentioned representation in terms of diagrams, equations (8.1) and (8.2) also define an equivalence relation between diagrams. Two diagrams are equivalent if and only if the associated strings in $\mathbb{R}^3$ are connected by smooth deformations in $\mathbb{R}^3$ performed by keeping the two planes, to which the endpoints of the strings belong, fixed. Thus, conditions (8.1) and (8.2) give an algebraic realization of a somewhat restricted form of ambient isotopy; its associated equivalence classes are described precisely by the elements of B_n.

Note that the generator g_i is associated to the overcrossing configuration L_+, occurring between the i-th and the $(i+1)$-th strings, whereas g_i^{-1} is associated to

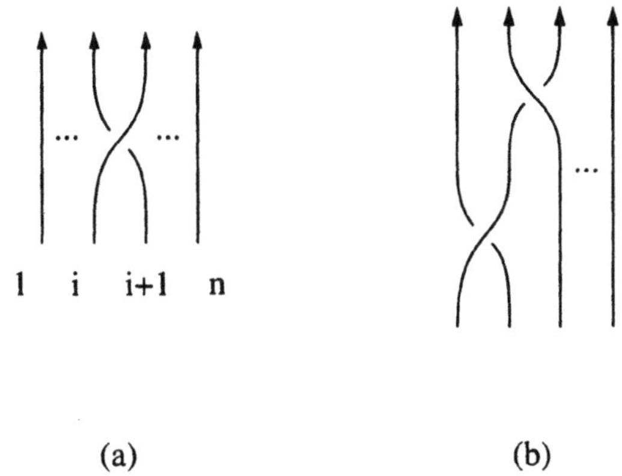

Figure 8.1.

the undercrossing configuration L_-. Let us consider the Reidemeister moves (RM). Invariance under RM of type II is automatically satisfied because, by definition, $g_i g_i^{-1} = 1 = g_i^{-1} g_i$. Invariance under RM of type III is also satisfied because of the defining relation (8.2). RM of type I only are excluded from this scheme and the reason is very simple: RM of type I are described in terms of diagram configurations in which the string orientations necessarily go upwards and downwards simultaneously. Only by taking into account the closure of braids can RM of type I be formulated.

The closure $\widehat{\sigma}$ of an element $\sigma \in B_n$ is a diagram obtained by connecting (in a ordering-preserving way) the endpoints of the strings of a graph associated to σ, as shown in Fig. 8.2(a). The closure $\widehat{\sigma}$ of the element σ will be represented as in Fig. 8.2(b). Closures of braids describe diagrams of oriented links in $\mathbb{R}^3$.

The fact that any link diagram can be represented as the closure of a braid may seem a rather trivial result; yet, it has important consequences. For example, this means that any link is a satellite of the unknot with a pattern link made of oriented strings. So, if one finds a general formula for satellites, one simply needs to determine the values of the unknots in order to compute the link invariants $\{E(L)\}$.

Given two generic elements σ_1 and σ_2 of B_n, the closure of σ_1 and the closure of $\sigma_2^{-1}\sigma_1\sigma_2$ are related by RM of type II, whereas in general $\sigma_1 \neq \sigma_2^{-1}\sigma_1\sigma_2$. Therefore, in order to have invariance under RM of type II, we must identify the closure of conjugate elements of B_n, i.e. elements related by the adjoint action of the group. For the conjugacy classes of B_n, invariance under RM of type III is satisfied. The conjugacy class of $\sigma \in B_n$ will be denoted by $\langle \sigma \rangle_n$.

Let us denote by B_∞ the union of the braid groups B_n for arbitrary n equipped with the natural inclusion $B_n \subset B_{n+1}$. The functions defined on the conjugacy

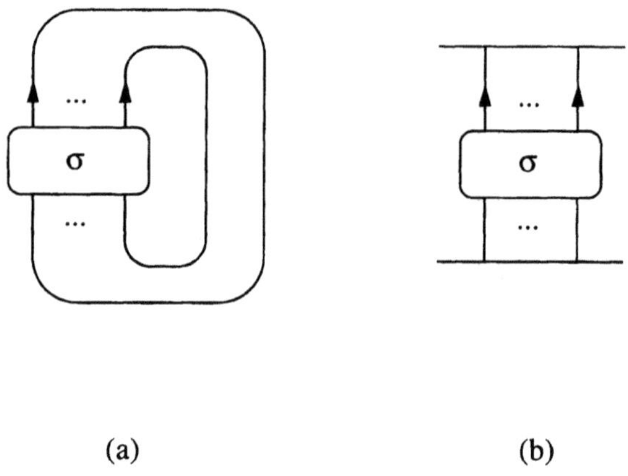

Figure 8.2.

classes of B_∞ and "covariant" under RM of type I represent link invariants of regular isotopy. An example of regular isotopy invariant can be obtained by considering a homomorphism ϕ of B_n into Z. Since the set Z of the integer numbers is an Abelian group with respect to the usual sum operation, conjugate elements of B_n have the same image under ϕ. Therefore, ϕ defines a function of the conjugacy classes of B_n into Z. Note that the generators of B_n are conjugate elements; so, it is sufficient to specify ϕ on a single generator only. When

$$\phi(g_i) = 1, \tag{8.3}$$

the regular isotopy invariant described by ϕ is the writhe number mentioned in Chapter 2. In order to recover ambient isotopy invariance, we must include RM of type I. It is easy to see that each RM of type I connects $\langle \sigma \rangle_n$ with $\langle \sigma g_n^{\pm 1} \rangle_{n+1}$. Therefore, the equivalence classes of ambient isotopic links are described by the set $\mathcal{D}$ of the conjugacy classes of B_∞ modulo the identification

$$\sigma \in B_n \leftrightarrow \sigma g_n^{\pm 1} \in B_{n+1} . \tag{8.4}$$

Of course, any function defined in $\mathcal{D}$ describes a link invariant of ambient isotopy. Summarizing, any oriented link diagram can be represented as the closure $\widehat{\sigma}$ of an element $\sigma \in \bigcup_n B_n$. The diagrams $\widehat{\sigma}_1$ and $\widehat{\sigma}_2$ describe ambient isotopic oriented links if σ_1 and σ_2 differ by a sequence of elementary moves (***Markov moves***) of two types [35]:

M1) σ_1 and σ_2 are conjugate elements of B_n;
M2) if $\sigma_1 \in B_n$, then $\sigma_2 = \sigma_1 g_n^{\pm 1} \in B_{n+1}$.

Any function invariant under Markov moves is called a Markov trace and defines a link invariant of ambient isotopy. Even if the definition of link invariants can be expressed in quite general terms, we usually use some specific representation of the braid group in order to construct such functions. Hecke algebra representations play a prominent role in connection with the classical link polynomials and the theory of Lie algebra representations.

8.2 Hecke algebra

Consider the algebra H_n generated by a set of $(n-1)$ elements $\{T_1, T_2, \ldots, T_{n-1}\}$ with defining relations

$$T_i\, T_j \;=\; T_j\, T_i\,, \qquad |i-j| > 1\,, \tag{8.5}$$

$$T_i\, T_{i+1}\, T_i \;=\; T_{i+1}\, T_i\, T_{i+1}\,, \qquad i < n-1\,, \tag{8.6}$$

$$T_i^{\,2} \;=\; z_1\, T_i \,+ z_2\,, \quad i = 1, 2, \ldots n-1\,, \tag{8.7}$$

where z_1 and z_2 are, for example, complex parameters. In general, z_1 and z_2 can be functions of some indeterminate with values in a given field K and, accordingly, H_n is defined on K. By comparing equations (8.1) and (8.2) with equations (8.5) and (8.6) it is clear that the homomorphism

$$g_i \;\mapsto\; T_i \tag{8.8}$$

defines a representation of B_n in H_n.

The quadratic relation (8.7) satisfied by the $\{T_i\}$ generators greatly simplifies the construction of link invariants. In fact, because of equations (8.5)-(8.7), each element of H_{n+1} can always be written as a sum of products of generators in which the generator T_n appears once at the most. More precisely, H_{n+1} admits a presentation as a (H_n, H_n)-bimodule

$$H_{n+1} \;\simeq\; H_n \oplus H_n \otimes_{H_{n-1}} H_n\,; \tag{8.9}$$

the natural map

$$\varphi \;:\; H_n \oplus H_n \otimes_{H_{n-1}} H_n \;\to\; H_{n+1} \tag{8.10}$$

is given by

$$\varphi\,(a + \sum_i b_i \otimes c_i\,) \;=\; a + \sum_i b_i\, T_n\, c_i\,, \tag{8.11}$$

and φ is an isomorphism [36].

Consider then $\sigma \in B_{n+1}$ and let $\tau \in H_{n+1}$ be the image of σ under the homomorphism (8.8). From equation (8.11), it follows that

$$\tau = a + \sum_i b_i \, T_n \, c_i \,, \tag{8.12}$$

where $\{ a, b_i, c_i \}$ are elements of the subalgebra H_n naturally included in H_{n+1}. The homomorphism (8.8) defines a linear relation (induced by equation (8.7)) between the diagrams associated with the elements of B_n; consequently, equation (8.12) can be represented as shown in Fig. 8.3.

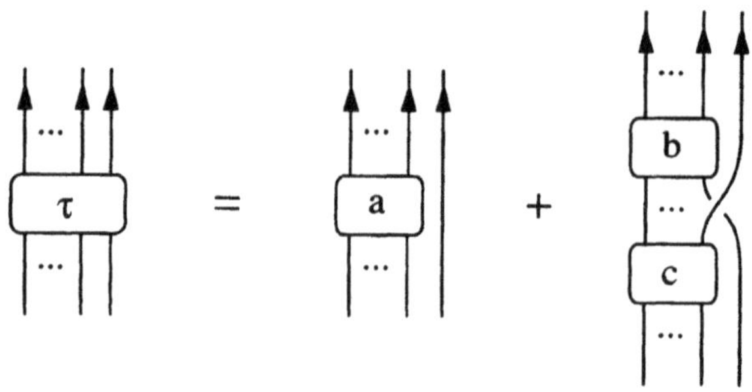

Figure 8.3.

The existence of the isomorphism (8.11) implies, by induction, that any element of H_n can be written as a linear combination of $(n!)$ fixed elements. The set of these elements can be chosen [36] to be in correspondence with a basis of the group algebra $(K\,P_n)$ of the permutation group P_n. For example, any braid diagram with three strings can always be written, in a given Hecke algebra representation of B_3, as a linear combination of the six diagrams shown in Fig. 8.4. Summarizing, the algebra H_n is of finite order; in fact, as a K-linear space, H_n has dimension $(n!)$ and the elements of a basis in H_n are in one-to-one correspondence with the elements of P_n.

Actually, the Hecke algebra H_n itself admits a faithful representation in $\mathrm{End}_K(K\,P_n)$. Let us recall that the generators $\{ s_1, s_2, \ldots, s_{n-1} \}$ of the permutation group P_n satisfy the relations

$$s_i \, s_j = s_j \, s_i \,, \qquad |i - j| > 1 \,, \tag{8.13}$$

$$s_i \, s_{i+1} \, s_i = s_{i+1} \, s_i \, s_{i+1} \,, \qquad i < n - 1 \,, \tag{8.14}$$

and

$$s_i^{\,2} = 1, \quad i = 1, 2, \ldots n - 1 \,. \tag{8.15}$$

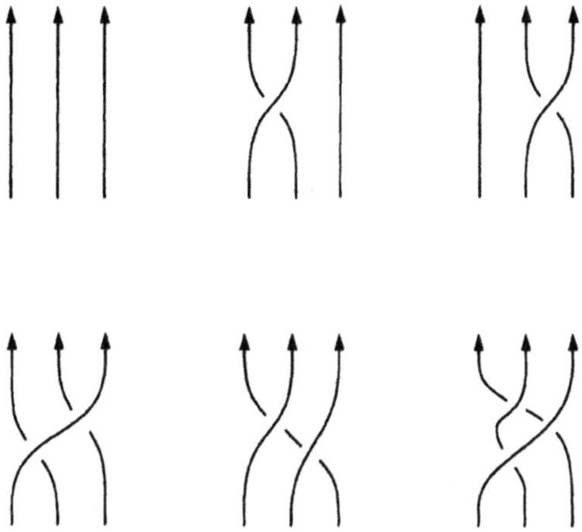

Figure 8.4.

We shall denote by $\ell : P_n \rightarrow N$ the "word length" in P_n with respect to the generators $\{ s_i \}$. For $i = 1, 2, \ldots, n-1$, let $L_i \in \mathrm{End}_K (K\, P_n)$ be defined by

$$L_i(\pi) = \begin{cases} s_i\, \pi & \text{if} \quad \ell(s_i\pi) > \ell(\pi); \\ z_2\, s_i\, \pi + z_1\, \pi & \text{if} \quad \ell(s_i\pi) < \ell(\pi); \end{cases} \tag{8.16}$$

for every $\pi \in P_n$. Then, the algebra map [36]

$$L(T_i) = L_i \tag{8.17}$$

defines the mentioned representation of H_n in $\mathrm{End}_K (K\, P_n)$.

Let us now consider the construction of a link invariant F naturally associated with a given Hecke algebra representation of the braid group. Further details concerning the definition of the Ocneanu trace can be found for instance, in Refs.[6,15,36]. In view of the linear relations between the elements of H_n, it is convenient to chose F to be K-linear on H_n. Therefore, from equation (8.7) it follows that F must satisfy

$$F(L_+) - z_2\, F(L_-) = z_1\, F(L_0), \tag{8.18}$$

where L_+, L_- and L_0 are shown in Fig. 2.5. The isomorphism (8.11) can then be used to simplify link diagrams. At this stage, in order to remove the remaining crossings of the link diagrams associated with the elements of a basis of H_n, we shall impose the condition

$$F(L^{(\pm)}) = \alpha^{\pm 1}\, F(L^{(0)}), \tag{8.19}$$

where $L^{(\pm)}$ and $L^{(0)}$ are shown in Fig. 2.6 and α is a given variable with values in K. In this way, the value taken by F on each link diagram L can be written as a linear combination of the value associated with the unknot. Equations (8.18) and (8.19) together with a given normalization for the value of the unknot specify the values of F on the link diagrams completely. Apart from some suitable normalization and redefinition of the variables, equations (8.18) and (8.19) are precisely the defining relations of the $S(L)$ polynomial introduced in Chapter 2. This construction clearly shows that the $S(L)$ polynomial is in fact the fundamental invariant from which all the other classical polynomials can be derived (for Hecke algebra representations of B_n).

Let us conclude this chapter by mentioning the coloured (or pure) braid group C_n which is a particular subgroup of B_n. The relation between B_n, C_n and the permutation group P_n can be understood as follows. Each element $\sigma \in B_n$ defines a permutation $\overline{\sigma} \in P_n$ of the n end-points of the strings of the associated diagrams. The map ν from B_n to P_n

$$\nu : \sigma \mapsto \nu(\sigma) = \overline{\sigma} \tag{8.20}$$

defines a homomorphism of B_n into P_n for which

$$\text{Ker}\, \nu = C_n \,, \tag{8.21}$$

where C_n is the coloured braid group. Clearly, the closures of elements of C_n describe links with n components. In general, the number of components of the link associated with the closure of $\sigma \in B_n$ is equal to the number of cycles contained in $\overline{\sigma} \in P_n$.

Different examples of B_n representations are known to be related to the classical polynomials. On the other hand, we know that in the CS model the gauge group structure is one of the fundamental aspects of the theory. Therefore, we are interested in the particular B_n representations whose construction is based on some Lie algebra structure.

Any link invariant satisfying a linear skein relation is clearly associated with a Hecke algebra realization of the braid group. Moreover, the elements of the algebra H_n, understood as a K-linear space, can be classified by means of the states of the irreducible representations of the permutation group. Indeed, the group algebra $(K P_n)$ defines precisely the regular representation of P_n. In conclusion, any linear skein invariant should be related (through the permutation group) with the Lie algebra structure of the unitary groups. Unfortunately, such a general connection is rather subtle. Up to now, it has not been elaborated in a form which can immediately be used in the context of the CS field theory.

As we shall see, the properties of the CS field theory will suggest how to connect, in general terms, the Lie algebra structure with the skein invariants.

On the other hand, particular examples of braid group representations, constructed by means of Lie algebras, have indeed been discovered: the so-called R-matrix representations. The R-matrix method is not directly related to the physics of the CS theory. Still, the link invariants obtained in the field theory approach are essentially the same as those constructed by means of the R-matrix formalism. The real reason for this observed coincidence is that these two different methods are both based on the same underlying Lie algebra structure.

Chapter 9

R – matrix and braids

In this chapter, we shall briefly review certain results concerning braid group representations associated with ordinary simple Lie algebras. The material presented here will not be used to solve the CS model. The purpose of this chapter is simply to recall some general features of the quasi-tensor category of quasi-triangular quasi-Hopf algebras which are useful for identifying the $E(L)$ polynomial obtained in the CS theory.

9.1 Quantum group approach

Suppose that a finite dimensional linear space V_i is associated with the i-th string of the diagrams representing the B_n elements. For the moment, we shall assume that all the spaces $\{V_i\}$ have the same dimension D. Let us try to find a matrix representation of B_n in the resulting space $V(n) = V_1 \otimes \cdots \otimes V_n$ of the form

$$g_i \mapsto G_i = \mathbb{1} \otimes \cdots \otimes R \otimes \cdots \otimes \mathbb{1} , \tag{9.1}$$

where R acts on $V_i \otimes V_{i+1}$. The defining relations (8.1) are obviously satisfied, whereas equation (8.2) implies

$$R_{12}\, R_{23}\, R_{12} = R_{23}\, R_{12}\, R_{23} , \tag{9.2}$$

where the matrices

$$\begin{aligned} R_{12} &= R \otimes \mathbb{1} , \\ R_{23} &= \mathbb{1} \otimes R , \end{aligned} \tag{9.3}$$

act on $V_1 \otimes V_2 \otimes V_3$. Equation (9.2) is called the quantum Yang-Baxter equation (QYBE). Each non-singular solution of equation (9.2) provides an explicit representation of B_n given in equation (9.1).

Assuming that R admits an enhancement μ, Turaev has shown [37] how to construct link invariants. The matrix μ, acting on the space associated to a single string, is called an enhancement for R if, in $V_1 \otimes V_2$, the following conditions are satisfied:

$$R\, \mu_1 \otimes \mu_2 \;=\; \mu_1 \otimes \mu_2\, R\,, \tag{9.4}$$

$$\mathrm{Tr}\Big|_{V_2} (R\, \mu_2) \;=\; \alpha \cdot \mathbb{1}\,\Big|_{V_1}\,, \tag{9.5}$$

$$\mathrm{Tr}\Big|_{V_2} \left(R^{-1} \mu_2\right) \;=\; \alpha^{-1} \cdot \mathbb{1}\,\Big|_{V_1}\,, \tag{9.6}$$

where α is a (complex) parameter. In equations (9.5) and (9.6), the matrices $R\,\mu_2$ and $R^{-1}\,\mu_2$ act on $V_1 \otimes V_2$ but the trace is performed in V_2 only. On the RHS of equations (9.5) and (9.6), the identity operator is defined in V_1. In general, μ_i denotes the enhancement matrix acting on V_i and the product

$$\mu^{\otimes n} \;=\; \mu_1 \otimes \mu_2 \otimes \cdots \otimes \mu_n \tag{9.7}$$

is called the measure in $V(n)$. Consider now a link diagram L described by the closure $\widehat{\sigma}$ of an element $\sigma \in B_n$ and let $M(\sigma)$ be the matrix representing σ in $V(n)$. The function

$$Y(L) \;=\; \mathrm{Tr}\left(M(\sigma)\, \mu^{\otimes n}\right) \tag{9.8}$$

is a regular isotopy invariant for link diagrams. Indeed, equation (9.4) implies that $Y(L)$ is invariant under Markov moves $M1$ of the first type. Moreover, from equations (9.5) and (9.6) it follows that

$$\begin{aligned} Y(L^{(+)}) &= \alpha\, Y(L^{(0)})\,, \\ Y(L^{(-)}) &= \alpha^{-1}\, Y(L^{(0)})\,, \end{aligned} \tag{9.9}$$

where $L^{(\pm)}$ and $L^{(0)}$ are shown in Fig. 2.6. Equations (9.9) mean that $Y(L)$ is a generalized $S(L)$ polynomial, in the sense that the covariance properties of $Y(L)$ under RM of type I (or Markov moves $M2$) coincide with those of the $S(L)$ polynomial. Finally, assume that the R matrix satisfies the polynomial equation

$$\sum_{p=a}^{b} c_p\, R^p \;=\; 0\,, \tag{9.10}$$

for some given coefficients $\{c_p\}$. Then, $Y(L)$ obviously satisfies

$$\sum_{p=a}^{b} c_p\, Y(L_p) \;=\; 0\,, \tag{9.11}$$

where the link diagrams $\{L_p\}$ coincide except for a small part of the diagrams in which L_p has a sequence of p consecutive crossings, as shown in Fig. 9.1.

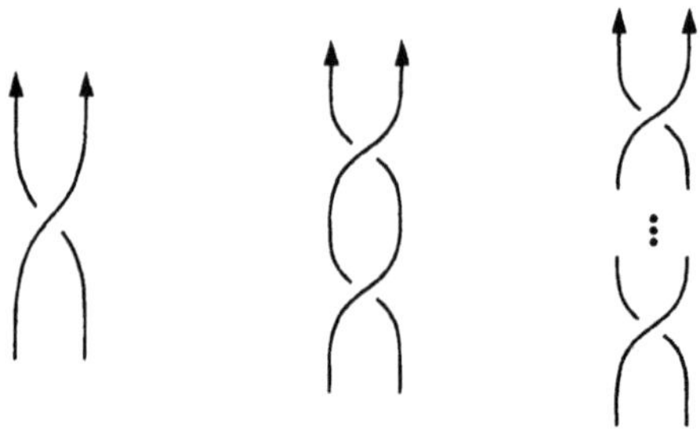

Figure 9.1.

Equation (9.11) is the generalization of the skein relation (iv) appearing in equation (2.11).

Given some explicit representation of the R matrix and of the enhancement μ, the direct computation of the trace entering the definition (9.8) of $Y(L)$ may become rather laborious. The computation of $Y(L)$ is very simple when the R matrix has only two different eigenvalues. In this case, the polynomial equation (9.10) becomes quadratic and equation (9.1) defines a Hecke algebra representation of B_n. Therefore, by means of the Ocneanu trace, $Y(L)$ can easily be reconstructed because, with a given normalization of the unknot, $Y(L)$ is completely fixed by the two eigenvalues of the R matrix and by the parameter α.

Unfortunately, when the polynomial equation (9.10) is of order higher than two, the normalization of the unknot, the value of α and the eigenvalues of the R matrix are no longer sufficient [38] for determining $Y(L)$ uniquely. In these conditions, the direct computation of the trace (9.8) has (in general) to be performed.

Of course, once $Y(L)$ has been constructed, one can easily define a new invariant of ambient isotopy simply by multiplying $Y(L)$ by the factor $\alpha^{-w(L)}$.

The problem now is to solve the QYBE (9.2). A wide class of solutions can be obtained by means of the so-called "quantum group" method. It has been shown [39] that a quasi-triangular Hopf (QTH) algebra structure, defined in terms of the so-called q-deformations of a simple Lie algebra $\mathcal{G}$, is naturally associated to the QYBE. As a result, for any D-dimensional irreducible representation ρ of $\mathcal{G}$, a particular solution of the QYBE can be constructed [39 - 44]. The associated B_n representation, defined by equation (9.1), is called an R-matrix representation. In the quantum group method, some representation of a given algebra A acts on the space $V(n)$ and the R matrix represents an invariant of this algebra.

Let A be an associative unital algebra with a homomorphism

$$\Delta \ : A \rightarrow \ A \otimes A\,, \tag{9.12}$$

called co-multiplication. We shall assume that Δ is co-associative; this means that, $\forall a \in A$,

$$(\mathrm{id} \otimes \Delta)(\Delta(a)) = (\Delta \otimes \mathrm{id})(\Delta(a)). \tag{9.13}$$

In terms of the permutation mapping $\sigma(a_1 \otimes a_2) = a_2 \otimes a_1$, one can construct another co-product Δ', namely the composition $\Delta' = \sigma \circ \Delta$. We are interested in the case in which the homomorphism Δ is not co-commutative, $\Delta \neq \Delta'$. Suppose an invertible element $\mathcal{R} \in A$ exists such that

$$\Delta'(a) = \mathcal{R}\, \Delta(a)\, \mathcal{R}^{-1}. \tag{9.14}$$

When the relations

$$(\Delta \otimes \mathrm{id})(\mathcal{R}) = \mathcal{R}_{13} \mathcal{R}_{23}, \tag{9.15}$$

$$(\mathrm{id} \otimes \Delta)(\mathcal{R}) = \mathcal{R}_{13} \mathcal{R}_{12}, \tag{9.16}$$

are satisfied, the Hopf algebra A is called quasi-triangular. In equations (9.15) and (9.16), if $\mathcal{R}_{12}$ is given by $\mathcal{R}_{12} = \sum_i a_i \otimes b_i \otimes \mathbb{1}$, then $\mathcal{R}_{13} = \sum_i a_i \otimes \mathbb{1} \otimes b_i$ and $\mathcal{R}_{23} = \sum_i \mathbb{1} \otimes a_i \otimes b_i$. The element $\mathcal{R}$ is also called the ***universal*** $\mathcal{R}$-matrix (of course, $\mathcal{R}$ is represented by a matrix in any matrix representation of A) and, from equations (9.14)-(9.16), it follows that

$$\mathcal{R}_{12} \mathcal{R}_{13} \mathcal{R}_{23} = \mathcal{R}_{23} \mathcal{R}_{13} \mathcal{R}_{12}. \tag{9.17}$$

Equation (9.17) represents the QYBE in its standard form. Given a matrix representation of A, equation (9.17) implies that the matrix

$$R = \Pi\, \mathcal{R}, \tag{9.18}$$

where Π is the permutation operator, satisfies precisely equation (9.2).

Interesting non-trivial examples of QTH algebras have been defined in terms of the so-called q-deformations of ordinary Lie algebras. For instance, starting from the Lie algebra associated with $SU(2)$

$$[J_3, J_\pm] = \pm J_\pm, \tag{9.19}$$

$$[J_+, J_-] = 2 J_3, \tag{9.20}$$

one can define the QTH algebra $SU(2)_q$ as the algebra whose elements are power series in some parameter q with coefficients belonging to the universal enveloping algebra of the $SU(2)$ Lie algebra. As generators, one can take $q^{\pm X_3/2}$ and $X\pm$ with relations

$$q^{X_3/2}(X_\pm)q^{-X_3/2} = q^{\pm 1/2}(X_\pm), \tag{9.21}$$

$$[X_+, X_-] = \frac{q^{X_3} - q^{-X_3}}{q^{1/2} - q^{-1/2}}. \tag{9.22}$$

Since in the $q \to 1$ limit equations (9.21) and (9.22) coincide with equations (9.19) and (9.20), q is also called the ***deformation parameter***. The co-product is given by

$$\Delta(q^{\pm X_3/2}) = q^{\pm X_3/2} \otimes q^{\pm X_3/2}, \tag{9.23}$$

$$\Delta(X_\pm) = q^{X_3/2} \otimes X_\pm + X_\pm \otimes q^{-X_3/2}, \tag{9.24}$$

and it is easy to verify that Δ defines a homomorphism of $SU(2)_q$ into $SU(2)_q \otimes SU(2)_q$.

The representation of $SU(2)_q$, associated with the fundamental representation of $SU(2)$, is given by 2×2 matrices in which

$$q^{X_3/2} = \begin{pmatrix} q^{1/4} & 0 \\ 0 & q^{-1/4} \end{pmatrix}, \tag{9.25}$$

and

$$X_+ = (X_-)^t = \begin{pmatrix} 0 & 1 \\ 0 & 0 \end{pmatrix}. \tag{9.26}$$

In this case, the braid group representation (9.1) is equivalent to the representation in which

$$R = \begin{pmatrix} q^{1/4} & 0 & 0 & 0 \\ 0 & q^{1/4} - q^{-3/4} & q^{-1/4} & 0 \\ 0 & q^{-1/4} & 0 & 0 \\ 0 & 0 & 0 & q^{1/4} \end{pmatrix}. \tag{9.27}$$

With respect to the standard $\mathcal{R}$-matrix normalization, the matrix (9.27) has been rescaled by a factor $q^{1/4}$. The enhancement μ is given by

$$\mu = q^{X_3}; \tag{9.28}$$

and the parameter α, appearing in equations (9.5) and (9.6), turns out to be

$$\alpha = q^{3/4}. \tag{9.29}$$

The minimal (characteristic) polynomial equation satisfied by R is

$$R^2 - (q^{1/4} - q^{-3/4})R - q^{-1/2} = 0, \tag{9.30}$$

and, therefore, we have a Hecke algebra representation of the braid group. For generic irreducible representations of $SU(2)$ and of the other simple Lie groups, the expression of R becomes rather complicated.

The universal $\mathcal{R}$-matrix for several examples of Lie algebras has been produced. Unfortunately, the form of the explicit representations of the $\mathcal{R}$ matrix is not very illuminating. In general, the direct computation of the Markov trace, even for simple links, is not a practical method of computing the link invariants.

Nevertheless, quantum groups represent one of the most important recent discoveries. They provide explicit nontrivial examples of QTH algebras. For our purposes, the crucial aspect of the quantum group construction lies in the (quite hidden) structure of the associated Lie algebra. The important point is that, for generic values of the deformation parameter q, the representation rings of quantum groups coincide with the representation rings of ordinary simple Lie algebras.

9.2 Lie algebras and monodromy representations

A simple physical interpretation of the quantum group construction has been discovered by Kohno [45]. Consider a set of n point-like particles moving on a plane. The positions of these particles can be parametrized by complex coordinates $\{z_i\}$ with $i = 1, \dots, n$. Assuming that any two particles never occupy the same position, the configuration space of the system is $X_n = \{(z_1, \dots, z_n) \in C^n : z_i \neq z_j \text{ for } i \neq j\}$. The world-lines of the particles, corresponding to a closed path in X_n, provide a geometric description of an element of the pure braid group C_n. If these particles are identified under permutations, the resulting configuration space is $Y_n = X_n / P_n$ and the fundamental group of Y_n is $\pi_1(Y_n) \cong B_n$. Suppose now that a linear space of dimension D, in which an irreducible representation ρ of the simple Lie algebra $\mathcal{G}$ is defined, is associated to each particle. Let $\{T_i^a\}$ be the generators of $\mathcal{G}$ in the ρ representation acting on the space H_i associated to the i-th particle. The total space associated to the n particles is $H(n) = H_1 \otimes \cdots \otimes H_n$. The one-form

$$\omega = \frac{\lambda}{\pi} \sum_{i \neq j} \Omega_{ij} \ \mathrm{d} \ln(z_i - z_j) , \tag{9.31}$$

where

$$\Omega_{ij} = T_i^a \otimes T_j^a \tag{9.32}$$

and λ is a free parameter, represents a flat connection in X_n and, since ω is symmetric under permutations of the particles, ω extends to a flat connection $\widetilde{\omega}$ in Y_n. For each element $\gamma \in \pi_1(Y_n)$, the associated holonomy

$$\Phi(\gamma) = \mathrm{P} \exp\left(i \oint_\gamma \widetilde{\omega} \right) \tag{9.33}$$

gives a matrix representation Φ of B_n in End$(H(n))$. Quite remarkably, Φ is equivalent [45] to the representation (9.1) obtained by means of the standard algebraic quantum group construction when the ***deformation parameter*** q is given by

$$q = \exp(-i\lambda) . \tag{9.34}$$

In the case of two particles only, the generator of B_2 is represented by

$$\Phi(g_1) = \Pi_{12}\, q^{\Omega_{12}} \tag{9.35}$$

where Π_{12} is the permutation operator acting on $H_1 \otimes H_2$.

The particular structure of the Kohno connection (9.31) implies that, in the generic case of n ($n \geq 2$) particles, each generator of B_n is represented in Φ by a matrix which is equivalent to the expression (9.35) but acts on a certain appropriate subspace of $H(n)$. The intertwining operator depends on the particular generator under consideration. Since Φ is equivalent to the R-matrix representation (9.1), this means that the matrix R is equivalent to the expression (9.35).

It should be noted that Kohno's construction is entirely based on the structure of the ordinary Lie algebra $\mathcal{G}$. Unfortunately, in this approach the inclusion $B_n \subset B_{n+1}$ is not realized in a natural way, because the intertwining operators connecting the generators have to be modified when going from B_n to B_{n+1}. On the other hand, in the R-matrix algebraic approach the naive inclusion $B_n \subset B_{n+1}$ is imposed by construction. However, the price to be paid is that the fundamental role of $\mathcal{G}$ is somewhat obscured. If one insists on requiring that the B_n generators have the form shown in equation (9.1), then the q-deformations of $\mathcal{G}$ seem to play a natural role.

For generic values of the deformation parameter q, whichever method one decides to adopt is a matter of taste. Both these methods can naturally be extended to the case in which different representations $\{\rho_i\}$ of $\mathcal{G}$ are associated to the different particles, or, to the different strings of the graphical presentation of B_n.

9.3 Quasi-Hopf algebra

Drinfeld has shown [5] that the QTH algebra associated to the QYBE actually provides a particular realization of a more general structure called the quasi-triangular quasi-Hopf (QTQH) algebra.

Consider the associative algebra $\mathcal{A}$ with co-product $\Delta : \mathcal{A} \to \mathcal{A} \otimes \mathcal{A}$ which is not necessarily co-associative. Equation (9.13) is now replaced by

$$(\mathrm{id} \otimes \Delta)(\Delta(a)) = \phi\, (\Delta \otimes \mathrm{id})(\Delta(a))\, \phi^{-1} , \tag{9.36}$$

where $\phi \in \mathcal{A}\otimes\mathcal{A}\otimes\mathcal{A}$ defines a (in general) non-trivial associativity isomorphism. It is assumed [5] that ϕ satisfies the pentagon-relation

$$(\mathrm{id}\otimes\mathrm{id}\otimes\Delta)(\phi) = (\mathbb{1}\otimes\phi)(\mathrm{id}\otimes\Delta\otimes\mathrm{id})(\phi)(\phi\otimes\mathbb{1}). \tag{9.37}$$

The triple $(\mathcal{A}, \Delta, \phi)$, satisfying equations (9.36) and (9.37) and other axioms concerning unit, co-unit and antipode, defines a quasi-Hopf algebra. Again, the existence of a commutativity isomorphism,

$$\Delta'(a) = \mathcal{R}\,\Delta(a)\,\mathcal{R}^{-1}, \tag{9.38}$$

is assumed. The presence of a non-trivial associativity isomorphism ϕ now implies that the quasi-triangular conditions (9.15)-(9.16) must be replaced by the hexagon-conditions [5]

$$(\Delta\otimes\mathrm{id})(\mathcal{R}) = \phi_{312}\,\mathcal{R}_{13}\,(\phi_{132})^{-1}\,\mathcal{R}_{23}\,\phi_{123}, \tag{9.39}$$

$$(\mathrm{id}\otimes\Delta)(\mathcal{R}) = (\phi_{231})^{-1}\,\mathcal{R}_{13}\,\phi_{213}\,\mathcal{R}_{12}\,(\phi_{123})^{-1}, \tag{9.40}$$

where $\phi_{123}\equiv\phi$.

The quadruple $(\mathcal{A}, \Delta, \phi, \mathcal{R})$ is called a QTQH algebra and, from its defining relations (9.36)-(9.40), the quasi-Yang-Baxter equality

$$\mathcal{R}_{12}\,\phi_{312}\,\mathcal{R}_{13}\,(\phi_{132})^{-1}\,\mathcal{R}_{23}\,\phi_{123} = \phi_{321}\,\mathcal{R}_{23}\,(\phi_{231})^{-1}\,\mathcal{R}_{13}\,\phi_{213}\,\mathcal{R}_{12} \tag{9.41}$$

can easily be derived. Given a QTQH algebra and an invertible element $F\in\mathcal{A}\otimes\mathcal{A}$, one can define a new QTQH algebra $(\mathcal{A}, \Delta_F, \phi_F, \mathcal{R}_F)$ where

$$\Delta_F(a) = F\;\Delta(a)\;F^{-1}, \tag{9.42}$$

$$\phi_F = F_{23}\;(\mathrm{id}\otimes\Delta)(F)\;\phi\;(\Delta\otimes\mathrm{id})(F^{-1})\;(F_{12})^{-1}, \tag{9.43}$$

and

$$\mathcal{R}_F = F_{21}\,\mathcal{R}\,F^{-1}. \tag{9.44}$$

Clearly, a QTH algebra is simply a QTQH algebra with a trivial associativity isomorphism ϕ. We have seen that the QTH algebra is relevant for the construction of braid group representations. The importance of the QTQH algebra is that, even if the QYBE is in general modified (as shown in equation (9.41)), the braid group representations described by the quasi-tensor category defined by the QTQH algebra still remain the same. The advantage of using a QTQH algebra is that the "generalized" R-matrix (9.18) can be taken [5] to be precisely

$$R = \Pi_{12}\, q^{T^a_{(p_1)}T^a_{(p_2)}}, \tag{9.45}$$

where $\{T^a_{(\rho_1)}\}$ and $\{T^a_{(\rho_2)}\}$ are the $\mathcal{G}$ generators in the representations ρ_1 and ρ_2. Of course, in these conditions, a non-trivial associativity isomorphism ϕ must be introduced. Drinfeld has constructed ϕ by using [5] solutions of the Knizhnik-Zamolodchikov [46] equation with appropriate boundary conditions. The link invariants obtained from the braid group representations associated with this quasi-tensor category are called by Drinfeld the universal link invariants [5].

The QTQH algebra is the algebraic context which naturally fits the more geometrical approach of Kohno. The generalized Kohno construction, defined for arbitrary representations of $\mathcal{G}$, provides the general structure of the B_n representations constructed by means of ordinary Lie algebras. Up to equivalence, the form of the B_n generators is shown precisely in equation (9.45). In conclusion, the quasi-tensor category associated with the Lie algebra structures is characterized by braiding matrices which are equivalent to the matrix (9.45) (up to possible abelian factors which trivially modify the braid representations).

As we have already stated, for q generic, the standard algebraic quantum group method and Kohno's method are equivalent. When q is a root of unity, Kohno's construction is still well defined, whereas the definition and the meaning of the quantum group itself present some problems. In this sense, and as far as braid group representations are concerned, Kohno's approach seems to be more simple and in fact more fundamental than the standard quantum group approach. On the other hand, quantum groups are very useful because they provide explicit matrix representations of the braid group.

A few possible methods for computing the link invariants (9.8) can in principle be used. As we have already mentioned, in the case of Hecke algebra B_n representations, the construction of the $S(L)$ polynomial can easily be performed by using, for instance, the skein relation. For B_n representations which are not of the Hecke type, one should find the explicit expression of the R matrix and of the enhancement μ. The direct computation of the trace (9.8) is then straightforward but, in practice, quite tedious. Kirillov and Reshetikhin [42,43] have developed a clever method of computing the link invariants in terms of the q-deformed $6j$-symbols of the quantum groups. This method can be practical or not, depending on how able you are in computing these $6j$-symbols. Particular examples of link invariants which cannot be obtained in Hecke algebra representations have been considered for instance in [47]. New examples of this kind have also been produced in [48]. What is still missing in all the different approaches is a systematic and simple method of producing these new invariants in such a way that the properties of the ordinary Lie algebras have a transparent and fundamental role. The quantum CS field theory naturally suggests how to solve this problem.

Chapter 10

Chern-Simons monodromies

We have seen that, in order to construct link invariants, it is important to specify the structure of the braiding matrix associated with a crossing between two strings of a braid. For Hecke algebra representations of B_n, the precise expression of the braiding matrix is actually irrelevant; only the knowledge of its eigenvalues is essential.

Let us now consider the CS field theory. Clearly, the link invariants defined by the expectation values of the Wilson line operators also admit an abstract interpretation in the braid group formalism described in the previous chapters. We would now like to derive information concerning the structure of the resulting braiding matrix.

10.1 Schrödinger picture

The expectation value $\langle W(L) \rangle$ can be interpreted as a scalar product between two states, just like in any field theory. Given a surface which separates $\mathbb{R}^3$ into two parts P_1 and P_2, the computation of the functional integral in $\mathbb{R}^3$ can be divided into two steps. First, the integral is performed separately in P_1 and P_2 with given boundary conditions on the surface for the fields. In this way, one obtains the wave functions of one "bra" and one "ket" state vectors. Secondly, the scalar product between these two states gives the result of the whole functional integral in $\mathbb{R}^3$. In this sense, one can consider the space of the states of the CS theory in the presence of Wilson line operators.

Let us consider the case of a surface intersecting a link L on which a Wilson line operator W(L) is defined. The states of the system are associated with a punctured surface; the positions of the punctures are defined by the intersection points of the surface with the oriented link L. Each puncture is characterized by the quantum numbers of a representation ρ_i of G inherited from $W(L)$. As it has been observed in [4], the non-trivial information carried by the states of the theory is represented by the monodromy matrices acting on the state wave functions when the punctures are exchanged according to the action of the braid group. This method of computing the monodromy matrices looks very similar to Kohno's construction described in

Chapter 9. In fact, finding the form of the monodromy matrices obtained in the CS theory is equivalent to determining the structure of the R matrix entering the braid representations realized in the state space of the CS model.

The direct computation of the state wave functions associated to a generic link and a generic punctured surface by means of the functional integral is rather complicated. On the other hand, our real interest does not concern the wave functions themselves but the monodromy matrices; our wish is to compute them exactly, to all orders in $\lambda = (2\pi/k)$. The crucial observation [4] which permits us to simplify and actually solve this problem is that the structure of the monodromy matrices is independent of the particular link under consideration and also of the measure giving the pairing between bra and ket vectors. This is just a consequence of the properties of the transition amplitudes in quantum field theory. This being the case, the computation of the monodromy matrices can consistently be performed in a particularly simple situation in which all the complications due to the pairing and the details of the links are absent.

Consider the Hamiltonian approach to the CS theory based on a $(2+1)$ decomposition of $\mathbb{R}^3$ [30,49]. The CS action is

$$S_{CS} = \frac{k}{4\pi} \int d^3x \, \left(A_2^a \dot{A}_1^a + A_0 F_{12}^a \right) , \tag{10.1}$$

where

$$F_{12}^a = \partial_1 A_2^a - \partial_2 A_1^a - f^{abc} A_1^b A_2^c . \tag{10.2}$$

In our notations $x^\mu = (x^0, \vec{x}\,) = (x^0, x^i)$ with $i = 1, 2$. The components $A_0^a(x)$ appear in the action without "time" derivatives acting on them and can be interpreted as Lagrange multipliers enforcing the Gauss Law which, in the absence of sources, reads

$$F_{12}^a = 0 . \tag{10.3}$$

The quantization can be performed in the usual way: first, one ignores the presence of the Lagrange multipliers and, as a result, from the expression (10.1) the canonical commutation relations

$$\left[A_1^a(\vec{x}\,), A_2^b(\vec{y}) \right] = i\, \delta^{ab} \frac{4\pi}{k} \delta(\vec{x} - \vec{y}) , \tag{10.4}$$

are derived. We can use the standard Schrödinger representation of (10.4) in which A_1^a play the role of the coordinates and A_2^a of the conjugate momenta. The composite operator F_{12}^a is well defined because, from equation (10.4), it follows that the expression (10.2) has no ordering ambiguities. The operator

$$G(\Lambda) = -\frac{k}{4\pi} \int d^2x \, F_{12}^a(\vec{x}\,)\, \Lambda^a(\vec{x}\,) , \tag{10.5}$$

where $\Lambda^a(\vec{x})$ have compact support (or vanish at infinity in a sufficiently rapid way), generates the residual (time-independent) gauge transformations. Then, the Gauss Law (10.3) takes the form of a constraint

$$G(\Lambda)\,\Psi = 0\,, \tag{10.6}$$

which selects the physical (gauge-invariant) subspace. Of course, in this formalism the physical state wave functions $\{\Psi\}$ are not normalizable: the group volume of time-independent gauge transformations should be factorized. Since we are interested in the monodromy matrices only, however, we can safely ignore the pairing. More generally, it is well known that the expectation values $\{\langle W(L)\rangle\}$, being vacuum-to-vacuum amplitudes, cannot be obtained by means of a naive Hamiltonian formulation. But in the Hamiltonian approach we do not want to compute $\{\langle W(L)\rangle\}$ at all; we only want to find the form of the monodromy matrices. So, we shall consider the state wave function associated with a fixed-time plane $\mathcal{P}$ cutting two oriented parallel straight lines running along the time direction, as shown in Fig. 10.1.

One can imagine that two Wilson line operators, in the irreducible representations ρ_1 and ρ_2 of the gauge group G, are associated with these lines so that the two punctures on $\mathcal{P}$ are characterized by the quantum numbers of these representations. Since the Wilson operators are functions of the Lagrange multipliers A_0^a, the Gauss Law (10.6) is actually modified and takes the form

$$\begin{aligned} iG(\Lambda)\,\Psi_{IJ} &= i\,\Lambda^a(\vec{x}_1)\,\left(T^a_{(\rho_1)}\right)_{KI}\,\Psi_{KJ} \\ &\quad + i\,\Lambda^a(\vec{x}_2)\,\left(T^a_{(\rho_2)}\right)_{KJ}\,\Psi_{IK}\,, \end{aligned} \tag{10.7}$$

where $T^a_{(\rho_1)}$ are the generators of G in the representation ρ_1 associated with the first puncture which has coordinates $\vec{x}_1$; similar notations have been used for the second puncture.

The wave function Ψ_{IJ} can easily be determined. Since we are adopting the Schrödinger picture, Ψ_{IJ} must be a functional of $A_1^a(\vec{x})$. Consider now the operator

$$U_\rho(\vec{x}) = U_\rho(x^1, x^2) = \mathrm{P}\exp\left(i\int_{-\infty}^{x^1} ds\, T^a_{(\rho)}\, A_1^a(s, x^2)\right)\,. \tag{10.8}$$

Since $G(\Lambda)$ is the generator of gauge transformations, one has

$$i\left[G(\Lambda)\,,\,U_\rho(\vec{x})\right] = i\,U_\rho(\vec{x})\,T^a_{(\rho)}\,\Lambda^a(\vec{x})\,. \tag{10.9}$$

Moreover $U_\rho(\vec{x})$, being function of A_1^a only, acts on the wave functions by multiplication. Let Ψ_0 be the "vacuum" wave function; i.e., the functional satisfying $G(\Lambda)\Psi_0 = 0$. For our purposes, the explicit expression [30,49] of Ψ_0 is not

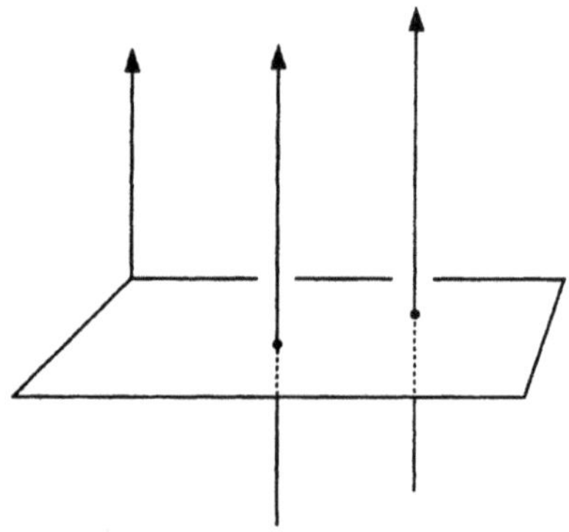

Figure 10.1.

needed. From equation (10.9), it follows that

$$\Psi_{IJ} = \Psi_{IJ,MN} = \left[U_{\rho_1}(\vec{x}_1) \right]_{MI} \left[U_{\rho_2}(\vec{x}_2) \right]_{NJ} \Psi_0 , \tag{10.10}$$

where the indices M and N are free.

We are now interested in the monodromy matrix, acting on the state wave functions, which describes the effect of a parallel transport of one of the punctures around a loop C enclosing the other puncture, as shown in Fig. 10.2.

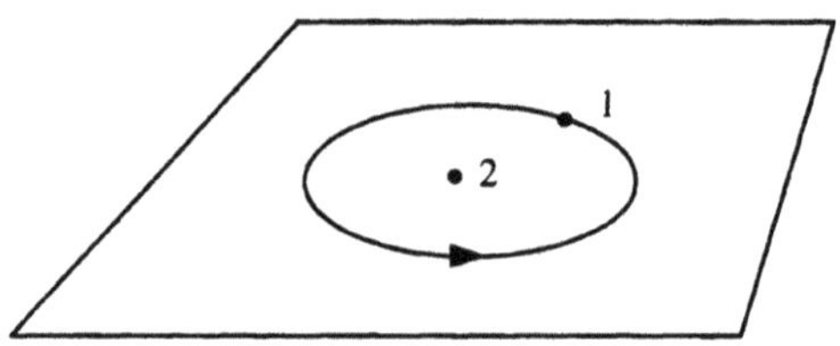

Figure 10.2.

The new wave function is given by

$$\Psi'_{IJ,MN} = \left[U_{\rho_1}(\vec{x}_1) \right]_{MK} [V(C)]_{KI} \left[U_{\rho_2}(\vec{x}_2) \right]_{NJ} \Psi_0 , \tag{10.11}$$

where

$$V(C) = \mathrm{P} \exp\left(i \oint_C dx^i \, A_i^a(x) \, T^a_{(\rho_1)} \right) , \tag{10.12}$$

is the quantum holonomy operator associated with the closed path C on the plane and with the orientation shown in Fig. 10.2. In the Schrödinger picture, $A_2^a(\vec{x})$ is represented by

$$A_2^a(\vec{x}) = -i\,\frac{4\pi}{k}\,\frac{\delta}{\delta A_1^a(\vec{x})}\,. \tag{10.13}$$

Therefore, in the expression (10.11), $A_2^a(\vec{x})$ acts necessarily on $U_{\rho_2}(\vec{x}_2)$ and on Ψ_0. We finally obtain [30]

$$\Psi'_{IJ,MN} = \Psi_{IJ,SQ}\,\mathcal{B}_{M,N}^{S,Q}\,, \tag{10.14}$$

where $\{\mathcal{B}_{M,N}^{S,Q}\}$ are the matrix elements of

$$\mathcal{B} = \exp\left(-i\,\frac{4\pi}{k}\,T^a_{(\rho_1)}T^a_{(\rho_2)}\right)\,. \tag{10.15}$$

It is quite remarkable that, in the canonical formalism considered in this section, the monodromy matrix $\mathcal{B}$ admits such a direct and clear derivation. In fact, this is the only method I know where the computation of the monodromy matrix $\mathcal{B}$ is so simple.

A particular prescription for dealing with the self-interaction effects related to the expression (10.12) has been used in the derivation of equation (10.14). This prescription is called "constant framing" in Ref.[30]. In any case, it should be noted that the self-interaction contributions (framing effects) are completely under control because of the general properties of the Wilson line operators derived in Chapter 6. This point will be discussed in Chapter 13.

10.2 Universality of the link invariants

Equation (10.15) gives us the missing information needed in order to reconstruct the universal link polynomial $E(L)$. From the knowledge of the monodromy matrix (10.15), it follows that the structure of the braiding matrix (or "half-monodromy"), associated with the exchange of the two punctures shown in Fig. 10.3, is fixed and is given by

$$M = \Pi_{12}\,q^{T^a_{(\rho_1)}T^a_{(\rho_2)}}\,, \tag{10.16}$$

where Π_{12} is the permutation operator and the ***deformation parameter*** q is given by

$$q = \exp\left(-\frac{2\pi i}{k}\right)\,. \tag{10.17}$$

The presence of the permutation operator in the expression (10.16) is clear and the fact that the "square root" of the monodromy $\mathcal{B}$ must be inserted in equation (10.16) is also quite obvious (a formal derivation of equation (10.16) from the expression (10.15) can be obtained by following the arguments presented in Ref.[45]).

Equations (10.16) and (10.17) can also be obtained by considering the meaning of the Gauss Law constraint in the CS theory in the presence of punctures on a plane, equation (10.7). The generator $G(\Lambda)$ of gauge transformations is proportional to the field strength F^a_{12}, equation (10.5); so, equation (10.7) states that the curvature of the connection on the surface is concentrated at the positions of the punctures.

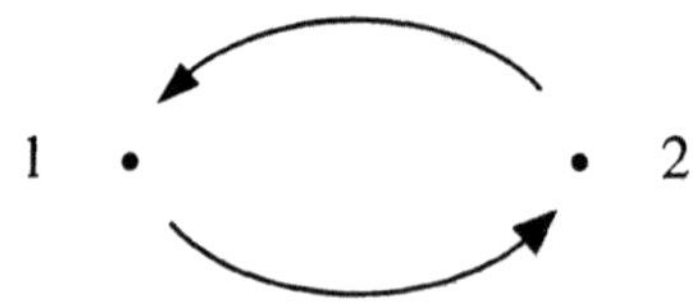

Figure 10.3.

The parallel transport of one puncture along a given path on the plane can be interpreted as the motion of a particle in the presence of a (non-Abelian) external vector field generated by the other particles. The effect produced by this external field is locally trivial because in any given open domain of the plane, which does not contain particles, the field can be written as a pure gauge configuration (the curvature is vanishing). However, when we consider a closed path enclosing some other particle, the global effect of the external field is non-trivial. Clearly, this can be understood as some kind of dynamical non-Abelian Bohm-Aharonov effect. We are now interested in the monodromy matrices which describe the amplitudes associated with closed paths of the particles. In a slightly different context, this problem has already been considered and solved by Kohno [45]. Indeed, the Kohno connection (9.31), with the parameter λ equal to $(2\pi/k)$, reproduces precisely the CS Gauss Law in the presence of non-Abelian point-like sources. The monodromy matrices associated with closed paths in the configuration space of the system are the pure braid group matrices of the monodromy representation considered in Chapter 9. As far as the braid group B_n is concerned, the structure of the resulting braiding matrix was given in equations (9.34) and (9.35), which coincide precisely with the equations (10.16) and (10.17) derived in the canonical field theory approach. Even if different methods have been used for analysing the problem, the final conclusions agree. This is the way it should be.

The matrix (10.16) cannot be taken to represent directly the braid generators, in the case of several punctures, because the braiding relations (8.2) are not satisfied. This is simply because the canonical approach, considered in the previous section, is not manifestly covariant. It should also be noted that the states, associated with the punctures on the plane, are not gauge invariant (by definition), therefore all the given expressions should be understood modulo equivalence relations (gauge transformations). As in ordinary gauge theories, each of these two drawbacks cures the other, of course, and in the covariant BRS approach both gauge invariance and general covariance are manifestly maintained. Similarly, in the Kohno construction the form of the braiding matrix is given by the expression (10.16) modulo some non-trivial equivalence relation. And in the QTQH algebra approach also, the generalized R matrix can be taken to coincide with the expression (10.16) at the cost of introducing a non-trivial associativity isomorphism.

Finally, we have reached the crucial point. In fact, Kohno's theorem [45] states that the monodromy representations of the braid group, constructed with the connection (9.31), are equivalent to the R-matrix representations obtained with the quantum group method. Equivalently, the braiding matrix (10.16) of the CS theory has precisely the structure of the generalized R matrix (9.45) entering the definition of the quasi-tensor category associated to the QTQH algebras in which the deformation parameter q is given in equation (10.17). The conclusion is that the link invariant $E(L)$, obtained in the CS theory, is nothing but the universal link invariant considered by Drinfeld.

Summarizing, the quantum group method, the monodromy representation construction of Kohno, the QTQH algebra approach of Drinfeld and the quantum CS theory represent different aspects of the same underlying mathematical structure. One of the great advantages of the topological field theory formalism is that one has an explicit three-dimensional framework for the construction of the knot invariants. Apart from all these different (and complementary) aspects, what are, after all, the essential ingredients entering the universal link polynomial ?
The answer to this question can again be formulated in several different ways. Each author has his favourite point of view and, in fact, all the different methods are interesting and instructive. As far as the construction of the link invariants is concerned, the fundamental ingredients are:

(1) the existence of a "tensor structure". By this I mean the existence of a well defined characterization of the satellites. As we shall see, this is given by the representation ring of the simple Lie algebras;
(2) the numerical "input". One has to specify some minimal numerical information (for example, the "eigenvalues" of the braiding matrix associated with the crossings) which is compatible with the satellite structure given in point (1).

Presumably, for certain satellite structures, the numerical input is essentially fixed. In these conditions only point (1) is really important. This seems to be the case for the tensor structure naturally associated with the Lie algebras of simple compact groups. At least this is what the CS theory and the results obtained by Drinfeld [5] suggest.

It is well known [50,-,56] that braid group representations associated to the quasi-tensor category of QTQH algebras can be also found in several two-dimensional systems. For instance, monodromy representations of B_n appear: in the critical Ising model, in WZNW models, in generalized conformal Toda systems and in rational conformal field theories in general. In this sense, the three-dimensional CS theory describes the universal structure of the braid group representations underlying all these different models. This property is also called the universality of the CS theory.

10.3 The inexistent shift

The results (10.16) and (10.17) are in agreement with the general properties derived in Chapter 6. In particular, under a mirror reflection each overcrossing is transformed into an undercrossing and, consequently, the braiding matrix M must be transformed into M^{-1}. On the other hand, a mirror reflection modifies the orientation of $\mathbb{R}^3$ and, as a result, the sign of the action is modified. This is equivalent to exchanging the coupling constant k with $-k$. From equation (10.17) it follows that, under this transformation, the braiding matrix M is transformed precisely into M^{-1}, as it should be.

In general, with the given structure (10.16) of the braiding matrix, this simple mirror-reflection argument implies that the deformation parameter q must satisfy

$$q(-k) = q^{-1}(k) . \qquad (10.18)$$

The explicit result (10.17), obtained with a direct computation, is in perfect agreement with the condition (10.18), of course.

Equations (10.16) and (10.17) are also in agreement with the perturbative computations of the expectation values of the Wilson line operators. In fact, equations (10.16) and (10.17) imply that, in the case in which the components of the links are associated with the fundamental representation of $G = SU(N)$, the parameters entering the skein relation are precisely those given in equations (5.16)-(5.19).

In Ref.[4] it was conjectured that the deformation parameter q, obtained in the CS theory in $\mathbb{R}^3$, is given by

$$\text{W conjecture} \quad : \qquad q = \exp\left(-i \, \frac{2\pi}{k + c_v} \right) , \qquad (10.19)$$

where c_v is the value of the quadratic Casimir operator in the adjoint representation. The conjectured expression (10.19) is similar to the real result (10.17). In fact, it can be obtained from (10.17) simply by replacing k with $k + c_v$. This *ad hoc* replacement has been called "shift".

The problem of determining how q depends on the physical (or renormalized) coupling constant k is also known as the shift-problem. This problem is well posed and, in order to solve it, one has to compare two quantities: the renormalized coupling constant k and the value of the deformation parameter q.

Several arguments can be used to show that the conjecture (10.19) is false. The expression (10.19) is in disagreement with several of the general properties derived in Chapter 6. Conjecture (10.19) is not in agreement with unitarity and CPT invariance of the CS field theory. In particular, equation (10.19) does not satisfy the condition (10.18). A simple second order computation of the expectation values of the Wilson line operators in perturbation theory is sufficient for showing that the shift conjecture is false. The result (10.15) also confirms that the conjectured shift is actually inexistent.

Some authors have misunderstood the problem of the shift and have considered a different problem: they have studied how the renormalized coupling constant k depends on the "bare" coupling constant k_0. This second problem has nothing to do with the problem of the shift. In fact, this second problem has nothing to do with physics at all. As we know from any textbook, the dependence of k on k_0 can be arbitrarily modified simply by choosing different regularization schemes; an explicit example, at one loop, has been given in Chapter 4. A finite shift in the "bare" coupling constant is equivalent to a change of the name of the coupling constant; this will never produce any observable effect. These authors are kindly invited to compute a transition amplitude in quantum electrodynamics and compare their result with the experiments. This exercise would be much more instructive than looking for some scheme in which, in the CS theory, k is equal to $k_0 + c_v$. If you find a regularization scheme in which $k = k_0 + c_v$, this does not mean that you have found the shift. In order to complete the exercise, you should also compute the deformation parameter q. If you find $q = \exp[-2\pi i/(k_0 + c_v)] = \exp[-2\pi i/k]$, then there is no shift. On the other hand, if you find $q = \exp[-2\pi i/(k_0 + 2c_v)] = \exp[-2\pi i/(k + c_v)]$, then you have found the shift.

We have already proved that the conjecture (10.19) is false. In the quantum CS theory there is no shift and the deformation parameter is given by equation (10.17).

As we have already mentioned, the monodromy properties of certain two-dimensional conformal models are described by the braid representations determined by the CS theory. The physical parameters of the different theories must be correctly identified, however. In particular, the CS coupling constant k should not be confused with the level ℓ of the WZWN model. For example, the deformation parameter determined by the CS theory with $G = SU(2)$ and $k \geq 3$ is equal to the deformation parameter obtained in the $SU(2)_\ell$ WZWN model with level $\ell = k - 2$.

Chapter 11

Defining relations

In this chapter, the information obtained from the three-dimensional CS field theory is organized in a set of rules permitting the reconstruction of $E(L)$. These rules are formulated for link diagrams corresponding to oriented framed links in $\mathbb{R}^3$. We use the vertical framing convention in which the writhe of each component of a link diagram is equal to the linking number of the corresponding component of the link in $\mathbb{R}^3$ and of its framing. The link diagrams are represented by the closure of elements of the braid group. Each string of the diagrams is oriented and associated with an irreducible representation of G.

11.1 Calculus rules

The calculus of the universal link polynomial $E(L)$ is based on a very simple idea: we want to find the relations connecting the values of $E(L)$ for different links. This means that we use the same strategy when we construct the "classical" polynomials by means of a skein relation.

In order to compute $E(L)$, the simple use of a skein relation is not sufficient, of course. As we have already seen, any generalized skein relation, associated with a braiding matrix with more than two different eigenvalues, does not identify uniquely a link invariant. Moreover, the normalization of the unknots cannot be arbitrarily fixed: in the CS theory the normalization of $E(L)$ is determined by the field theory definition (5.7) and (5.8).

Actually, by analysing the topological field theory, a great deal of information has been obtained. In addition to the generalized skein relations, which can easily be derived from equation (10.16), we know that all the general properties discovered in Chapter 6 must hold. In fact, now we have all the ingredients for the complete reconstruction of $E(L)$.

The relations following from the discrete symmetries of the field theory, the satellite formulae of Section 6.3 and the explicit numerical result (10.16) uniquely determine $E(L)$. It may seem a little surprising, but the solution of the quantum CS field theory in $\mathbb{R}^3$ is already contained in the material presented in the previous chapters. What is more, the topological nature of the CS theory implies that, in reality, the solution of the quantum CS theory in any orientable three-manifold $\mathcal{M}$ is implicitly contained in the previous chapters as well.

At this stage, the problem of computing the expectation values of the Wilson line operators in any three-manifold $\mathcal{M}$ admits a unique solution. Of course, this problem can be solved but the construction of the solution is not trivial. So, I will now try to explain the method I use for computing the expectation values of the Wilson line operators.

In the computation of $E(L)$, the recipe [9,57] for connecting different link diagrams is based on a finite number of rules which are in order.

(1) Projection decomposition

We have seen in Chapter 6 that one of the main properties of the Wilson line operators is represented by the fusion property (6.26), which can be represented as in Fig. 6.4. The elementary blocks entering the decomposition (6.28) are the inequivalent irreducible representations of G. So, given two irreducible representations ρ_1 and ρ_2, the tensor product $\rho_1 \otimes \rho_2$ is decomposed

$$\rho_1 \otimes \rho_2 = \oplus_t \, \rho(t) \tag{11.1}$$

in terms of the irreducible components $\{\rho(t)\}$ of G; t represents a set of parameters which are used to label the inequivalent representations. A given representation may happen to appear in the decomposition (11.1) several times. In this case, we simply introduce in the set t a new index which is used to distinguish these different but equivalent irreducible representations.

We now want to consider the local properties of the link diagrams. Given two oriented strings associated with the irreducible representations ρ_1 and ρ_2, we introduce the projection decomposition shown in Fig. 11.1. This rule means that the elementary blocks, which are potentially associated with the two given strings, are precisely those described by equation (11.1). The boxes introduced in Fig. 11.1 are called projectors because we assume that the "orthogonality" property shown in Fig. 11.2 holds. The projection decomposition rule can be interpreted as a completeness relation and will be used for analysing the crossings.

Of course, the insertion of projectors in a given link diagram drastically modifies the picture of the diagram and, in fact, what we get is no longer a link diagram at all. The point is that, by means of the rules that we are describing, each projector can also be eliminated and replaced with a suitable combination of ordinary string configurations. So, the use of projectors is just a convenient method of deriving relations between link diagrams.

The strategy of our construction is the following: we start with a given link diagram. In the intermediate steps of the whole procedure we introduce and use projectors. These projectors are then eliminated and, at the end, we get ordinary

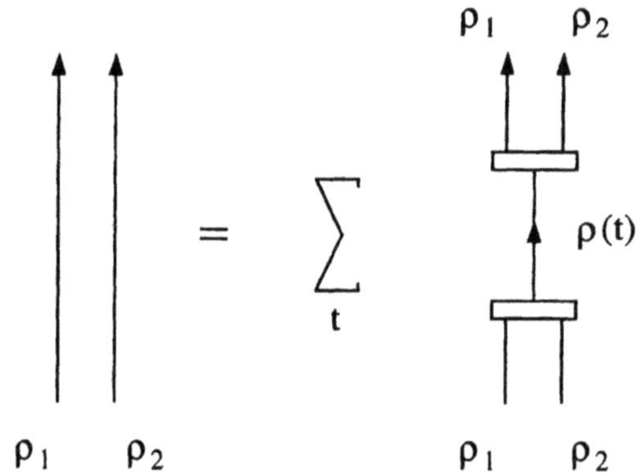

Figure 11.1.

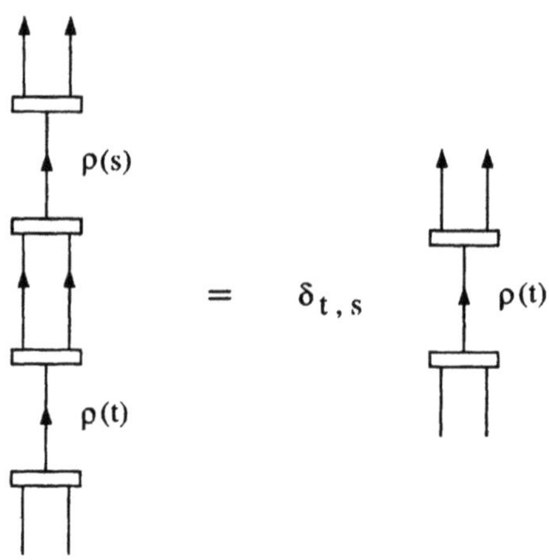

Figure 11.2. Orthogonality property of projectors

link diagrams again. As a result, the value taken by the universal link polynomial on the starting link diagram is expressed in terms of the values taken on a set of different link diagrams.

(2) Crossing

We know that a braiding matrix, which is equivalent to the expression (10.16), is associated to each crossing between strings. On the other hand,

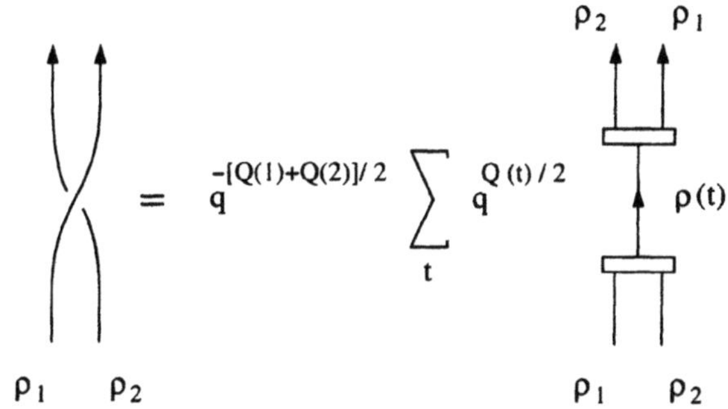

Figure 11.3. Overcrossing

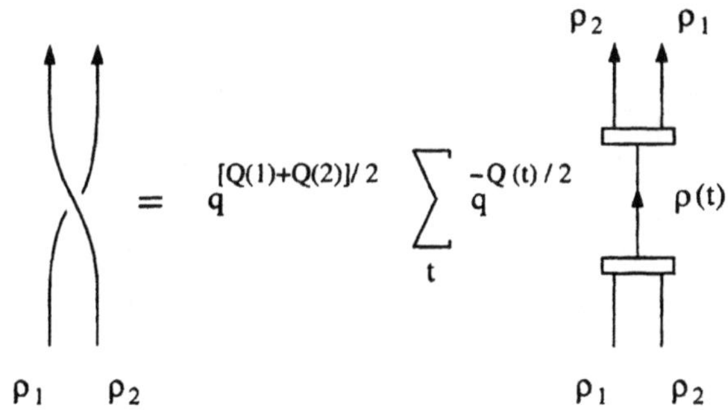

Figure 11.4. Undercrossing

$$
\begin{aligned}
T^a_{(\rho_1)} \otimes T^a_{(\rho_2)} &= \frac{1}{2}\left[\left(T^a_{(\rho_1)} + T^a_{(\rho_2)}\right)^2 - T^a_{(\rho_1)}T^a_{(\rho_1)} - T^a_{(\rho_2)}T^a_{(\rho_2)}\right] \\
&= \frac{1}{2}\left[Q(\rho_1 \otimes \rho_2) - Q(\rho_1) - Q(\rho_2)\right] ,
\end{aligned}
\tag{11.2}
$$

where $Q(\rho)$ is the value of the quadratic Casimir operator in the representation ρ. Therefore, by making use of the projection decomposition (**1**), the crossing relations take the form shown in Fig. 11.3 and Fig. 11.4. At this stage, it is clear that the introduction of projectors is a very convenient method of analysing the crossings because the irreducible components, appearing in the decomposition of the tensor product, "diagonalize" the braiding operations (or, the braiding matrices).

The presence of the permutation operator in equation (10.16) has been taken into account; in fact, in the crossing relations of Fig. 11.3 and Fig. 11.4, the two

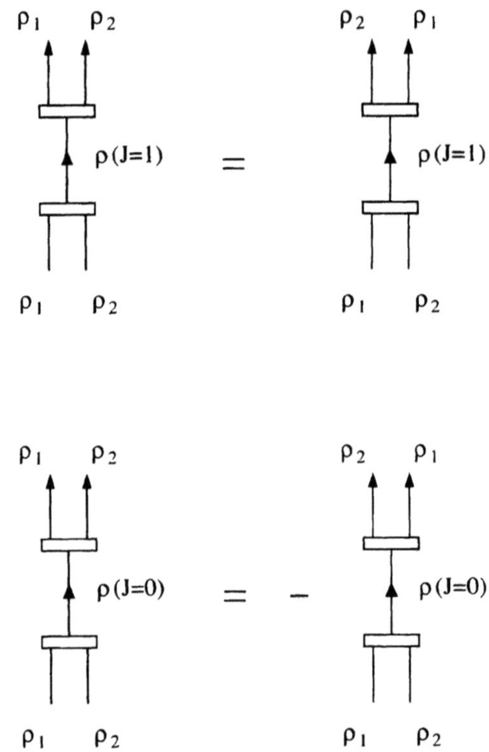

Figure 11.5. Symmetry property of projectors when $\rho_1 \cong \rho_2 \cong (J = 1/2)$

representations ρ_1 and ρ_2 appear in the projectors with the right positions. When ρ_1 and ρ_2 are equivalent, $\rho_1 \cong \rho_2$, the positions of ρ_1 and of ρ_2 in each projector can be exchanged; in doing this, one gets a factor $(+1)$ if $\rho(t)$ is symmetric and a factor (-1) if $\rho(t)$ is antisymmetric under permutations of ρ_1 and ρ_2.
For example, let us consider the case in which $G = SU(2)$ and both ρ_1 and ρ_2 are equivalent to the fundamental (spin $J = 1/2$) representation of $SU(2)$. The decomposition reads $(1/2) \otimes (1/2) = (1) \oplus (0)$. The adjoint (1) representation is symmetric, whereas the trivial (0) representation is antisymmetric. In this case, the relations shown in Fig. 11.5 hold.

The introduction of "projectors" in the crossing relations looks similar to the Kirillov-Reshetikhin construction [42,43]. However, the meaning and the use of the projectors in the two cases are completely different. In the method presented here, the explicit form of these projectors, interpreted as operators acting on the tensor product of the group representation spaces associated with the strings, will never be used. So, we never pose the problem of finding either the explicit matrix form of these projectors (understood as operators acting on the tensor product of vector spaces) or the matrix form of the associated representations of the braid group. In fact, in the procedure that we are considering the existence of a tensor product of the group representation spaces is not assumed at all. In our construction, the strings of the link diagrams are not associated with vector spaces: on the contrary, they are considered as "scalar" quantities. Our projectors are not operators acting on the tensor product of the the group representation spaces but represent the independent components of the braid conjugacy classes. Our idea is that only the "tensor structure" defined by the ordinary Lie algebra is relevant for the construction of the link invariants. This structure is described exclusively by the representation ring of the simple Lie algebra; only for this reason is the labelling of the strings made in terms of the irreducible representations of the group.

Rules **(1)** and **(2)** are used for analysing the local properties of the link diagrams and, clearly, the closure of the braids has not been considered. We shall now introduce the remaining three rules in which the partial or total closure of braids is involved.

(3) Twisting

A change of framing of a given component of the link L is represented by a change in the writhe of the same component of the link diagram. In the braid description of the link diagrams, each elementary operation $\Delta w = \pm 1$ on an arbitrary component can be performed by means of a Markov move of the second type (M2). For this reason, the covariant behaviour of a link invariant under a change of framing of one component gives, in general, information on the partial closure of the braids. To be more precise, the variable α which describes the effect of an elementary modification of the framing characterizes, in equations (9.5) and (9.6), the enhancement which is used in Turaev's method of computing the Markov trace.

The perturbative computations performed in [22] show that a $\Delta w = \pm 1$ modification of the framing of a string associated with a representation ρ of G results in the presence of a multiplicative factor $\alpha^{\pm 1}$ which (at order λ^2 included) is consistent with the expression

$$\alpha(\rho) = q^{Q(\rho)} . \tag{11.3}$$

As we shall see in Chapter 13, this is the only possible value of the ***twist variable*** α (with $\alpha \to 1$ when $k \to \infty$) which is consistent with the form (10.15) of the monodromy matrix and with the satellite formula (6.35). So, we assume that the relations shown in Fig. 11.6 hold.

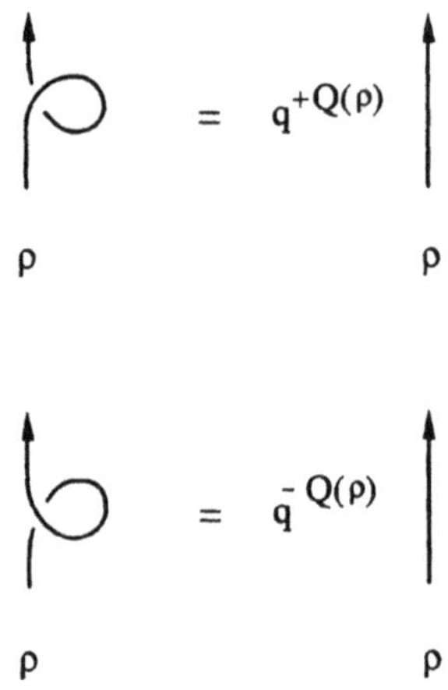

Figure 11.6.

(4) Projection compatibility

The satellite formulae derived in Chapter 6 and, in particular, equation (6.35) show that the irreducible representations of G appearing in the decomposition (11.1) and entering the projection decomposition rule can be used to construct satellites. Remember that our definition of satellites is based on a specific choice of the homeomorphism $h^\diamond$; $h^\diamond$ is determined by the framing of the link component which has to be removed and replaced with $h^\diamond(P)$ where P is the pattern link. In Chapter 6, we have considered the case in which the pattern link is equal to the link B shown in Fig. 6.6. On the other hand, rule **(1)** tells us that two parallel strings can be understood in terms of projectors as shown in Fig. 11.1. At this stage, we must specify how the projection decomposition is related to the satellite structure described by equation (6.35). The natural requirement is that there is a complete compatibility between satellites and projection decomposition. That is, we assume that (in the construction of satellites based on our choice of the homeomorphism $h^\diamond$) the generalized "pattern link" defined by means of a projector associated with the representation ρ is equivalent to the pattern link made of a single component labelled by the representation ρ. So, we assume that the relation shown in Fig. 11.7 holds.

The meaning of the projection compatibility rule can easily be understood in terms of link diagrams. Suppose that, in the intermediate steps of our analysis of a link diagram, we have introduced and used projectors. At this point, we have a set of pictures which do not represent link diagrams because, somewhere in each picture, a projector appears. Suppose now that the two strings, which originate from (and come back to) a given projector, are parallel on the plane of a given picture.

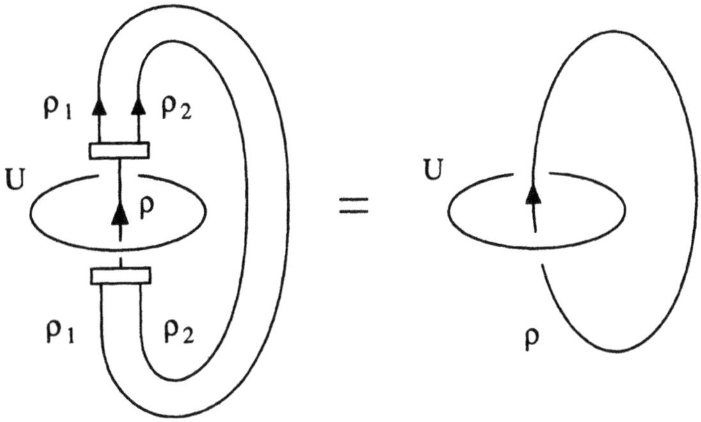

Figure 11.7.

These two components may produce overcrossings and undercrossings with several other components but, in any case, they remain parallel. As we have seen in Chapter 2, this means that, if we ignore the presence of the projector, the picture would represent the diagram of a satellite constructed with the pattern link B. In these conditions, equation (6.35) states that this couple of parallel strings can be substituted by a single string and we have to sum over the irreducible representations contained in the tensor product of the original representations associated with the two strings. Now, what happens if the two parallel strings originate from and come back to a given projector ? Rule **(4)** simply means that there is no longer a sum over the irreducible representations; only the component ρ, selected by the projector, must be considered. In other words, the relation shown in Fig. 11.7 means that the companion of a generalized diagram, representing a satellite obtained with a pattern link which contains a projector (as shown in Fig. 11.7), is just the link in which the relevant string is associated with the representation selected by the projector.

Rule **(4)** actually specifies the meaning of the projectors introduced in **(1)** and **(2)**. Among the five defining relations, condition **(4)** represents the essential point which characterizes the whole construction.

(5) Factorization

We have seen in Chapter 6 that, because of the peculiar properties of the quantum CS field theory, for the link $L = L_1 \cup L_2$, which is the distant union of L_1 and

L_2, the following relation

$$E(L_1 \cup L_2) = E(L_1)\, E(L_2) \tag{11.4}$$

must hold.

This concludes the list of the five defining relations of the universal link polynomial; in the remaining chapters, I will illustrate how to use them in order to reconstruct $E(L)$.

It would be useful to add some final comments. First of all, the structure of the defining relations **(1)**-**(5)** does not depend on the particular value taken by the deformation parameter q. This is in agreement with the fact that, in the perturbative computations of $\{ \langle W(L) \rangle \}$ in $\mathbb{R}^3$, the particular value of the coupling constant k does not play any relevant role. We shall return to this point when we consider surgery on three-manifolds.

According to the results of the previous chapters, equation (10.17) in particular, the deformation parameter q entering these defining relations is actually given by

$$q = \exp\left(-\frac{2\pi i}{k}\right) . \tag{11.5}$$

Note that our formulae for the "eigenvalues" of the braiding matrix, equations (10.16) and (10.17), and for the value of the twist variable (11.3) have been obtained with the normalization convention (4.1). If a different notation is adopted, the same quantities have different expressions but the final numerical values agree, as they should.

Rules **(1)**-**(5)** represent a generalization of the standard rules of computation based on the use of a simple skein relation. In the computation of the values of the universal link polynomial by means of rules **(1)**-**(5)**, the invariance under ambient isotopy transformations of the framed links is assumed. This is similar to the case of the ordinary "skein calculus". In other words, the information derived from **(1)**-**(5)** is perfectly compatible with the regular isotopy invariance of the link diagrams (or ambient isotopy invariance of the framed links). This is why the calculus based on **(1)**-**(5)** is simpler and more practical than any other known method. The formal proof that the rules **(1)**-**(5)** are consistent with the regular isotopy invariance of the link diagrams can be obtained by means of the known methods used in the case of the ordinary skein invariants. Such proof is not at all trivial, however, and will not be produced here. There is a simple argument which can be used to show the internal consistency of rules **(1)**-**(5)**. This argument is the fact that the quantum CS field theory exists. Indeed, rules **(1)**-**(5)** summarize exactly the whole information obtained in the field theory context; this will be explained in Chapter 13.

Finally, the numerical coefficients entering the defining relations are determined exclusively by the values of the quadratic Casimir operator of the given simple Lie algebra. The structure of the particular Lie algebra is used in the decomposition (11.1) of the tensor products into a sum over the irreducible components. As we have already mentioned, an extra index may be introduced to count the different but equivalent representations appearing in the decomposition (11.1). However, since all the results ultimately depend on the values of the quadratic Casimir operator only, this extra index will never appear in the final expression of $E(L)$. In fact, this extra index is used in connection with the projectors and these projectors are completely eliminated at the end of the computations. So, only the multiplicities of the inequivalent representations and their symmetry properties under permutations are relevant.

Chapter 12

The extended Jones polynomial

In this chapter we shall consider the case in which $G = SU(2)$ and the irreducible representations associated with the link components are arbitrary. In this case, the universal link polynomial $E(L)$ is called the extended Jones polynomial. The irreducible $SU(2)$ representations are labelled by J with $2J = 0, 1, 2, \ldots$. In our notations

$$Q(J) = J(J+1), \tag{12.1}$$

and the dimension $D(J)$ of the J-representation is

$$D(J) = 2J + 1. \tag{12.2}$$

12.1 The values of the unknots

The fundamental blocks, which are needed to reconstruct $E(L)$, are the values $\{ E_0[J] \}$ of the unknot U_0 (with zero writhe) associated with the representation J; let us compute them.

In the case in which $\rho_1 \simeq \rho_2 \simeq (J = 1/2)$, the projection decomposition relation of Fig. 11.1 takes the form shown in Fig. 12.1. Consider now the ***closure*** of the pictures appearing in both sides of this equation. Here, closure means that we connect the endpoints of the strings in the standard ordering-preserving way, as in the case of ordinary braid group elements. On the left-hand-side of the new equation that we obtain by taking the closure, we get a link diagram representing the distant union of two unknots with preferred framings (zero writhe), both associated with the $J = 1/2$ representation. Factorization **(5)** then implies that the associated value of $E(L)$ is just

$$E_0[1/2]\, E_0[1/2]. \tag{12.3}$$

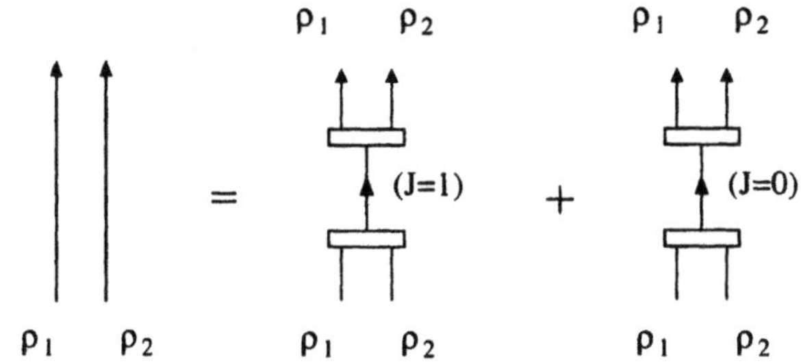

Figure 12.1.

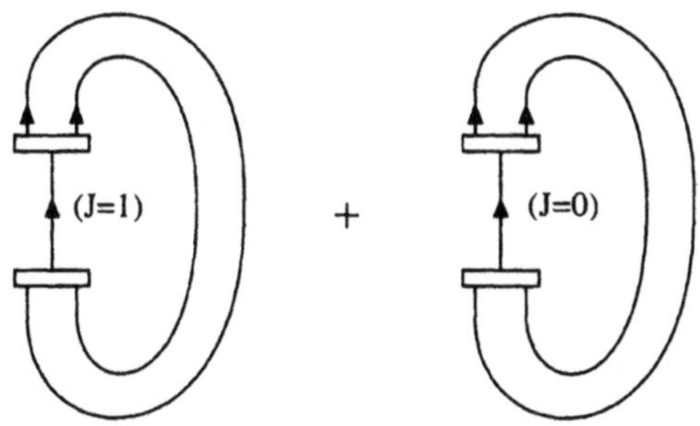

Figure 12.2.

On the right-hand-side of the equation, we get the sum of the pictures shown in Fig. 12.2. According to the projection compatibility rule **(4)**, these pictures are equivalent to those shown in Fig. 12.3.

Summarizing, by means of the rules introduced in Chapter 11, we have obtained the relation shown in Fig. 12.4. Note that, at this stage, the projectors have been eliminated and the pictures appearing in Fig. 12.4 are ordinary link diagrams.

Of course, the relation shown in Fig. 12.4 refers to the values of the extended Jones polynomial taken on the links appearing in the picture. In other words, the equality in Fig. 12.4 means

$$E_0[1/2]\, E_0[1/2] \;=\; E_0[1] + E_0[0]\,. \tag{12.4}$$

Equation (12.4) is just a particular case of equation (6.29). We now have three variables $E_0[0]$, $E_0[1/2]$ and $E_0[1]$ and only one relation; so, we shall look for two new equations involving these variables.

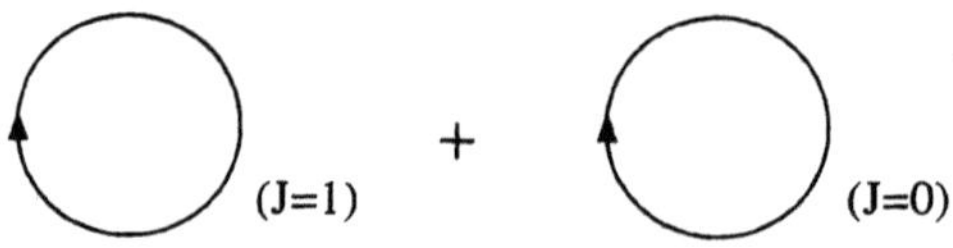

Figure 12.3.

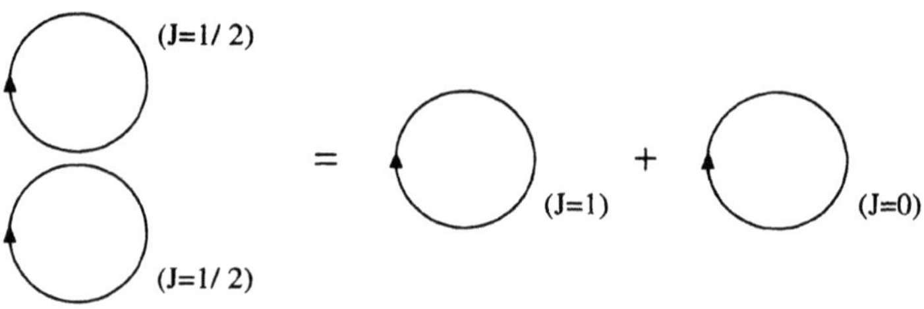

Figure 12.4.

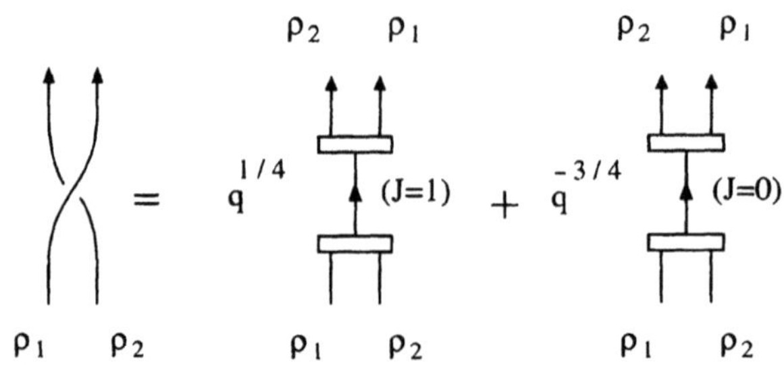

Figure 12.5.

When $\rho_1 \simeq \rho_2 \simeq (J = 1/2)$, the overcrossing picture of Fig. 11.3 takes the form shown in Fig. 12.5. By using the symmetry properties of the projectors illustrated in Fig. 11.5, the overcrossing relation can be written as in Fig. 12.6.
At this stage, we consider the closure of both sides of this equation. Note that, in order to make use of the projection compatibility rule **(4)**, the permutation of

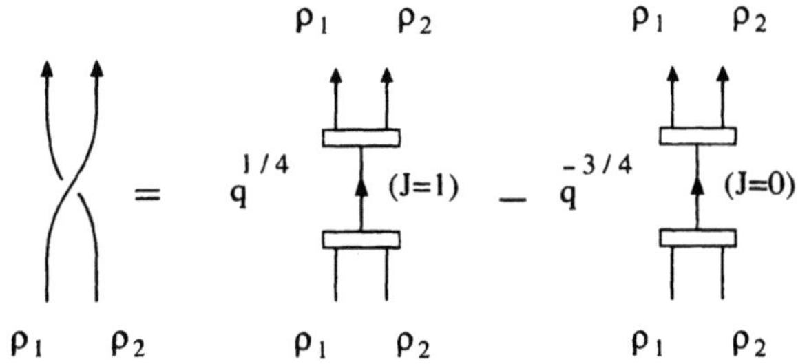

Figure 12.6.

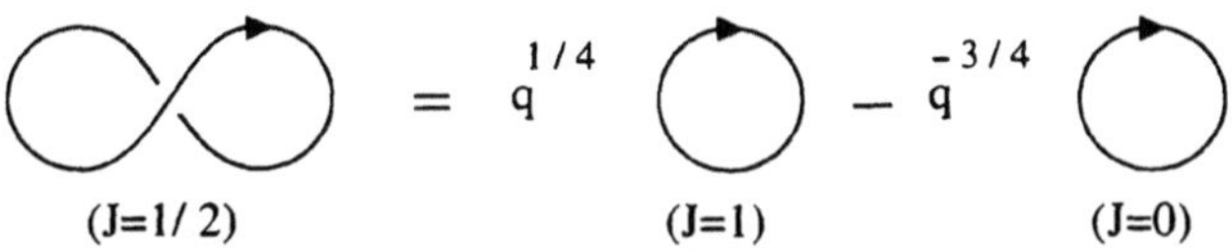

Figure 12.7.

the positions of ρ_1 and ρ_2 in the projectors was necessary. Finally, we obtain the relation shown in Fig. 12.7.
In Fig. 12.7, ordinary link diagrams appear. (The projectors have been eliminated). By taking into account the twisting rule (**3**), the unknot with writhe $w = +1$ is easily seen to correspond to $q^{3/4} E_0[1/2]$ and, therefore, the relation shown in Fig. 12.7 reads

$$q^{3/4}\, E_0[1/2] \;=\; q^{1/4}\, E_0[1] - q^{-3/4}\, E_0[0]\,. \tag{12.5}$$

The third equation connecting $E_0[0]$, $E_0[1/2]$ and $E_0[1]$ can be obtained by considering the undercrossing relation of Fig. 11.4. When $\rho_1 \simeq \rho_2 \simeq (J = 1/2)$, the picture of Fig. 11.4 takes the form shown in Fig. 12.8.
The same argument, which has been used in the case of overcrossing, now leads to

$$q^{-3/4}\, E_0[1/2] \;=\; q^{-1/4}\, E_0[1] - q^{3/4}\, E_0[0]\,. \tag{12.6}$$

By combining equations (12.5) and (12.6), one gets

$$E_0[1/2] \;=\; (q^{1/2} + q^{-1/2})\, E_0[0]\,, \tag{12.7}$$

$$E_0[1] \;=\; (q + 1 + q^{-1})\, E_0[0]\,. \tag{12.8}$$

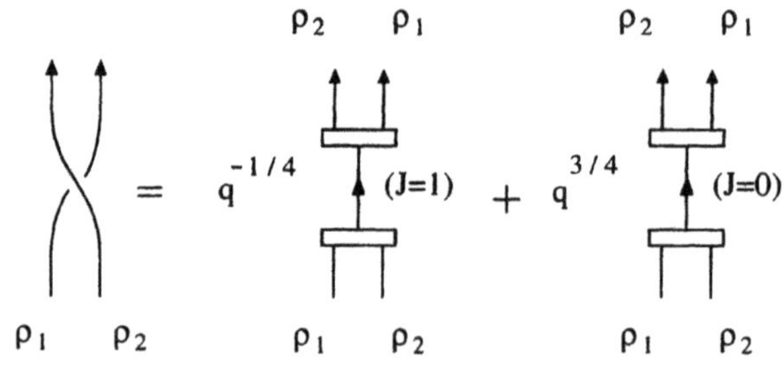

Figure 12.8.

If one inserts equations (12.7) and (12.8) into equation (12.4), one finds a quadratic equation for $E_0[0]$ and the two solutions are given by

$$\text{solutions of equation (12.4)} : \begin{cases} E_0[0] = 0 & ; \\ E_0[0] = 1 & . \end{cases} \tag{12.9}$$

The first possibility is that $E_0[0]$ vanishes. This would imply that $E_0[1/2]$ and $E_0[1]$ also vanish and, in fact, that $E(L)$ is vanishing for any link L. The existence of this trivial solution was expected because the solution $E(L) = 0$ (for any L) satisfies the five defining relations introduced in Chapter 11. Since all the computations are based exclusively on the relations **(1)**-**(5)**, it is clear that the trivial solution must be found.

In the field theory context, however, the possibility $E_0[0] = 0$ is excluded by definition, see equations (5.7) and (5.8). So, only the second solution

$$E_0[0] = 1 \tag{12.10}$$

is acceptable. Equation (12.10) gives precisely the right answer because we already know that $E_0[0]$ must be equal to unity. Summarizing, by using all the five defining relations **(1)**-**(5)** of Chapter 11, we have found equation (12.10) and

$$E_0[1/2] = q^{1/2} + q^{-1/2} , \tag{12.11}$$

$$E_0[1] = q + 1 + q^{-1} . \tag{12.12}$$

How to determine $E_0[3/2]$ should be clear now; we only need to consider the projection decomposition of Fig. 11.1 in the case in which $\rho_1 \simeq (J = 1/2)$ and $\rho_2 \simeq (J = 1)$. One finds the relation shown in Fig. 12.9.
The closure of the pictures shown in Fig. 12.9 gives

$$E_0[1/2]\, E_0[1] = E_0[3/2] + E_0[1/2] , \tag{12.13}$$

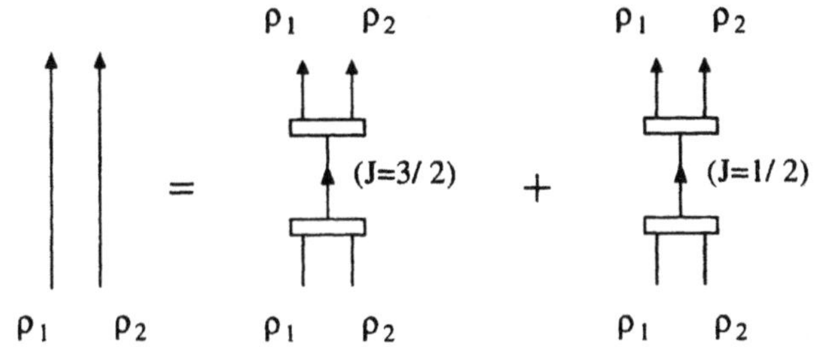

Figure 12.9.

and, by means of the expressions (12.11) and (12.12), we get

$$E_0[3/2] \;=\; q^{3/2} + q^{1/2} + q^{-1/2} + q^{-3/2} \;=\; \frac{q^2 - q^{-2}}{q^{1/2} - q^{-1/2}} \,. \tag{12.14}$$

At this point, we can give the general result concerning the values of the unknots:

Property E1. *The value $E_0[J]$ of the unknot with preferred framing and associated with the J-representation of $SU(2)$ is given by*

$$E_0[J] \;=\; \frac{q^{(2J+1)/2} - q^{-(2J+1)/2}}{q^{1/2} - q^{-1/2}} \,. \tag{12.15}$$

Proof. We have already seen that equation (12.15) is true when J is equal to 0, 1/2, 1 and 3/2. Assume now, by induction, that equation (12.15) holds for $0 \leq J \leq j$ with $j \geq 1/2$. Consider then the projection decomposition in the case in which $\rho_1 \simeq (j)$ and $\rho_2 \simeq (1/2)$. The decomposition into the irreducible components reads

$$(j) \otimes (1/2) \;=\; (j + 1/2) \oplus (j - 1/2) \,. \tag{12.16}$$

By taking the closure of the pictures entering the associated projection decomposition, one finds

$$E_0[j]\, E_0[1/2] \;=\; E_0[j + 1/2] + E_0[j - 1/2] \,. \tag{12.17}$$

The value of $E_0[j + 1/2]$ can be obtained from this equation and the resulting expression is in agreement with equation (12.15). So, assuming that equation (12.15) holds for $J \leq j$, we have shown that equation (12.15) is true also for $J = j + 1/2$. □

12.2 Hopf link

The diagram representing the Hopf link is shown in Fig. 12.10. We shall now compute the associated value of the extended Jones polynomial when the two components C_1 and C_2 are characterized by the representations $\rho_1 \simeq (J_1)$ and $\rho_2 \simeq (J_2)$.

The diagram shown in Fig. 12.10 is described by the closure of the braid shown in Fig. 12.11. Since we have two consecutive crossings, we only need to consider the crossing relation of Fig. 11.3 twice and make use of the orthogonality property of the projectors shown in Fig. 11.2.

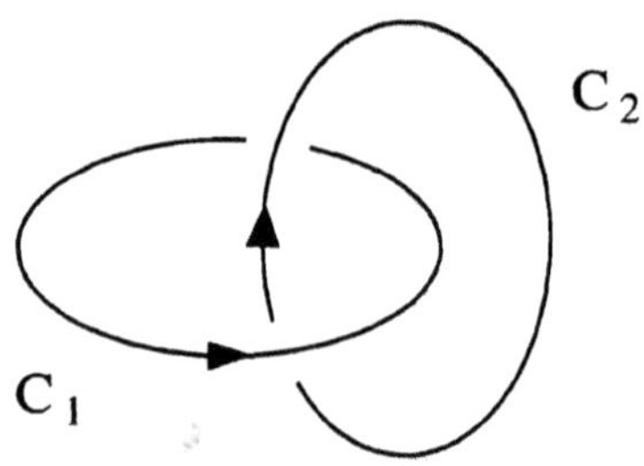

Figure 12.10.

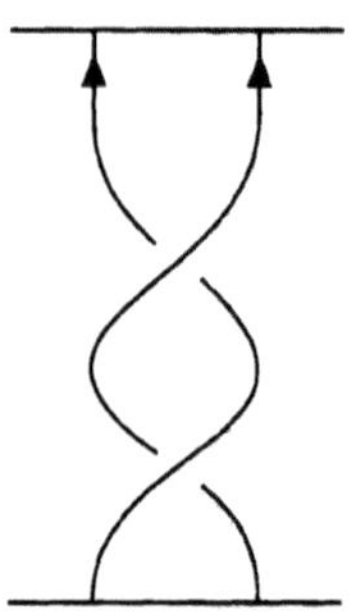

Figure 12.11.

First of all, the tensor decomposition reads

$$(J_1) \otimes (J_2) = \oplus_{J=|J_1-J_2|}^{J_1+J_2} (J) , \qquad (12.18)$$

where of course the sum is performed on the values of J which differ by unity. It should be noted that equation (12.18) and the values (12.1) of the quadratic Casimir operator represent the whole information coming from the group $SU(2)$. All the rest of the calculus is based on the (group-independent) topological properties of the links.

By means of rule **(2)**, we obtain the relation shown in Fig. 12.12, where the coefficients $d(J)$ are given by

$$d(J) = q^{-[J_1(J_1+1)+J_2(J_2+1)]/2} q^{J(J+1)/2} . \tag{12.19}$$

The crossing, appearing on the top of each projector in Fig. 12.12, is again decomposed according to the crossing relation **(2)**. As a result, we obtain a double sum of two consecutive projectors: the first projects on the representation (J) and the second projects on (J').

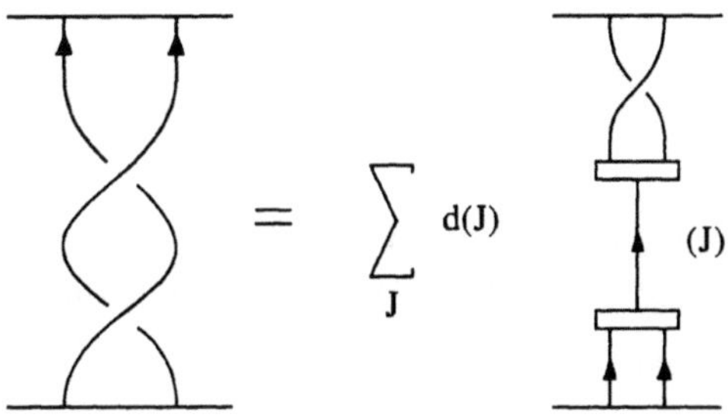

Figure 12.12.

By using the orthogonality property of the projectors, Fig. 11.2, the double sum is reduced to a single sum on J of single projectors with coefficients $\{d^2(J)\}$. Finally, rule **(4)** gives the relation shown in Fig. 12.13. Again, all the projectors have been eliminated and ordinary link diagrams are obtained.

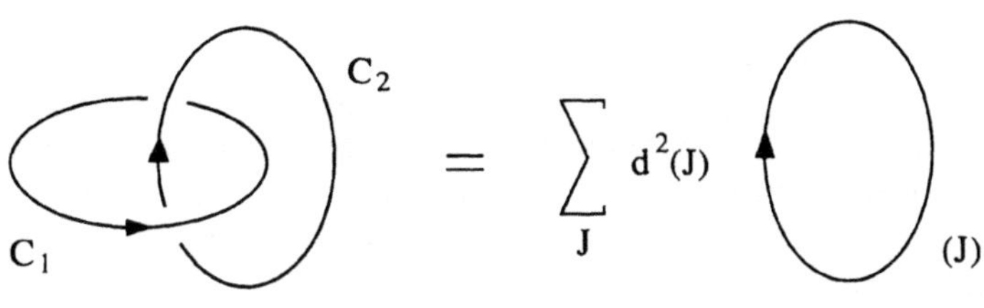

Figure 12.13.

So, for the value of $E(L)$ on the Hopf link, we have found

$$E(C_1, C_2; J_1, J_2) = \sum_{J=|J_1-J_2|}^{J_1+J_2} d^2(J)\, E_0[J] . \tag{12.20}$$

The sum appearing in equation (12.20) can easily be performed; indeed, by means of equation (12.15), one finds

$$q^{Q(J)}\, E_0[J] = \frac{1}{1-q}\left[q^{J^2} - q^{(J+1)^2} \right] . \tag{12.21}$$

Therefore,

$$\sum_{J=|J_1-J_2|}^{J_1+J_2} q^{Q(J)}\, E_0[J] = \frac{1}{1-q}\left[q^{(J_1-J_2)^2} - q^{(J_1+J_2)^2} \right] . \tag{12.22}$$

Finally, we have

Property E2. *The value of the Hopf link, whose components C_1 and C_2 have preferred framings and are associated with the representations J_1 and J_2 of $SU(2)$, is given by*

$$E(C_1, C_2; J_1, J_2) = \frac{q^{(2J_1+1)(2J_2+1)/2} - q^{-(2J_1+1)(2J_2+1)/2}}{q^{1/2} - q^{-1/2}} . \tag{12.23}$$

Proof. Equation (12.23) is simply a consequence of equations (12.19), (12.20) and (12.22). □

12.3 Trefoil knot

The right-handed trefoil diagram T, shown in Fig. 12.14(a), can be described by the closure of the braid represented in Fig. 12.14(b). When the knot T is characterized by the $(J = 1/2)$ representation, the computation of $E(T; 1/2)$ can be directly performed by means of the crossing relation shown in Fig. 12.6. Since the three crossings of Fig. 12.14(b) are consecutive, the orthogonality property of the projectors eliminates all the terms apart from the "diagonal" contributions in which two single projectors, selecting the (J=1) and (J=0) representation respectively, remain. Each associated numerical coefficient is equal to the third power of the coefficient appearing in Fig. 12.6. Therefore,

$$\begin{aligned} E(T;1/2) &= q^{3/4}\,E_0[1] - q^{-9/4}\,E_0[0] \\ &= (q+1)\,(q^2+q-1)\,q^{-9/4}\,. \end{aligned} \tag{12.24}$$

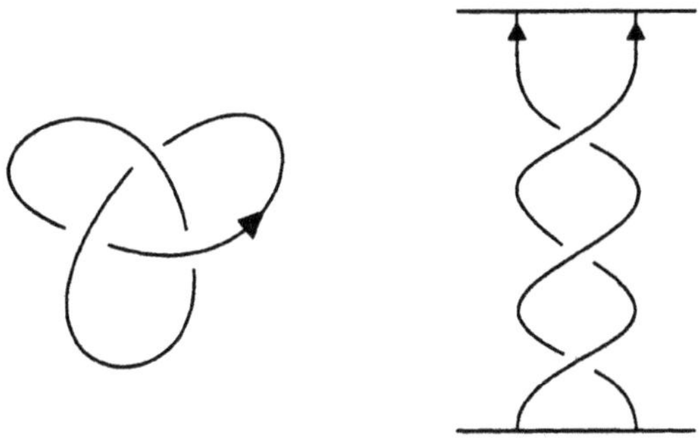

(a) (b)

Figure 12.14.

When $J = 1$, the tensor decomposition is

$$(1)\otimes(1) = (2)\oplus(1)\oplus(0)\,, \tag{12.25}$$

where (1) is antisymmetric and (2) and (0) are symmetric under a permutation of the two factors associated with the two equivalent ($J = 1$) representations. Therefore, the analogue of equation (12.24) reads

$$E(T;1) = q^3\,E_0[2] - q^{-3}\,E_0[1] + q^{-6}\,E_0[0]\,. \tag{12.26}$$

Let us now consider the case in which the trefoil knot is associated with a generic representation J. The tensor decomposition is

$$(J)\otimes(J) = (2J)\oplus(2J-1)\oplus\cdots\oplus(0)\,, \tag{12.27}$$

and, under a permutation of the two equivalent representations (J), each irreducible component keeps a factor

$$\text{representation}\quad (2J-n) \quad\Longrightarrow\quad \text{factor}\quad (-1)^n\,. \tag{12.28}$$

As usual, the closure of the three consecutive crossings, shown in Fig. 12.14, is analysed by using the crossing relation, the orthogonality property of the projectors and the projection compatibility rule. The final result can be summarized as:

Property E3. *The value* $E(T;\, J)$ *for the trefoil knot* T, *with writhe* $w = +3$ *and associated with the representation* J *of* $SU(2)$, *is given by*

$$E(T;\, J) = \frac{q^{3J^2}}{q^{1/2} - q^{-1/2}} \sum_{n=0}^{2J} (-1)^n\, q^{3/2n(n-4J-1)} \left[q^{(2J-n+1/2)} - q^{-(2J-n+1/2)} \right] . \tag{12.29}$$

Proof. Equation (12.29) is a consequence of the five defining relations introduced in Section 11. □

It is clear that the value of $E(L)$, associated with a knot or a link described by the closure of a braid of the type g_1^m in B_2, can immediately be obtained. In the next example, we shall consider a more "complicated" knot. The method that we shall use to determine the value of the extended Jones polynomial will give a more precise idea of how the defining relations **(1)**-**(5)** can conveniently be applied.

12.4 Figure-eight knot

The figure-eight knot diagram K_{F8}, shown in Fig. 12.15, can be described by the closure of the braid $g_1 g_2^{-1} g_1 g_2^{-1} \in B_3$, shown in Fig. 12.16, and the associated conjugacy class is denoted by $\langle g_1 g_2^{-1} g_1 g_2^{-1} \rangle_3$.

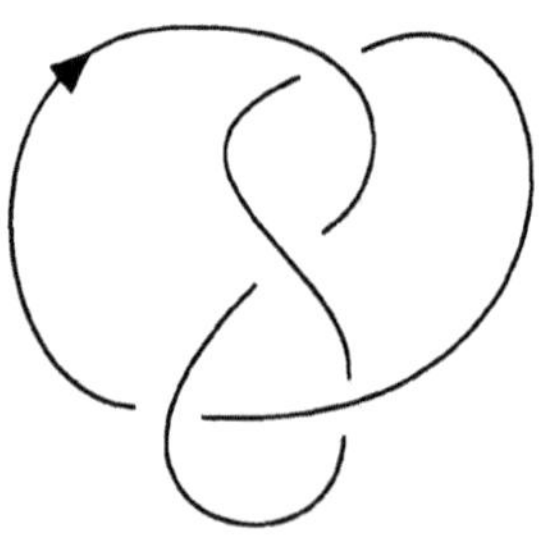

Figure 12.15.

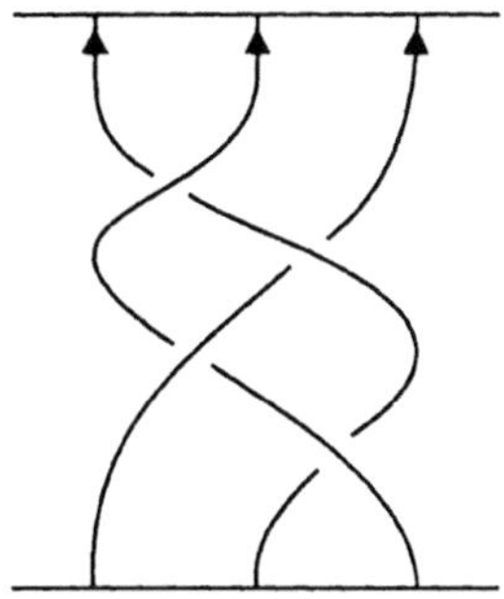

Figure 12.16.

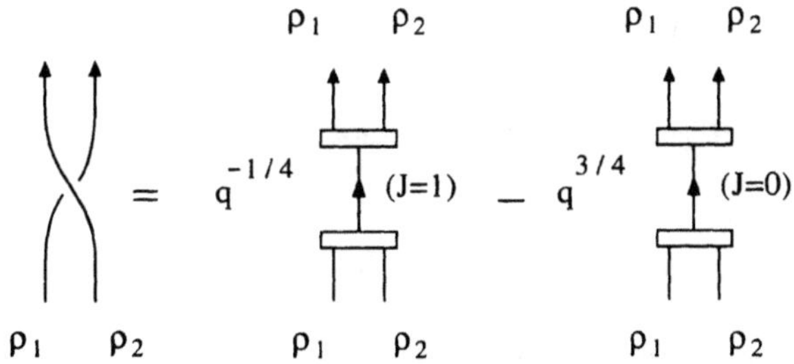

Figure 12.17.

In general, in the computation of the link invariants there is a simple "experimental" rule which states that the most convenient way of analysing a link diagram is to introduce a minimal number of modifications. In our case it is clear that, if we could exchange an overcrossing for an undercrossing, the job would be done. We shall now try to follow this strategy.

Let us consider the case in which K_{F8} is labelled by the fundamental ($J = 1/2$) representation of $SU(2)$. By taking into account the symmetry property of the projectors, the undercrossing relation of Fig. 12.8 can be written in the form shown in Fig. 12.17.

On the other hand, the two projectors on ($J = 1$) and ($J = 0$) also appear in the overcrossing relation of Fig. 12.6 and in the projection decomposition of Fig. 12.1. So it is clear that the undercrossing can be written as a suitable linear combination of an overcrossing and a "no-crossing". In fact, by eliminating the projectors, the relation shown in Fig. 12.18 is immediately obtained.

The result shown in Fig. 12.18 is precisely the skein relation which holds for the fundamental representation of $SU(2)$. Consequently, the generator g_2^{-1} of the braid group can be substituted by

$$g_2^{-1} \Longrightarrow q^{1/2}\, g_2 - \left(q^{3/4} - q^{-1/4}\right), \tag{12.30}$$

and, instead of the class $\langle g_1 g_2^{-1} g_1 g_2^{-1} \rangle_3$, we can consider the linear combination

$$q^{1/2}\, \langle g_1 g_2^{-1} g_1 g_2 \rangle_3 - \left(q^{3/4} - q^{-1/4}\right) \langle g_1 g_2^{-1} g_1 \rangle_3 \,. \tag{12.31}$$

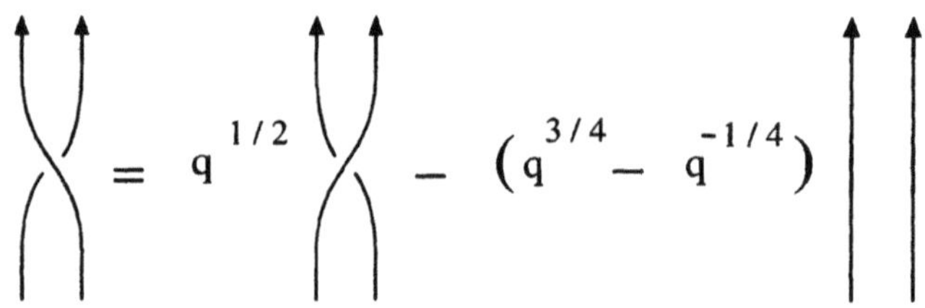

Figure 12.18. Skein relation for ($J = 1/2$)

Now, for the first class appearing in the expression (12.31), one has

$$\langle g_1 g_2^{-1} g_1 g_2 \rangle_3 = \langle g_2^{-1} g_1 g_2 g_1 \rangle_3 = \langle g_2^{-1} g_2 g_1 g_2 \rangle_3 = \langle g_1 g_2 \rangle_3 \,. \tag{12.32}$$

Clearly, the closure of $\langle g_1 g_2 \rangle_3$ simply describes the unknot with writhe $w = +2$; therefore

$$\langle g_1 g_2 \rangle_3 \Longrightarrow q^{3/2}\, E_0[1/2] \,. \tag{12.33}$$

On the other hand, the closure of $\langle g_1 g_2^{-1} g_1 \rangle_3$ describes a Hopf link in which one of the components has writhe $w = -1$ and so

$$\langle g_1 g_2^{-1} g_1 \rangle_3 \Longrightarrow q^{-3/4} \frac{q^2 - q^{-2}}{q^{1/2} - q^{-1/2}} \,. \tag{12.34}$$

In conclusion, by combining equations (12.31)-(12.34), we find, for the figure-eight knot diagram K_{F8},

$$E(K_{F8}\,;\, 1/2) = q^{5/2} + q^{-5/2} \,. \tag{12.35}$$

In the computation of $E(K_{F8};\, 1/2)$ we have shown: first, how to derive a skein relation from the defining relations **(1)-(5)** and, secondly, we have followed the standard method for the construction of link invariants based on the use of the skein relation. The case in which the knot K_{F8} is labelled by $J = 1$ is more interesting because it presents some novel aspects.

As before, in order to eliminate an undercrossing, we first derive a skein relation. Consider an overcrossing between two strings whose associated representations are $\rho_1 \simeq \rho_2 \simeq (J = 1)$. From the tensor decomposition (12.25), it follows that the overcrossing relation takes the form shown in Fig. 12.19. Note that the three projectors, appearing in Fig. 12.19, are those which also enter the projection decomposition (1).

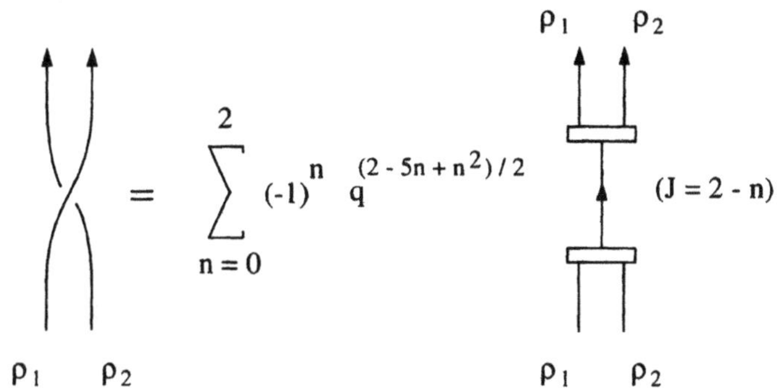

Figure 12.19.

Now we shall consider several consecutive crossings; in particular, let us denote by L_p (p is a positive integer) the configuration with p consecutive crossings of the two strings, as shown in Fig. 9.1. Because of the orthogonality property of the projectors, each L_p configuration can be written again as a linear combination of the three projectors mentioned above. Moreover, each numerical coefficient, entering the linear combination L_p, is equal to the p-th power of the corresponding coefficient shown in Fig. 12.19.

Of course, in the case of no-crossing at all, each coefficient is equal to the identity; i.e., the 0-th power must be considered. So, we can denote the no-crossing configuration by L_0. An undercrossing is represented by L_{-1} because, in this case, each coefficient entering the undercrossing relation coincides with the inverse of the corresponding coefficient entering L_1. Accordingly, p consecutive undercrossings will be denoted by L_{-p}.

In conclusion, all the configurations $\{L_p\}$, L_0 and $\{L_{-p}\}$ are expressed as linear combinations of three projectors only. Consequently, there is a "minimal" linear combination of these configurations telling us that, in reality, only three inequivalent projectors are used. This means that a generalized skein relation has finally been obtained.

Needless to say, our derivation of the skein relation is similar to the standard construction based on the minimal characteristic equation satisfied by the matrix representations of the generators of the braid group. In our method, we have not a matrix representation of the braid generators and our skein relations describe directly the properties of the values taken by $E(L)$ on the link diagrams which differ only by the presence of a different number of consecutive crossings.

The skein relation for $J = 1$, which follows from the overcrossing of Fig. 12.19, can be written as

$$q^2 L_2 - (1 - q + q^3) L_1 - (q^{-1} - q + q^2) L_0 + L_- = 0 . \tag{12.36}$$

In terms of the braid generators, equation (12.36) takes the form

$$g^{-1} = -q^2 g^2 + (1 - q + q^3) g + (q^{-1} - q + q^2) . \tag{12.37}$$

Therefore, instead of the class $\langle g_1 g_2^{-1} g_1 g_2^{-1} \rangle_3$, we can consider the linear combination

$$\begin{aligned}(1 - q + q^3) \langle g_1 g_2^{-1} g_1 g_2 \rangle_3 + (q^{-1} - q + q^2) \langle g_1 g_2^{-1} g_1 \rangle_3 \\ - q^2 \langle g_1 g_2^{-1} g_1 g_2^2 \rangle_3 ,\end{aligned} \tag{12.38}$$

which has been obtained from $\langle g_1 g_2^{-1} g_1 g_2^{-1} \rangle_3$ by substituting one generator g_2^{-1} according to equation (12.37). The first two classes appearing in the expression (12.38) have already been analysed. Since we are considering the case in which $J = 1$, the analogue of equation (12.33) now reads

$$\langle g_1 g_2^{-1} g_1 g_2 \rangle_3 \Longrightarrow q^4 E_0[1] , \tag{12.39}$$

and the analogue of equation (12.34) is

$$\langle g_1 g_2^{-1} g_1 \rangle_3 \Longrightarrow q^{-2} \frac{q^{9/2} - q^{-9/2}}{q^{1/2} - q^{-1/2}} . \tag{12.40}$$

Finally, $\langle g_1 g_2^{-1} g_1 g_2^2 \rangle_3$ must be considered. Clearly,

$$\langle g_1 g_2^{-1} g_1 g_2^2 \rangle_3 = \langle g_2^{-1} g_1 g_2^2 g_1 \rangle_3 , \tag{12.41}$$

and the closure of $g_2^{-1} g_1 g_2^2 g_1 \in B_3$ is shown in Fig. 12.20.

At first sight, it seems that our construction is leading nowhere because the link diagram corresponding to the picture of Fig. 12.20 seems to be as complicated as the starting diagram of Fig. 12.16. However, if you consider the picture more carefully, you can see that the link described by the diagram of Fig. 12.20 is just a satellite of the Hopf link: the first string of the braid shown Fig. 12.20 represents one of the components of the Hopf link whereas the remaining two strings represent the image of the pattern link which takes the place of the second component of

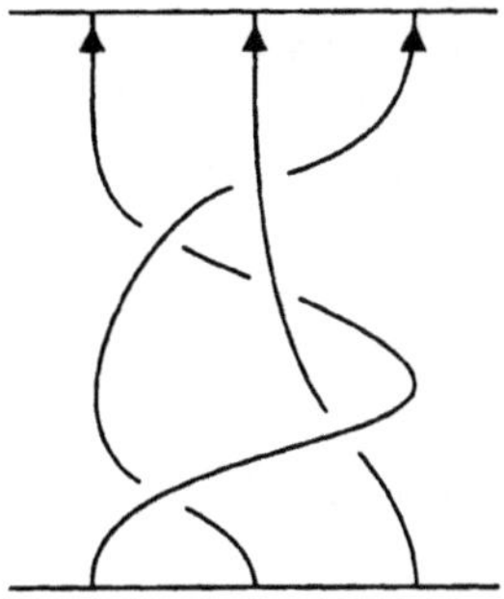

Figure 12.20.

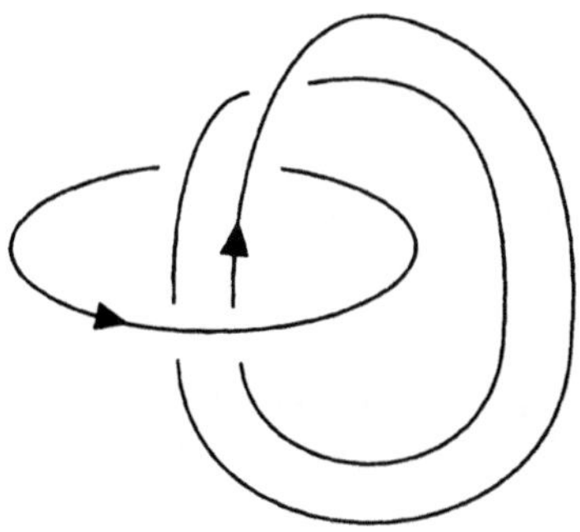

Figure 12.21.

the Hopf link. In other words, the link diagram H', corresponding to the closure of the braid of Fig. 12.20, is shown in Fig. 12.21.
One of the components of H' is a simple circle and the second component of H' is the image of the pattern link. In this particular case, the complement of a tubular neighbourhood of the first component of H' in S^3 is a solid torus standardly embedded in S^3 and, for this reason, the image of the pattern link is ambient isotopic with the pattern link itself. In conclusion, in order to compute $E(H')$, we only need to decompose the pattern link into a sum of its irreducible components and make use of the result (12.23) concerning the Hopf link.

We are interested then in the pattern link which contains a single undercrossing with $J = 1$. First of all, we use rule **(2)** which gives us a relation of the same type shown in Fig. 12.19. In the undercrossing relation, each coefficient is the inverse of the corresponding coefficient appearing in Fig. 12.19. This relation implies that the resulting pattern link is equivalent to a linear combination of generalized pattern links constructed by means of projectors. Secondly, we use the projection compatibility rule **(4)** and we obtain the relation shown in Fig. 12.22. Therefore,

$$E(H') = q^{-1}\, E(C_1, C_2; 1, 2) - q\, E(C_1, C_2; 1, 1) + q^2\, E(C_1, C_2; 1, 0)\,, \tag{12.42}$$

where $E(C_1, C_2; J_1, J_2)$ is the value of the Hopf link derived in Section 12.2.

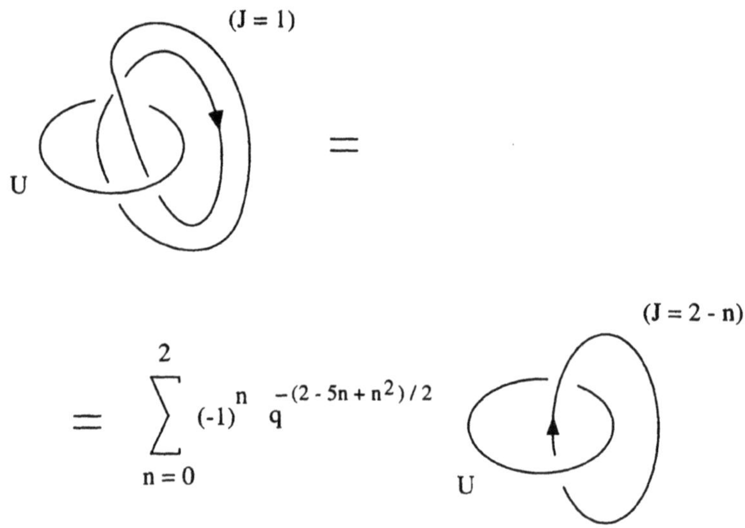

Figure 12.22.

Finally, from equations (12.38)-(12.42) it follows that the value of the extended Jones polynomial on the figure-eight knot diagram K_{F8} (with $J = 1$) is given by

$$E(K_{F8}\,; 1) = q^7 - q^5 + q + 1 + q^{-1} - q^{-5} + q^{-7}\,. \tag{12.43}$$

12.5 Connection with the Jones polynomial

A precise relation exists between the universal link polynomial $E(L)$ when $G = SU(2)$ and the Jones polynomial.

Suppose that all the components of the links are associated with the fundamental $J = 1/2$ representation of $SU(2)$. In this particular case, the projection decomposition takes the form shown in Fig. 12.1, whereas the crossing relations are shown in Figs.12.6 and 12.17. As we have already mentioned, we have three configurations (overcrossing, undercrossing and no-crossing) expressed in terms of two projectors only. Therefore, there is a linear relation between them, shown in Fig. 12.18, which is precisely the ordinary skein relation with particular values of the parameters. Taking into account the properties of $E(L)$ under twisting (or Reidemeister moves of the type I), we conclude that for $G = SU(2)$ and when all the link components are labelled by the representation $J = 1/2$ only, $E(L)$ satisfies the skein relation (2.11) of the $S(L)$ polynomial with

$$\alpha = q^{3/4}, \tag{12.44}$$

$$\beta = q^{1/4}, \tag{12.45}$$

$$z = q^{1/2} - q^{-1/2}, \tag{12.46}$$

and with the normalization of the unknot given in equation (12.11). Therefore, (for $J = 1/2$) the combination

$$V(L; q^{1/2}) = \frac{q^{-3w(L)/4}}{q^{1/2} + q^{-1/2}} \cdot E(L) \tag{12.47}$$

is an ambient isotopy invariant and coincides with the Jones polynomial $V(L)$ defined in Chapter 2. This is the reason why $E(L)$, for $G = SU(2)$, is called the extended Jones polynomial.

Equation 12.47 also shows that $E(L)$, for $G = SU(2)$ and when all the link components have $J = 1/2$, is actually a polynomial in $q^{\pm 1/4}$ because $V(L; q^{1/2}) \in Z[q^{\pm 1/2}]$. Of course, the same conclusion can be obtained by analysing directly the defining relations of the $S(L)$ polynomial.

It would be useful to recall our definition of the Jones polynomial. Let us denote by x the quantity

$$x = q^{1/4}. \tag{12.48}$$

Then, according to our notation, $V(L; x^2)$ is defined by

(i) ambient isotopy invariance,
(ii) $V(U; x^2) = 1$, for the unknot U,
(iii) $x^4 V(L_+) - x^{-4} V(L_-) = (x^2 - x^{-2}) V(L_0)$. (12.49)

In the standard conventions adopted in mathematics, however, the exchange relation (12.49) takes a different form. In order to clarify the relation between the two invariants, let us introduce the original Jones polynomial $V_J(L; x^2)$ defined by

(i) ambient isotopy invariance,
(ii) $V_J(U; x^2) = 1,$ for the unknot U,
(iii) $x^4 V_J(L_+) - x^{-4} V_J(L_-) = -(x^2 - x^{-2}) V_J(L_0)\,.$ (12.50)

Equation (12.50) represents the exchange relation in the form which is usually reported in connection with the original Jones polynomial. If we compare equation (12.49) with equation (12.50), it is clear that

$$V(L; x^2) = V_J(L; -x^2)\,, \tag{12.51}$$

and then (for $J = 1/2$)

$$V_J(L; -x^2) = \frac{x^{-3w(L)}}{x^2 + x^{-2}} \cdot E(L; x)\,. \tag{12.52}$$

12.6 Bracket connection

The Kauffman's bracket polynomial $V_B(L)$ is an interesting polynomial associated with unoriented link diagrams; $V_B(L)$ is strictly related to the Jones polynomial. The bracket polynomial is defined by [58]

(i) regular isotopy invariance,
(ii) $V_B(U_0; x) = -(x^2 + x^{-2}),$ for the circle U_0,
(iii) $V_B(L^{(+)}) = -x^3\, V_B(L^{(0)}), \quad V_B(L^{(-)}) = -x^{-3}\, V_B(L^{(0)})\,,$
(iv) $V_B(L_C) = x\, V_B(L_U) + x^{-1}\, V_B(L_V)\,.$ (12.53)

The configurations L_C, L_U and L_V are shown in Fig. 12.23, whereas $L^{(+)}$, $L^{(-)}$ and $L^{(0)}$ are those appearing in Fig. 2.6. Note that, since the writhe of a single component of a link diagram is invariant under a modification of the orientation

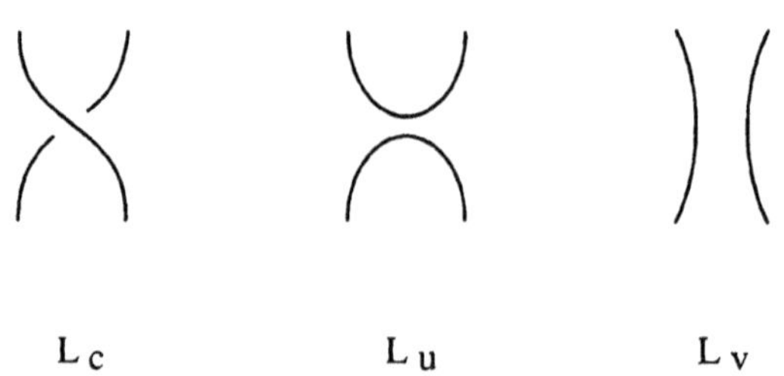

Figure 12.23.

of this component, the distinction between $L^{(+)}$ and $L^{(-)}$ is also meaningful for unoriented link diagrams.

It is clear that only for unoriented link diagrams can the configurations L_C, L_U and L_V be used in the exchange relation.

Now, any invariant I of unoriented link diagrams defines (by trivial extension) an invariant I_E for oriented links. Obviously, the value of I_E is the same on all the oriented link diagrams whose only difference is the orientation of some component. This trivial extension of the bracket polynomial on the set of oriented link diagrams will be denoted by the same symbol $V_B(L, x)$.

For any given oriented link diagram L, it is easy to see that the bracket polynomial is related to the original Jones polynomial by

$$V_B(L; x) = -(x^2 + x^{-2}) \left(-x^3\right)^{w(L)} V_J(L; x^2) . \tag{12.54}$$

By comparing equation (12.52) with equation (12.54), we get (for $J = 1/2$)

$$V_B(L; x) = (\pm i)^{w(L)} E(L; \mp ix) , \tag{12.55}$$

which can obviously be rewritten as

$$E(L; x) = \exp\left(\mp\frac{i\pi\, w(L)}{2}\right) V_B(L; \pm ix) . \tag{12.56}$$

In conclusion, when the gauge group is $SU(2)$, the CS vacuum expectation values of the Wilson line operators (associated with the framed links whose components are characterized by the fundamental $J = 1/2$ representation) are not equal to the values of the bracket polynomial. The precise relation between the two is shown in equation (12.56).

The reason why the bracket polynomial cannot describe directly the expectation values of the field theory is clear: the covariant behaviour of $V_B(L; x)$ under an elementary modification of the framing is not the correct behaviour of the Wilson line operators. In the defining relation (iii) of $V_B(L; x)$, there is an extra minus sign which is not present in the field theory context. The absence of this minus sign in the quantum CS theory can easily be checked, for instance, by considering the $k \to \infty$ "semi-classical" limit. In this limit, the expectation values of the Wilson line operators become equal to the product of the dimensions of the representations associated with the different components of the links. This quantity does not depend on framing; therefore, for $k \to \infty$, the twist variable must be equal to the identity.

12.7 Reconstruction theorems

In the previous sections we have seen how the defining relations **(1)**-**(5)** can be used to compute the values of $E(L)$ when $G = SU(2)$. Only a few significant examples have been considered and it is natural to ask, now, whether rules **(1)**-**(5)** completely determine the value of the extended Jones polynomial on a generic link. The answer is that they do.

We shall now present constructive proof of this claim; that is, we shall produce a direct and general method for the computation of $E(L)$. The proof consists of two parts. Firstly, we will show how to compute $E(L)$ in the case in which all the representations associated with the link components are equivalent to the fundamental ($J = 1/2$) representation. Secondly, we shall give a recursive relation which determines $E(L)$ for arbitrary representations.

Theorem R1. *For each oriented link diagram L, whose components $\{C_i\}$ are labelled by the fundamental ($J = 1/2$) representation of $SU(2)$, the associated non-trivial value*

$$E(L) = E(C_1, C_2, \ldots, C_m; 1/2, 1/2, \ldots, 1/2), \tag{12.57}$$

is uniquely determined by

$$\begin{aligned}
&\text{(i)} \quad \textit{regular isotopy invariance,} \\
&\text{(ii)} \quad E(U_0) = E_0[1/2] = (q^{1/2} + q^{-1/2}), \\
&\text{(iii)} \quad E(L^{(+)}) = q^{3/4}\, E(L^{(0)}), \quad E(L^{(-)}) = q^{-3/4}\, E(L^{(0)}), \\
&\text{(iv)} \quad q^{1/4}\, E(L_+) - q^{-1/4}\, E(L_-) = (q^{1/2} - q^{-1/2})\, E(L_0).
\end{aligned} \tag{12.58}$$

The configurations L_+, L_- and L_0 are shown in Fig. 2.5 *and $L^{(+)}$, $L^{(-)}$ and $L^{(0)}$ appear in* Fig. 2.6. *The value $E(L)$, shown in* equation (12.57), *can be computed by using the skein relation recursively. Therefore, $E(L)$ is a regular isotopy invariant which takes the form of a (finite) Laurent polynomial in the variable $q^{1/4}$ with integer coefficients. Furthermore, $E(L)$ is invariant under a modification of the orientation of a generic link component.*

Proof. From twisting **(3)**, it follows that $E(L)$ is a regular isotopy invariant and the value of the twist variable coincides with the value reported in (iii). The value $E_0[1/2]$ of the unknot U_0 with zero writhe has been derived, by means of **(1)**-**(5)**, in Section 12.1. Neglecting the trivial solution, which is not acceptable, $E_0[1/2]$ is unique and is given in (ii). The skein relation (iv), satisfied by $E(L)$ for ($J = 1/2$), has been derived in Section 12.4, see Fig. 12.18. It is well known that conditions (i)-(iv) are consistent and identify a unique regular isotopy invariant, so we will not insist on this point. The values of this invariant can easily be obtained by means of the standard methods of skein theory: the final result is that, for $(J = 1/2)$, $E(L) \in Z[q^{\pm 1/4}]$.

We shall now prove that $E(L)$ (shown in equation (12.57)) is invariant under a modification of the orientation of a generic single component, say C_1, of the link diagram L. Let L' be the link diagram obtained from L by a modification of the orientation of the component C_1. Then, equation (12.56) gives

$$E(L'; q^{1/4}) = \exp\left(\mp \frac{i\pi\, w(L')}{2}\right) V_B(L'; \pm i q^{1/4}), \tag{12.59}$$

where $w(L')$ is the writhe of L'. Since the bracket polynomial is an invariant for unoriented links,

$$V_B(L'; \pm i q^{1/4}) = V_B(L; \pm i q^{1/4}). \tag{12.60}$$

Let us now consider the writhe $w(L')$. Under a modification of the orientation of C_1, $w(C_1)$ does not change; only the writhe numbers of the crossings between C_1 and the remaining components of L are modified. The number of crossings of C_1 with each single remaining component of L is necessarily even. Therefore,

$$w(L') = w(L) + 4n \tag{12.61}$$

for some integer n. Equations (12.60) and (12.61), inserted in equation (12.59) give

$$E(L'; q^{1/4}) = E(L; q^{1/4}). \tag{12.62}$$

Equation (12.62) concludes the proof of theorem **R1**. □

It should be noted that $E(L)$, defined by theorem **R1**, is an ambient isotopy invariant of unoriented framed links in $\mathbb{R}^3$ and in S^3 as well. Indeed, any link in $\mathbb{R}^3$ (or in S^3) belongs to the interior of a three-ball and, within this three-ball, rules (i)-(iv) uniquely determine the value of the invariant. We shall now use satellites to analyse the remaining representations of the group.

Theorem R2. *For each oriented link diagram L, whose components $\{C_i\}$ are labelled by a set $\{J_i\}$ of irreducible representations of $SU(2)$, the associated non-trivial value*

$$E(L) = E(C_1, C_2, \ldots, C_m; J_1, J_2, \ldots, J_m), \tag{12.63}$$

is uniquely determined by (and can be reconstructed by means of) the five defining relations **(1)-(5)**. *Moreover, $E(L) \in Z[q^{\pm 1/4}]$ and represents the value of an ambient isotopy invariant of unoriented framed links in $\mathbb{R}^3$ and in S^3 as well.*

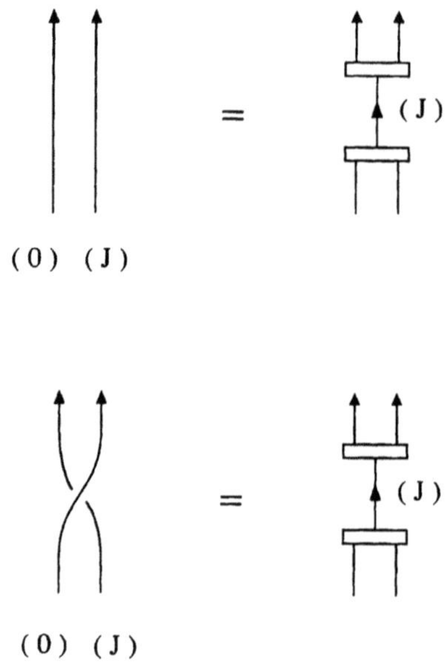

Figure 12.24.

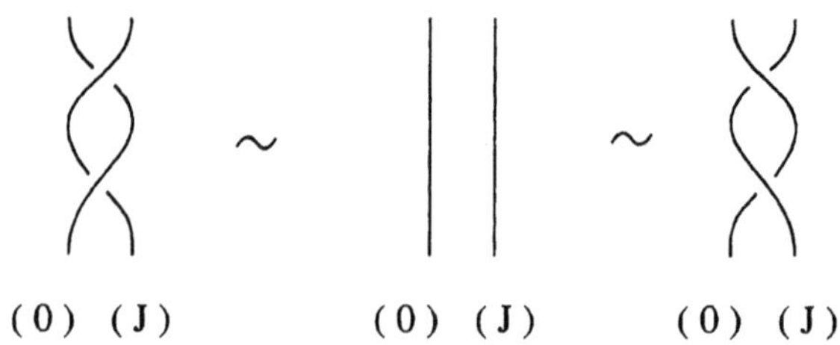

Figure 12.25.

Proof. When each representation J_i is equivalent to the fundamental ($J = 1/2$) representation of $SU(2)$, theorem **R1** implies that all the statements of theorem **R2** are true.

Let us analyse the trivial $J = 0$ representation. Consider the set of two string configurations in which one of the two strings is labelled by $\rho_1 \simeq (J = 0)$ and the other string is associated with $\rho_2 \simeq (J)$ for arbitrary J. In this case, the projection decomposition **(1)** and the crossing relation **(2)** are shown in Fig. 12.24. Consequently, the relations shown in Fig. 12.25 are valid. Consider now a generic link diagram L in which one of the components, say C_1, is characterized by the trivial $J = 0$ representation. Because of the relations of Fig. 12.25, the component

C_1 can be unlinked from all the remaining components of L. Then, C_1 can be unknotted and reduced to a simple circle. Finally, from factorization **(5)**, we get

$$E(L) = E_0[0]\, E(L^{(r)}) = E(L^{(r)}) \tag{12.64}$$

where $L^{(r)}$ is the link which is obtained from L by removing the component C_1. In conclusion, any link component associated with the trivial $J = 0$ representation can be removed.

We shall now present a general prescription for the computation of $E(L)$ shown in equation (12.63). The method we shall describe is called ***standard cabling***; it is based on

(a) a recursive use of the satellite relations;
(b) the fact that, for each representation (J) of $SU(2)$, one has

$$(1/2) \otimes (J) = (J + 1/2) \oplus (J - 1/2)\,. \tag{12.65}$$

Let $L' = \{C^{(2)}, C_2, \ldots, C_m\}$ be a satellite of a link $L = \{C_1, C_2, \ldots, C_m\}$, where $C^{(2)} = h^\circ(B)$ is the image of the pattern link B, shown in Fig. 2.11. The satellite L' is constructed by means of the homeomorphism h° defined in Chapter 2. All the remaining components $\{C_2, \ldots, C_m\}$ and their associated representations are kept fixed.

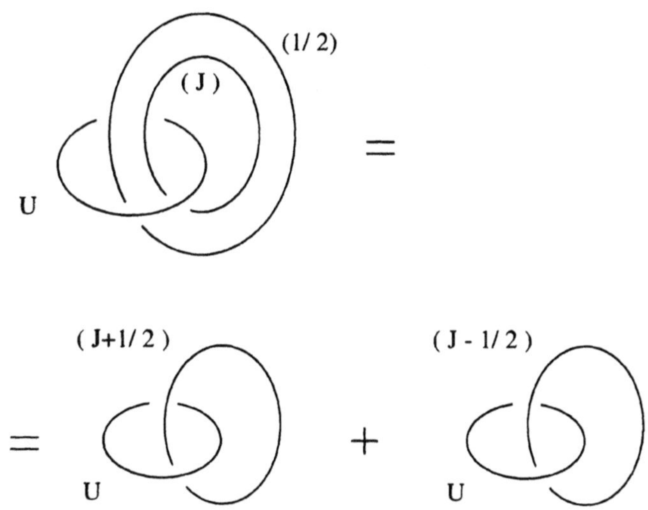

Figure 12.26.

Let the two components K_1 and K_2 of $C^{(2)}$ be associated with the representations $\rho_1 \simeq (J = 1/2)$ and $\rho_2 \simeq (J)$ for an arbitrary J. From rules **(1)** and **(4)**, it follows that, in the construction of satellites, the relation shown in Fig. 12.26 holds. This means

$$E(C^{(2)}, \ldots; 1/2, J, \ldots) = E(C_1, \ldots; J + 1/2, \ldots) + E(C_1, \ldots; J - 1/2, \ldots). \tag{12.66}$$

Equation (12.66) can be used recursively for the computation of the extended Jones polynomial in the case in which the representations $\{ J_i \}$, associated with the components $\{ C_i \}$ of the link L, are arbitrary.

Indeed, for each link component C labelled by (J) with $J > 1/2$, $E(L)$ can be written as a linear combination of two terms

$$\begin{aligned} E(\ldots, C, \ldots; \ldots, J, \ldots) = & - E(\ldots, C, \ldots; \ldots, J - 1, \ldots) \\ & + E(\ldots, C^{(2)}, \ldots; \ldots, 1/2, J - 1/2, \ldots). \end{aligned} \tag{12.67}$$

In the first term, the component C is labelled by $(J - 1)$. In the second term, this component is "doubled" ($C^{(2)}$ is the image of the pattern link B) and the two components of $C^{(2)}$ are associated with the representations $(1/2)$ and $(J - 1/2)$. This procedure can be used recursively and, at the end, only $(J = 1/2)$ or $(J = 0)$ representations remain. At this stage, the link components associated with $(J = 0)$ can be removed and all the remaining components are labelled by $(J = 1/2)$. Therefore the value of $E(L)$, shown in equation (12.63), is equal to a suitable combination, with integer coefficients, of contributions evaluated in the case in which all the link components are labelled by $(J = 1/2)$. In these conditions, we can use the skein relation of theorem **R1**.

Consequently, $E(L)$ is a regular isotopy invariant for unoriented link diagrams and $E(L) \in Z[q^{\pm 1/4}]$ for arbitrary representations of $SU(2)$. Finally, the whole construction illustrated above is valid for arbitrary framed links in $\mathbb{R}^3$ and in S^3 also. In conclusion, the reconstruction theorem **R2** is proved. □

Chapter 13

General properties

The introduction of the defining relations **(1)-(5)** has been "very strongly" suggested by the three-dimensional CS theory. We now want to prove that in the $E(L)$ polynomial, defined by **(1)-(5)**, all the general properties of the expectation values $\{ \langle W(L) \rangle \}$ obtained in Chapter 6 are indeed satisfied. In the present chapter, the gauge group G will be taken to be a generic compact simple Lie group. Several features of the universal link polynomial will be analysed. We shall concentrate, in particular, on the generalized satellite relations, on connected sums of links and on mutations.

13.1 Twist variable

The properties of the Wilson line expectation values, which have been derived in Chapter 6, are the most important features of the field theory construction. Among the results of Chapter 6, the ***satellite formula*** (6.35) plays a fundamental role because it implies the complete integrability of the CS theory.

We would now like to show that equation (6.35) can be derived from **(1)-(5)**. The proof is very simple. Indeed, for each irreducible representation $\rho(t)$ contained in the tensor product of ρ_1 and ρ_2, the associated projector appears in the projection decomposition **(1)**. On the other hand, from projection compatibility **(4)**, it follows that the satellites constructed with the closure of these projectors are equivalent to the satellites constructed with ordinary knots. Therefore, combining **(1)** and **(4)**, the relation shown in Fig. 13.1 is obtained; this relation implies equation (6.35). The fact that equation (6.35) is a direct consequence of the defining relations is not surprising, of course, because the defining relations themselves have been formulated in such a way that all the field theory properties are satisfied.

One of the consequences of the satellite formula (6.35) is represented by the deep connection which must exist between the twist variables and the eigenvalues of the monodromy matrix. Suppose that in the twisting relations **(3)** the twist variable $\alpha(\rho)$, associated with the representation ρ, is not specified as in equation (11.3) but is left undetermined. Then, the known asymptotic behaviour of $\alpha(\rho)$ together with the consistency between crossing **(2)** and projection compatibility **(4)** require

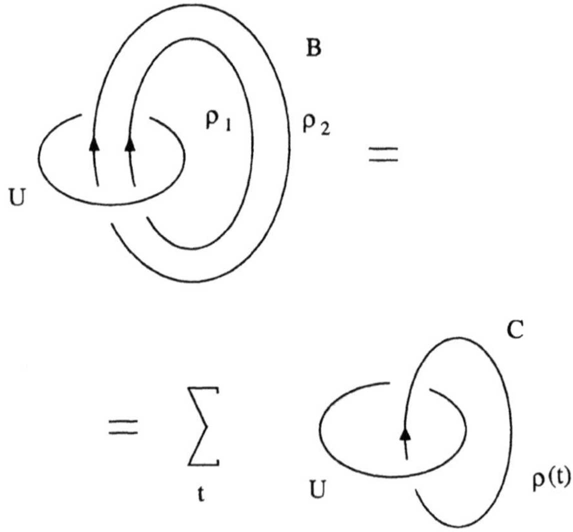

Figure 13.1.

that equation (11.3) must hold. The proof is based on the validity of the satellite relation shown in Fig. 13.1.

Let us denote by $\mathcal{V}$ the complement of a tubular neighbourhood of the circle U in S^3. $\mathcal{V}$ is a solid torus standardly embedded in S^3 and Fig. 13.1 can be interpreted in the following way. The two-component link B in $\mathcal{V}$, appearing on the left-hand-side of the relation, is a satellite of the unknot C, contained in $\mathcal{V}$, appearing on the right-hand-side. This framed unknot in $\mathcal{V}$ has standard framing (vertical framing convention); in other words, the unknot C shown in Fig. 13.1 has a vanishing self-linking number. Suppose now that the framing of C is modified in such a way that its new self-linking number is equal to $+1$. In this case, the framings and the linking number of the two components of the corresponding satellite will be modified accordingly. Therefore, the analogue of the relation shown in Fig. 13.1 will take the form shown in Fig. 13.2.

Let us denote by $B^{(1)}$ the two-component link in $\mathcal{V}$ shown in Fig. 13.2. $B^{(1)}$ represents the image of the link B of Fig. 6.6 under the action of a right-handed ***twist*** of the solid torus $\mathcal{V}$ which contains B. Twist homeomorphisms of solid tori will be analysed in detail in Chapter 16.

By means of twisting **(3)**, the relation of Fig. 13.2 can be written in the form shown in Fig. 13.3, where the values of the twist variable have not been specified.

On the left-hand-side of the relation appearing in Fig. 13.3, a monodromy matrix (equivalent to the expression (10.15)) is associated with the pure braid group crossing between the two components characterized by ρ_1 and ρ_2. For each irreducible component $\rho(t)$, appearing in the decomposition of the tensor product $\rho_1 \otimes \rho_2$, the

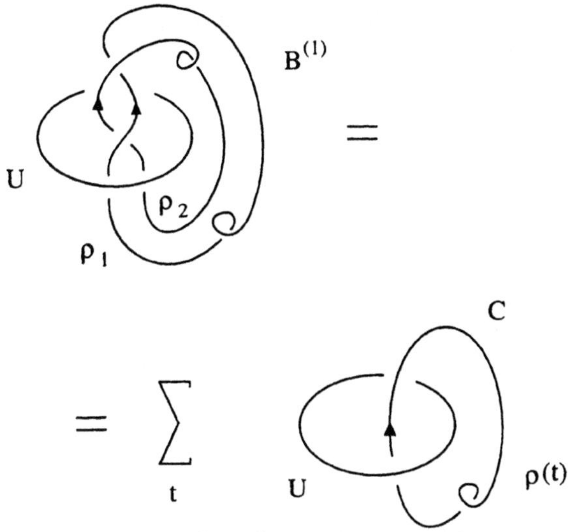

Figure 13.2.

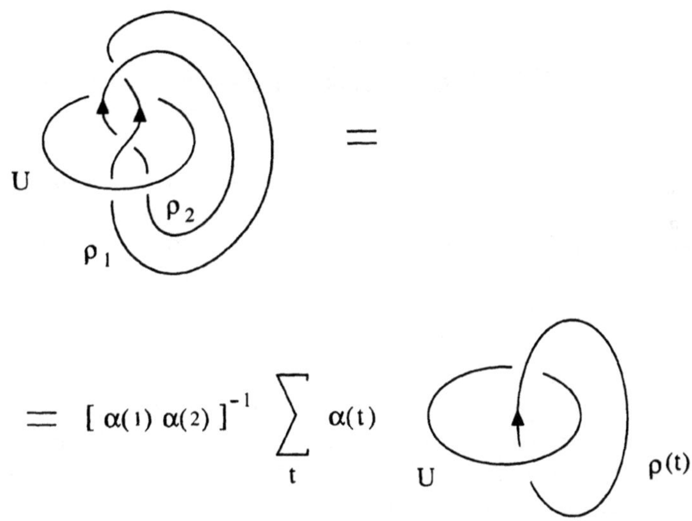

Figure 13.3.

corresponding eigenvalue of the monodromy matrix reads

$$m(t) = q^{-Q(\rho_1)}\, q^{-Q(\rho_2)}\, q^{Q(\rho(t))} \,. \tag{13.1}$$

On the other hand, the projectors have been introduced in order to diagonalize the monodromy and, in fact, on the right-hand-side of the relation of Fig. 13.3 the effect of the pure braid group crossing is expressed, on each single component with given $\rho(t)$, by the multiplicative factor

$$m(t) = [\alpha(\rho_1)\,\alpha(\rho_2)]^{-1}\,\alpha(\rho(t))\,. \tag{13.2}$$

If we compare equations (13.1) and (13.2), we finally obtain

Property G1. *The only possible value of the twist variable $\alpha(\rho)$, which is consistent with the satellite formulae and coincides with the identity in the semiclassical $k \to \infty$ limit, is given by*

$$\alpha(\rho) = q^{Q(\rho)}\,. \tag{13.3}$$

The value (13.3) of the twist variable is in agreement with the results of the perturbative computations [22], of course. In the defining relation **(3)**, the correct value (13.3) of $\alpha(\rho)$ has simply been assumed.

13.2 Recovered field theory

We shall now consider the behaviour of $E(L)$ under a modification of the orientation of the ambient space $\mathbb{R}^3$ (or S^3). Let $\widetilde{L}$ be the ***mirror image*** of the link diagram L; i.e. $\widetilde{L}$ is related to L by exchanging overcrossings with undercrossings (and vice-versa). Therefore, from **(2)** and **(3)**, it follows that $E(\widetilde{L})$ is simply obtained from $E(L)$ by replacing q with q^{-1}. Since $q = \exp(-2\pi i/k)$, this means that equation (6.15) is satisfied.

Another discrete symmetry which has to be considered is ***inversion***. Let L^{-1} be the link diagram obtained by reversing the orientations of all the components of L. Under the inversion operation $L \to L^{-1}$, the writhe number of each crossing is not modified. In other words, under inversion each overcrossing (undercrossing) remains an overcrossing (undercrossing). Consider now a link invariant in $\mathbb{R}^3$ (or S^3) whose construction is based on the analysis of link diagrams. If only the difference between over/under-crossing is relevant, this invariant is left unchanged under inversion. The universal link polynomial defined by **(1)-(5)** is precisely of this type and so

$$E(L) = E(L^{-1})\,. \tag{13.4}$$

This property was already known to hold for the Jones, HOMFLY and $S(L)$ polynomials and reflects the isotropy properties of the ambient space. On the other hand, the complex conjugation of $E(L)$ is equivalent to exchanging q^* for q. Since $q = \exp(-2\pi i/k)$, it is easy to see that equation (6.14) is satisfied.

We can also introduce a discrete symmetry which may be called ***charge conjugation***. Consider an oriented framed link L in $\mathbb{R}^3$ (or S^3) whose m components $\{C_i\}$ are characterized by the representations $\{\rho_i\}$ of the gauge group G. Let L^* be the link obtained from L by replacing ρ_i with ρ_i^* for $i = 1, 2, \ldots, m$. The projection decompositions of the couples (ρ_i, ρ_j) and (ρ_i^*, ρ_j^*) have the same structure and the corresponding crossing relations coincide because $Q(\rho) = Q(\rho^*)$. Therefore

$$E(L^*) = E(L). \tag{13.5}$$

Since the link L^{-1} (obtained from L by inversion) coincides with L^*, equation (13.5) is in agreement with equation (13.4).

Suppose now that C, which is one of the components of L, is associated with a real representation ρ. Reversing the orientation of C only ($C \to C^{-1}$ and all the other components of L are kept fixed) is equivalent to performing the substitution $\rho \to \rho^*$ in L. In our case since ρ is real, the projection decomposition **(1)** and all the crossing relations **(2)** are invariant under $\rho \to \rho^*$. Therefore, the universal link polynomial $E(L)$ is invariant under $C \to C^{-1}$.

When all the representations $\{\rho_i\}$ are real, $E(L)$ represents an invariant of regular isotopy for unoriented link diagrams, in perfect agreement with the conclusions we came to in Chapter 6. To give an example: when $G = SU(2)$, each irreducible representation is equivalent to its complex conjugate. In this case, we have explicitly shown in Chapter 12 that the extended Jones polynomial is an invariant of unoriented framed links.

In the previous chapter, we have also shown how the ***generalized skein relations*** can be derived from **(1)**-**(5)**. Consider the crossing relation, shown in Fig. 11.3, when ρ_1 is equivalent to ρ_2. On each irreducible component $\rho(t)$, appearing in the decomposition of the tensor product

$$\rho_1 \otimes \rho_2 = \oplus_t \, \rho(t), \tag{13.6}$$

the crossing operation is realized by multiplication with

$$\lambda(t) = (-1)^{s(t)} \, q^{[Q(\rho(t)) - Q(\rho_1) - Q(\rho_2)]/2}, \tag{13.7}$$

where $s(t) = 0$ (1) if $\rho(t)$ is symmetric (antisymmetric) under a permutation of ρ_1 and ρ_2. Let us denote by $\mathcal{K}$ the crossing operation between two strings and by $\mathcal{K}^p$ the presence of p consecutive crossings. Then $\mathcal{K}$ satisfies

$$\prod_t{}' \, [\mathcal{K} - \lambda(t)] = \sum_p c_p \mathcal{K}^p = 0, \tag{13.8}$$

where the product is performed on the different values of $\{\lambda(t)\}$. Consequently

$$\sum_p c_p \, E(L_p) \; = \; 0 \, , \tag{13.9}$$

which represents a generalized skein relation of the type considered in equation (9.11). An explicit example of a generalized exchange relation is shown in equation (12.36). Clearly, the generalized skein relations can also be written when the two strings are associated with inequivalent representations. In this case, the numbers of consecutive crossings must be even. As we have already mentioned, equation (13.9) is not sufficient for reconstructing $E(L)$. In fact, as we have seen in Chapter 12, in addition to equation (13.9) much more information can be derived from the defining relations **(1)**-**(5)**.

In conclusion, all the general features of the topological field theory are contained in the defining relations **(1)**-**(5)**. Only the particular values of the numerical inputs remain to be verified. But, as we have seen in Chapter 11 and in the previous section, the coefficients entering the crossing relations have been chosen to agree with the results of the field theory concerning the monodromy. So, rules **(1)**-**(5)** summarize all the information derived from the quantum CS field theory. As a result, it follows that

Property G2. *The result of any computation based on* **(1)**-**(5)** *coincides with the result obtained in the quantum CS field theory.*

Property **G2** is quite remarkable; the CS model is the first example of a nontrivial quantum field theory whose structure can unambiguously and equivalently be described in purely topological terms. Here, there are no "nice axioms" to be imposed, there aren't any physically inequivalent choices or ambiguities in the quantization and you cannot modify any single numerical factor: everything follows exclusively from an action principle.

In this model, the applicability of topological arguments is not restricted to a region in which the semiclassical approximation is valid. In fact, no kind of approximation is at all necessary. There is no low-energy limit to be considered, nor any other limit, in some given parameter, to be performed. The quantum CS theory is a "true" topological field theory in the sense that any observable and any result obtained in this model has exclusively a topological origin and a topological meaning.

Finally, let us consider the construction of an invariant, defined on link diagrams, which represents an ambient isotopy invariant for (not framed) links. In terms of link diagrams, we would like to connect $E(L)$, which is a ***regular isotopy*** invariant, with an ***ambient isotopy*** invariant naturally associated with it. Because of the covariance properties of $E(L)$ under Reidemeister moves of type I, it is very easy to obtain ambient isotopy invariants from $E(L)$. Let $\{C_1, \ldots, C_m\}$ be the components of the link diagram L and $\{\rho_1, \ldots, \rho_m\}$ be the associated representations of G. Then

$$X(L) = E(L)\, q^{-\sum_{i=1}^{m} w(C_i) Q(\rho_i)} \tag{13.10}$$

is clearly an ambient isotopy invariant. For equivalent representations, $\rho_i \simeq \rho \quad \forall i = 1, 2, \ldots, m$, one can define

$$X'(L) = E(L)\, q^{-w(L) Q(\rho)}, \tag{13.11}$$

which also represents an ambient isotopy invariant for (not framed) links.

13.3 Links in a solid torus

The quantum CS field theory provides an intrinsic three-dimensional description of the link invariants and, for this reason, several features of $E(L)$ admit a simple physical interpretation. In this section we shall use field theory arguments to derive the most important property of the universal link polynomial.

Let U be a circle in S^3, N a tubular neighbourhood of U and $\mathring{N}$ its interior. As we have mentioned in Chapter 2, $\mathcal{V} = S^3 - \mathring{N}$ is also a solid torus and we simply call it the complement of U in S^3. The circle $C \subset \mathcal{V}$, shown in Fig. 13.4, is essential in $\mathcal{V}$ and can be taken to represent the "core" of $\mathcal{V}$. We can also introduce an orientation for C and a preferred framing C_f, which has a linking number zero with C.

Suppose now that a Wilson line operator $W(C; \rho)$ is associated with the knot C. As we have shown in Chapter 7, a complete set of gauge invariant observables associated with the circle C is given by the set of Wilson line operators $\{ W(C; \rho) \}$, with ρ varying in the set of all the inequivalent irreducible representations of the gauge group G. Clearly, when ρ is the trivial representation, the Wilson operator is just the identity. Equivalently, you can imagine that in this case there is not a Wilson operator at all.

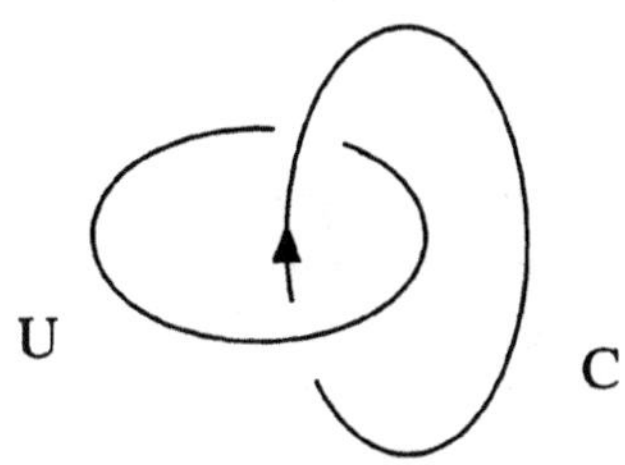

Figure 13.4.

If the representation ρ is not irreducible, we can decompose it into a sum of irreducible components

$$\rho = \oplus_{i=1}^{n} \rho(i) . \tag{13.12}$$

Since the Wilson line operator is the trace of the quantum holonomy, the decomposition (13.12) results in the sum of Wilson operators

$$W(C;\rho) = \sum_{i=1}^{n} W(C;\rho(i)) . \tag{13.13}$$

Thus, the definition of $W(C)$ can be extended, by linearity, to the general situation in which the knot C is associated with a linear combination of inequivalent irreducible representations. For example, let K be a generic oriented framed knot in S^3. The "colour" of K can be taken to be

$$\chi = x_1 \rho(1) + x_2 \rho(2) + \cdots + x_n \rho(n) , \tag{13.14}$$

where $\{x_i\}$ are complex numbers and $\{\rho(i)\}$ are irreducible representations; this simply means that the associated Wilson operator is

$$W(K;\chi) = x_1 W(K;\rho(1)) + x_2 W(K;\rho(2)) + \cdots + x_n W(K;\rho(n)) . \tag{13.15}$$

According to this definition, the vacuum expectation value of both sides of this equation will give

$$E(K;\chi) = x_1 E(K;\rho(1)) + x_2 E(K;\rho(2)) + \cdots + x_n E(K;\rho(n)) . \tag{13.16}$$

Similarly, let $L = \{C_1 \cup C_2\}$ be a two-component oriented framed link in S^3. When the two components of L have colours $\chi_1 = \sum_i x_i \rho(i)$ and $\chi_2 = \sum_j y_j \rho(j)$ respectively, the associated invariant is simply

$$E(C_1, C_2; \chi_1, \chi_2) = \sum_i \sum_j x_i y_j E(C_1, C_2; \rho(i), \rho(j)) . \tag{13.17}$$

Let us now reconsider the solid torus $\mathcal{V}$. Let P be an oriented framed link in $\mathcal{V}$ and let each component of P be characterized by an irreducible representation of G.

The product $W(P)$ of the Wilson line operators associated with the components of $P \subset \mathcal{V}$ represents a gauge-invariant observable defined in the solid torus $\mathcal{V}$. Let us imagine that $\mathcal{V}$ is a tube standardly embedded in S^3. Since the CS model is a topological general-covariant theory, the thickness of the tube is totally irrelevant. In the limit in which this thickness goes to zero, the tube degenerates to a simple circle C: the core of the tube. Consequently, $W(P)$ can be interpreted as a gauge-invariant observable associated with the knot C.

Since $\{ W(C;\rho) \}$ is a complete set of gauge-invariant observables associated with C, $W(P)$ necessarily admits an expansion of the form

$$W(P) = \sum_{\rho} \xi(\rho)\, W(C;\rho)\,, \tag{13.18}$$

where the sum is over the inequivalent irreducible representations of G and the complex coefficients $\{ \xi(\rho) \}$ encode the particular structure of $P \subset \mathcal{V}$. Equation (13.18) can be represented (symbolically) as shown in Fig. 13.5. In conclusion, the following property holds

Property G3. *Any link contained inside a solid torus, standardly embedded in* S^3, *is equivalent to (and can be substituted by) a single component* C, *which coincides with the core of solid torus. The colour associated with* C *is a suitable linear combination of the inequivalent irreducible representations of the gauge group. With a given choice of the framing of* C *(we shall always use preferred framing), the coefficients entering this combination are fixed.*

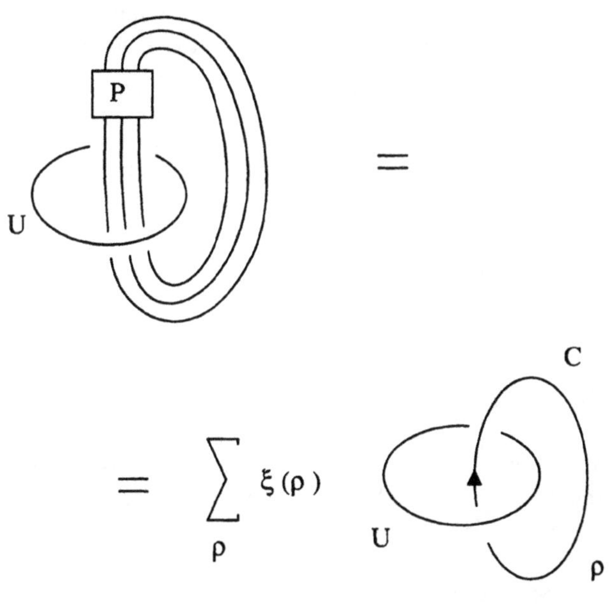

Figure 13.5.

Equation (13.18) represents the most important feature of the universal link polynomial and has to be understood as an equivalence relation which can be used in the computation of the universal link polynomial.

For any given link P in $\mathcal{V}$, the expression (13.18) contains a finite number of non-vanishing coefficients $\{\xi(\rho)\}$. The representations appearing in the sum (13.18) can be, at most, those entering the decomposition of the tensor product of the representations associated with the components of P which intersect an arbitrarily fixed meridinal disc of $\mathcal{V}$. Indeed, gauge invariance implies that the non-Abelian "electric" flux associated with a meridinal discs of $\mathcal{V}$ must be conserved; therefore, it must be the same for any choice of the meridinal disc.

From the physical point of view, equation (13.18) can be understood as a consequence of two basic symmetry properties of the CS theory. Firstly, gauge invariance implies that the non-Abelian flux associated with a meridinal disc of the solid torus must be conserved. Secondly, general covariance implies that this flux can always be imagined to be concentrated on a single knot which coincides with the core of the solid torus. It is interesting to compare this physical picture with the mathematical derivations of equation (13.18). In the algebraic approach [7] of Reshetikhin and Turaev, equation (13.18) is a consequence of the Hopf structure of the relevant ribbon algebra. In the general knot theory context, equation (13.18) can easily be derived when some skein relation can be used. Indeed suppose that, in this case, we analyse the links contained in a solid torus $\mathcal{V}$. By making use of the skein relation recursively, any given link in $\mathcal{V}$ can be decomposed in terms of a finite number of "elementary" links in $\mathcal{V}$; the associated modules label the conjugacy classes of the braid group. A detailed discussion of the mathematical derivations of equation (13.18) is beyond the purposes of the present lectures. We simply note that the quantum CS field theory provides a simple explanation of equation (13.18).

Let us consider a few examples. Suppose that $P \subset \mathcal{V}$ is actually contained inside a three-ball $\mathcal{B} \subset \mathcal{V}$. In this case, a meridinal disc of $\mathcal{V}$ which does not intersect any component of P exists. Consequently, the only representation which appears in the sum (13.18) is the trivial representation. The only (possibly) non-vanishing coefficient is just $\xi(\rho = \text{ trivial })$ and, by taking the vacuum expectation value of both sides of equation (13.18), it follows that $\xi(\rho = \text{ trivial })$ must coincide with $E(P)$. So, equation (13.18) reads

$$W(P) = E(P)\, W(C; \rho = \text{ trivial }) = E(P) \cdot \mathbb{1} = E(P) . \qquad (13.19)$$

In the second example, the link P is of the type shown in Fig. 13.6.
The box, appearing in Fig. 13.6, contains a ***tangle*** X (which is simply part of a link) and the component of P, which crosses the meridinal disc shown in the figure, is associated with a given (non-trivial) irreducible representation $\widetilde{\rho}$. Also in this case, the sum appearing in equation (13.18) contains only one term

$$W(P) = \xi(\widetilde{\rho})\, W(C; \widetilde{\rho}) . \qquad (13.20)$$

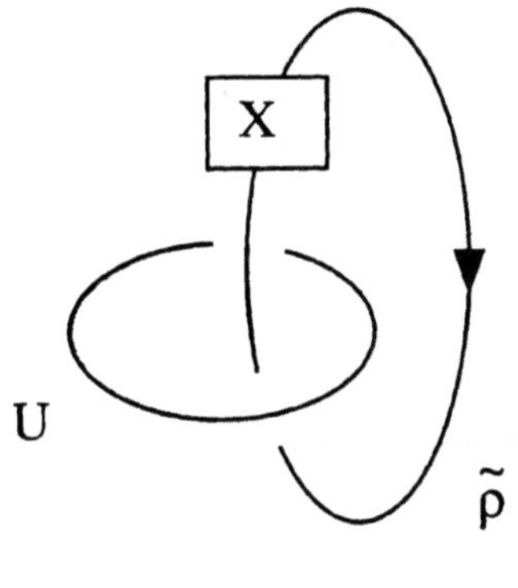

Figure 13.6.

The vacuum expectation value of both sides of equation (13.20) determines the value of the coefficient $\xi(\tilde{\rho})$

$$\xi(\tilde{\rho}) = \frac{E(P)}{E(C;\tilde{\rho})} = \frac{E(P)}{E_0[\tilde{\rho}]}. \tag{13.21}$$

The link $P \subset \mathcal{V}$ of the last example is shown in Fig. 13.7.

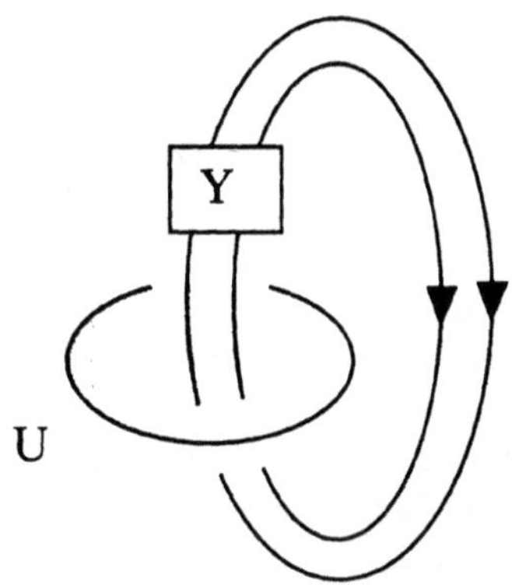

Figure 13.7.

In order to simplify our discussion, let us consider the gauge group $G = SU(2)$. Suppose now that the two link components, which intersect the meridinal disc, are associated with the representations ($J = 1/2$) and ($J = 1$) of $SU(2)$. Since

$$(1/2) \otimes (1) = (3/2) \oplus (1/2), \tag{13.22}$$

equation (13.18) reads

$$W(P) = \xi(3/2)\, W(C; 3/2) + \xi(1/2)\, W(C; 1/2). \tag{13.23}$$

The vacuum expectation value of both sides of equation (13.23) gives

$$E(P) = \xi(3/2)\, E_0[3/2] + \xi(1/2)\, E_0[1/2]. \tag{13.24}$$

In order to compute the coefficients $\xi(1/2)$ and $\xi(3/2)$, a second equation is necessary. We can consider, for example, the satellite $P^{(1)}$ constructed with P and whose companion is the unknot with writhe $+1$. Equivalently, $P^{(1)}$ represents the image of P under a right-handed twist of the solid torus $\mathcal{V}$ which contains P. In this case, by using the same argument presented in Section 13.1, we find

$$E(P^{(1)}) = q^{15/4}\, \xi(3/2)\, E_0[3/2] + q^{3/4}\, \xi(1/2)\, E_0[1/2]. \tag{13.25}$$

Clearly, the system of equations (13.24) and (13.25) admits a unique solution for $\xi(1/2)$ and $\xi(3/2)$. Once the values of the coefficients $\xi(1/2)$ and $\xi(3/2)$ are known, the computation of $E(L)$ for the satellites constructed with P can be simplified. Indeed, as we shall demonstrate in the next section, we only need to compute the value of the universal link polynomial for the companion.

13.4 Satellites

Let L be an oriented framed link in S^3 with m components. Each component C_i of L (with $i = 1, 2, \ldots, m$) has framing C_{if} and is labelled by a given irreducible representation ρ_i of the gauge group. Let us denote by N_i a tubular neighbourhood of the link component C_i.

We shall also consider the circle U in S^3 introduced in the previous section. The complement of U in S^3 is the solid torus $\mathcal{V}$ standardly embedded in S^3; the core of $\mathcal{V}$ is the oriented knot C, shown in Fig. 13.4, which has preferred framing C_f.

As we have seen in Chapter 2, for each component C_i of L there is a unique (up to ambient isotopy) homeomorphism

$$h_i^\circ : \mathcal{V} \Longrightarrow N_i\,, \tag{13.26}$$

so that $h_i^\circ(C) = C_i$ and $h_i^\circ(C_f) = C_{if}$. Thus, any link L in S^3 can be interpreted as the union of several (disjoint) solid tori $\{ N_i \}$ and, for each N_i, there is a specified homeomorphism $h_i^\circ : \mathcal{V} \to N_i$ which represents the framing of the considered component.

Since the image of the *framed* core C of $\mathcal{V}$ under h_i° is precisely the *framed* component C_i of L, we can write

$$\begin{aligned} L &= C_1 \cup C_2 \cup \cdots \cup C_m \\ &= h_1^\circ(C) \cup h_2^\circ(C) \cup \cdots \cup h_m^\circ(C) \,. \end{aligned} \tag{13.27}$$

The representations $\{ \rho_i \}$, associated with $\{ C_i \}$, are "internal" quantum numbers which are universal because they do not depend on the particular embeddings N_i in S^3. As we have explained in Section 13.3, the definition of the universal link polynomial can be extended by linearity to the general situation in which each component C_i is associated with a linear combination of inequivalent irreducible representations of the gauge group. So, in general, $E(L)$ is understood to be a multilinear function

$$E(L) = E(C_1, C_2, \ldots, C_m; \chi_1, \chi_2, \ldots, \chi_m) \,, \tag{13.28}$$

on the space of the possible colourings $\{ \chi_1, \chi_2, \ldots, \chi_m \}$ of the link components. For fixed irreducible representations $\{\chi_i = \rho_i \}$, equation (13.28) becomes

$$E(L) = E(C_1, C_2, \ldots, C_m; \rho_1, \rho_2, \ldots, \rho_m) \,. \tag{13.29}$$

Let us concentrate on a single solid torus, say N_1, and suppose that, instead of the core C of $\mathcal{V}$, we now consider a link $P \subset \mathcal{V}$. The substitution

$$C_1 = h_1^\circ(C) \Longrightarrow h_1^\circ(P) \,, \tag{13.30}$$

gives origin to a new link L', which is a satellite of L constructed by means of the pattern link P;

$$L' = h_1^\circ(P) \cup h_2^\circ(C) \cup \cdots \cup h_m^\circ(C) \,. \tag{13.31}$$

Of course, all the remaining components $\{ C_2, \ldots, C_m \}$ of L and their associated representations $\{ \rho_2, \ldots, \rho_m \}$ are kept fixed. In the standard definition of satellites, it is usually assumed that P is essential in $\mathcal{V}$. This restriction just simplifies the companionship relation between links. However, for the satellite relations that we are going to derive to be valid, this assumption is not necessary. Therefore, we shall consider the case in which no restriction is imposed on the link $P \subset \mathcal{V}$. The link L', shown in equation (13.31) and obtained from L by means of the substitution (13.30), is called a ***generalized satellite*** of L.

It is quite clear that $E(L')$ must be related to $E(L)$; this relation must depend on the pattern link $P \subset \mathcal{V}$. We would now like to understand the general structure of the satellite relations.

In order to solve this problem, let us consider the Wilson operator $W(P)$ corresponding to the pattern link P inside the solid torus $\mathcal{V}$. According to equation (13.18), $W(P)$ is equivalent to an appropriate linear combination of Wilson line operators associated with the framed core C of $\mathcal{V}$. The values of the ξ-coefficients, entering equation (13.18), are fixed and depend on the pattern link $P \subset \mathcal{V}$. In agreement with the statement of property **G3**, the pattern link P can be substituted therefore by the framed core C of $\mathcal{V}$; moreover, the colour χ associated with C is specified by the ξ-coefficients of equation (13.18)

$$\chi = \sum_{\rho} \xi(\rho)\, \rho \, . \tag{13.32}$$

Consequently, $h_1^\circ(P)$ is equivalent to

$$h_1^\circ(P) \Longleftrightarrow h_1^\circ(C) = C_1 \, , \tag{13.33}$$

with the "colour state" χ, given in equation (13.32), associated with C_1.

Thus, we have found the solution to the problem of the satellites. Indeed, from equations (13.31) and (13.33), it follows that

$$\begin{aligned} E(L') &= E(C_1, C_2, \dots, C_m; \chi\, , \rho_2\, , \dots, \rho_m) \\ &= \sum_{\rho} \xi(\rho)\, E(C_1, C_2, \dots, C_m; \rho\, , \rho_2\, , \dots, \rho_m)\, . \end{aligned} \tag{13.34}$$

In this expression, the representations $\{ \rho_2\, , \dots, \rho_m \}$ are fixed. The whole information concerning the pattern link P is contained in the coefficients $\{ \xi(\rho) \}$. These coefficients are universal because they are independent of the particular form of the companion.

As a consequence of property **G3**, it follows then:

Property G4. *Any generalized satellite is related to its companion by the same coefficients specified in* **G3**, *as shown in* equation (13.34).

Equation (6.35) represents a particular example of the general satellite relations derived in this section. As a final remark, let us note that the general satellite relations, expressed in the appropriately modified form, hold for any skein invariant of links. The improvement introduced by the topological quantum CS theory on this subject is remarkable; in the three-dimensional framework of the field theory, the derivation of the satellite formulae is very simple. The structure of the Lie algebra of the gauge group is encoded in the general satellite relations. We shall discuss this point in detail in Chapter 15.

13.5 Skein relation

The general satellite relations, discussed in the previous section, shed new light on the origin of the skein relations and on the meaning of projectors. With the vertical framing convention, any given link diagram L describes a framed link in S^3. Let us concentrate on a small part of the diagram contained inside some open disc on the plane of the diagram. Suppose that, inside this disc, the picture of the diagram represents part of two strings (these two strings may belong to a single component of the link L, of course). By modifying the two-string configuration inside the disc, and by keeping all the rest of the diagram fixed, we may obtain diagrams corresponding to skein related links.

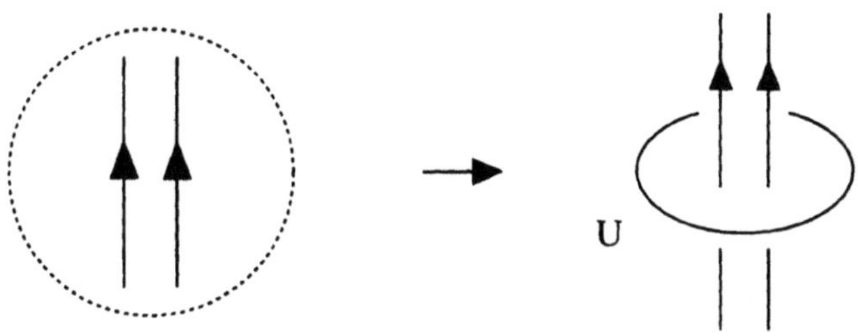

Figure 13.8.

Let us assume that, at the beginning, a no-crossing two-string configuration is contained inside the disc. In order to disentangle the structure of the two-string configurations from the rest of the link diagram, we shall introduce a circle U enclosing the two strings, as shown in Fig. 13.8. The circle U is not associated with a Wilson line operator and, consequently, it should not be interpreted as a new component of the link. U is just an imaginary circle in S^3 which is used to define a solid torus $\mathcal{V}$, which is the complement of U in S^3. Clearly, the whole link L is contained in $\mathcal{V}$ and, by construction, the two strings we are considering intersect a meridinal disc of $\mathcal{V}$. Consequently, the whole link L in S^3 can also be represented as shown in Fig. 13.9, where the tangle Y depends on the structure of the rest of the link diagram.
Since we will never modify the rest of the link diagram, the tangle Y is fixed. This being the case, if the starting no-crossing configuration is replaced for example by an overcrossing, the new link L_+ will be described by Fig. 13.10.

Instead of an overcrossing, one can insert an undercrossing or, in general, several consecutive over(under)-crossings; clearly, the associated pictures will contain the same tangle Y connected with the corresponding two-string configurations. All these links are inside the solid torus $\mathcal{V}$ and two strings intersect a meridinal disc

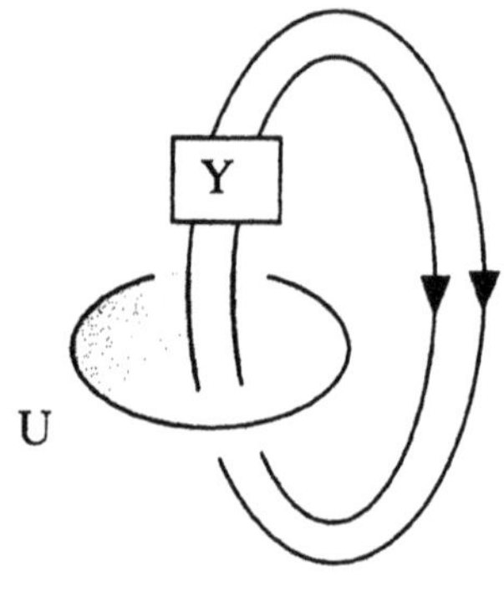

Figure 13.9.

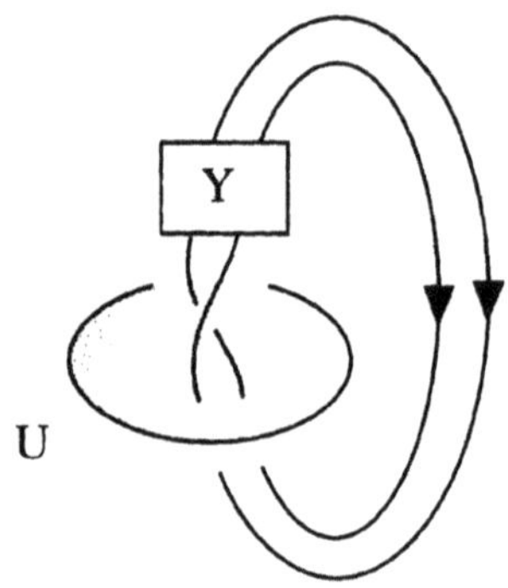

Figure 13.10.

of $\mathcal{V}$. Let ρ_1 and ρ_2 be the irreducible representations of the group associated with these two strings.
For each irreducible representation $\rho(t)$ entering the tensor product decomposition

$$\rho_1 \otimes \rho_2 = \oplus_t \rho(t), \tag{13.35}$$

a corresponding projector appears in the projection decomposition **(1)** and selects the representation $\rho(t)$ passing through a meridinal disc of the torus $\mathcal{V}$, as shown in Fig. 13.11. Consequently, rule **(1)** is strictly connected with the decomposition shown in equation (13.18). In fact, when the irreducible representations contained in the decomposition (13.35) are inequivalent (i.e. $\rho(t) \not\simeq \rho(t')$ for any $t \neq t'$), equation (13.18) becomes

$$W(L) = \sum_t \xi(\rho(t))\, W(C; \rho(t)), \tag{13.36}$$

where each term of this sum corresponds precisely to a representation of the decomposition (13.35). In our case, the coefficients $\{\xi(\rho(t))\}$ depend on the structure of the tangle Y, of course. The vacuum expectation value of both sides of equation (13.36) gives

$$E(L) = \sum_t \xi(\rho(t))\, E_0[\rho(t)]. \tag{13.37}$$

Therefore, the insertion of the projector associated with the representation $\rho(t)$, as shown in Fig. 13.11, selects the factor

$$\xi(\rho(t))\, E_0[\rho(t)]$$

from the sum (13.37).

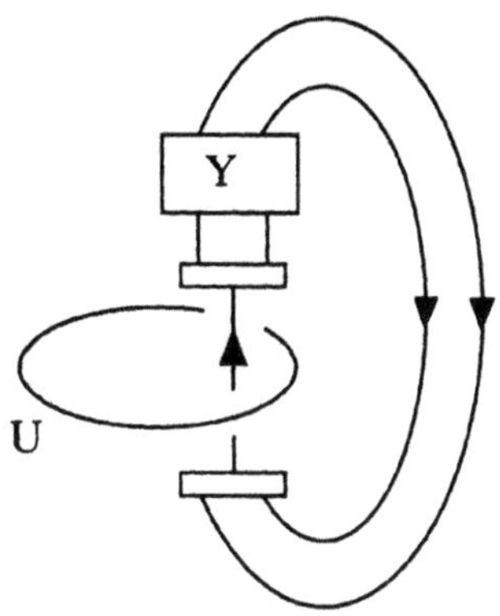

Figure 13.11.

When the two representations ρ_1 and ρ_2 are equivalent, the no-crossing two-string configuration can be substituted, for example, by an overcrossing and we obtain the link L_+ shown in Fig. 13.10. The universal link polynomial associated with L_+ can be written as

$$E(L_+) = \sum_t \xi'(\rho(t))\, E_0[\rho(t)], \tag{13.38}$$

and, according to the crossing rule (2), each coefficient $\xi'(\rho(t))$ is given by

$$\xi'(\rho(t)) \;=\; (-1)^{s(t)}\, q^{[Q(\rho(t))-Q(\rho_1)-Q(\rho_2)]/2}\, \xi(\rho(t))\,, \tag{13.39}$$

where $s(t) = 0$ (1) if $\rho(t)$ is symmetric (antisymmetric) under a permutation of ρ_1 and ρ_2.

Similarly, rules (1) and (2) can be used to analyse the case in which several consecutive crossings are present. The projection compatibility rule (4) together with (1) and (2) reproduce the generalized satellite formulae (13.34) and, as we have seen, the twisting rule (3) is fixed by consistency. Finally, factorization (5) ensures that the links associated with the trivial representation of the gauge group can be eliminated and correspond to the vacuum (absence of links).

In conclusion, the defining relations (1)-(5) form a consistent set of rules describing the general satellite relations satisfied by the universal link polynomial. The satellite relations, in turn, are uniquely determined by the decomposition rule of the tensor product of the representations of the gauge group. Therefore, only the representation ring of the ordinary Lie algebra is really relevant. This is the origin and the meaning of the universality of the link invariants.

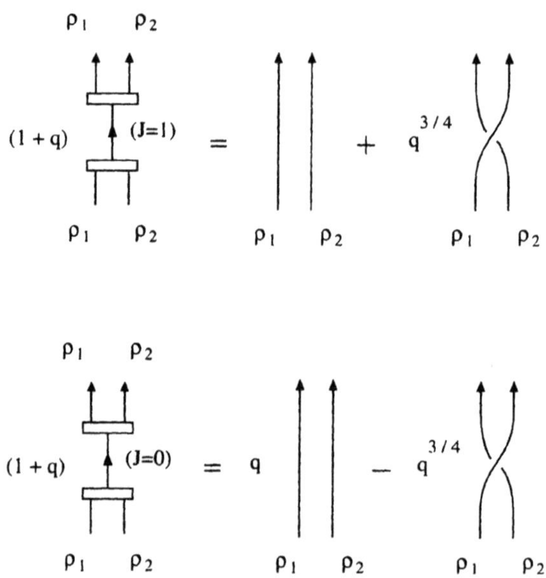

Figure 13.12.

13.6 Projectors

As we have already mentioned in Chapter 11, our projectors have an intrinsic meaning as "scalars", or invariants, under the action of the group. When the projectors are represented in terms of string configurations, the invariance of the projectors under the action of the group becomes evident.

Let us consider, for example, the case in which the gauge group is $G = SU(2)$. For two strings associated with the fundamental $J = 1/2$ representation, there are two inequivalent projectors. From the relations shown in Fig. 12.1 and in Fig. 12.6, it follows that the two projectors on $J = 1$ and $J = 0$ can be written in the form shown in Fig. 13.12. This example shows that our projectors can be understood as skein modules.

Similarly, any other projector can be written as a suitable combination of string configurations. This presentation of the projectors may be used for the computation of the ξ-coefficients entering equation (13.18). It should be clear that a generic skein module does not represent a projector. Projectors must diagonalize twist transformations or, equivalently, must diagonalize crossing operations as shown in Fig. 11.3 and Fig. 11.4.

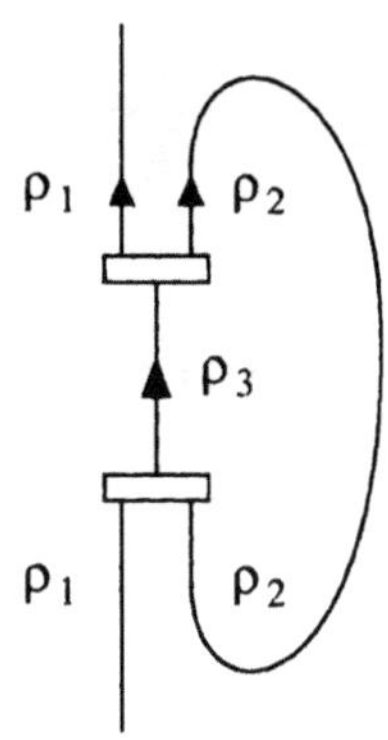

Figure 13.13.

At this stage, a useful computation rule concerning the partial closure of projectors will be derived. The ***partial closure*** of a projector is shown in Fig. 13.13.

Suppose that we are interested in the case in which, in some given diagram, the partial closure of a projector appears. We know that each projector admits a description in terms of string configurations; consequently, the partial closure shown in Fig. 13.13 can be understood as a single string, associated with the representation ρ_1, connected with a tangle X, as shown in Fig. 13.14.

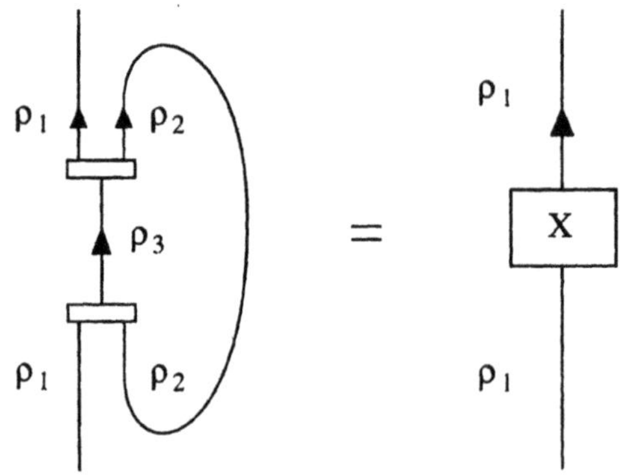

Figure 13.14.

The link that we are considering is therefore a generalized satellite: one of its components is the image of a pattern link of the type shown in Fig. 13.6 in which $\widetilde{\rho} = \rho_1$. Consequently, we can use equation (13.20) and the coefficient $\xi(\widetilde{\rho}) = \xi(\rho_1)$, shown in equation (13.21), in our case reads

$$\xi(\rho_1) = \frac{E_0[\rho_3]}{E_0[\rho_1]}. \tag{13.40}$$

This result can be represented as shown in Fig. 13.15. The relation shown in Fig. 13.15, where $\xi(\rho_1)$ is given in equation (13.40), is called [9] the ***partial closure rule***.

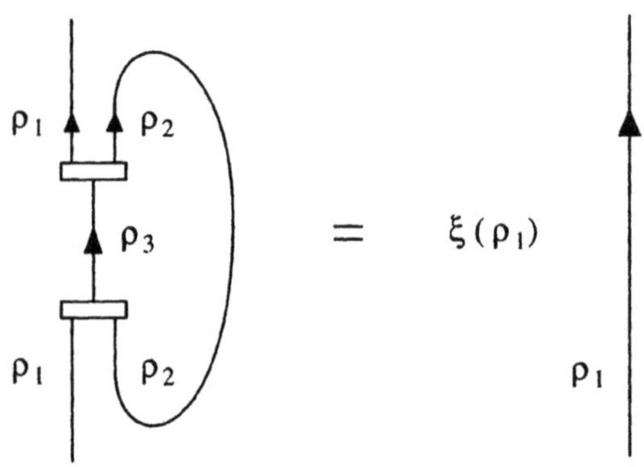

Figure 13.15.

13.7 Borromean rings

In Chapter 11, the defining relations **(1)**-**(5)** were formulated for link diagrams represented by the closure of oriented braids. It should be clear that the braid group itself does not play any essential or intrinsic role in the definition and in the computation of $E(L)$. The braid description of link diagrams has been used just because it represents a standard method of classifying the diagrams. In reality, the crucial properties which characterize the universal link polynomial are described by equations (13.18) and (13.34).

In this section, we shall present a simple example which illustrates how equations (13.18) and (13.34) can be used in the computation of the universal link polynomial.

Let us consider the link diagram B_R, shown in Fig. 13.16, which corresponds to the so-called Borromean rings.

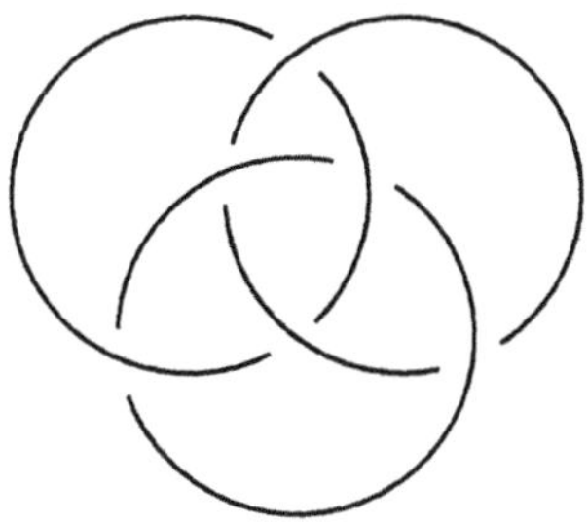

Figure 13.16.

To simplify matters, let the gauge group be $SU(2)$ and let each component of B_R be associated with the $J = 1/2$ representation. The link diagram B_R is described by the closure of the element $(g_2^{-1}\, g_1)^3$ of B_3. Instead of using the skein relation (12.58), let us compute $E(B_R)$ by means of a different method.

The diagram shown in Fig. 13.16 is regular isotopic with the diagram shown in Fig. 13.17 which shows that the link B_R is a satellite of the Hopf link. The complement of the component $C_1 \in B_R$ in S^3 is a solid torus (standardly embedded in S^3) which contains the two-component pattern link. Since the values of $E(L)$ for the Hopf link are known, we only need to determine the coefficients entering the decomposition (13.18) for the pattern link. In our case, the representations running along the core of the solid torus are those with $J = 1$ and $J = 0$ because

$$(1/2) \otimes (1/2) = (1) \oplus (0) \, ; \tag{13.41}$$

therefore, the (possibly) non-vanishing coefficients are $\xi(1)$ and $\xi(0)$. These two coefficients can be computed by means of several different methods. I shall now present a construction which is particularly useful for illustrating the meaning of projectors.

Let us concentrate on the possible two-string configurations inside a solid torus; we are interested in the case in which the two strings are both associated with the $J = 1/2$ representation of $SU(2)$ but are oriented along opposite directions, as shown in Fig. 13.17.

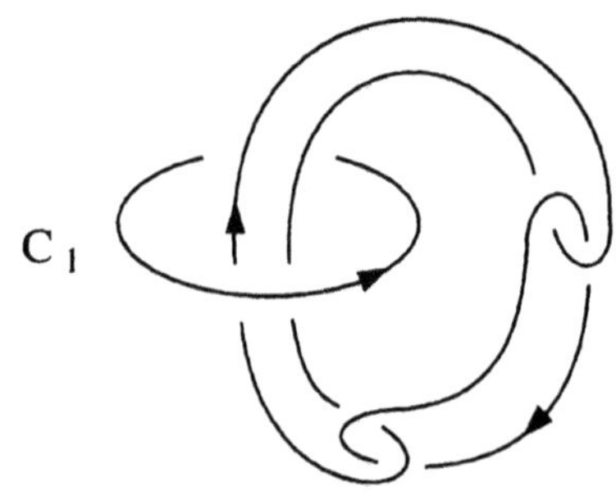

Figure 13.17.

Only two "global" irreducible representations are associated with a meridinal disc of the solid torus: one representation has $J = 1$ and the other has $J = 0$. So, any configuration can be expressed in terms of the two projectors shown in Fig. 13.18.

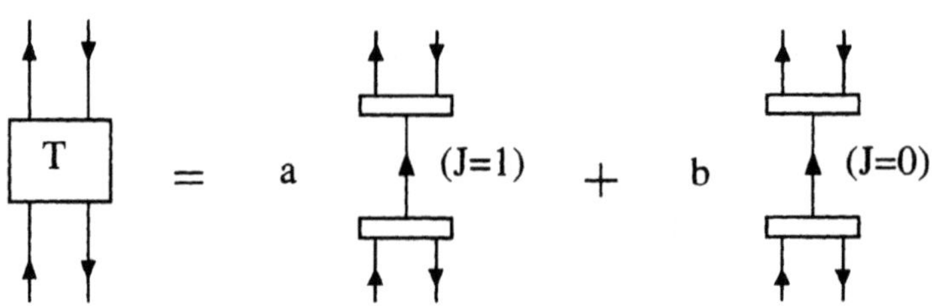

Figure 13.18.

Each tangle T can be interpreted as a suitable combination of the two projectors and is characterized by the two coefficients (a, b) shown in Fig. 13.18. If the tangle T is the result of the composition of two consecutive tangles T_1 and T_2, which have coefficients (a_1, b_1) and (a_2, b_2) respectively, then the orthogonality property of the projectors implies that the coefficients of T are $(a = a_1 a_2,\ b = b_1 b_2)$.

The correct normalization of the projectors is fixed by the completeness relation which is expressed in terms of the trivial (no-crossing) configuration as shown in Fig. 13.19.

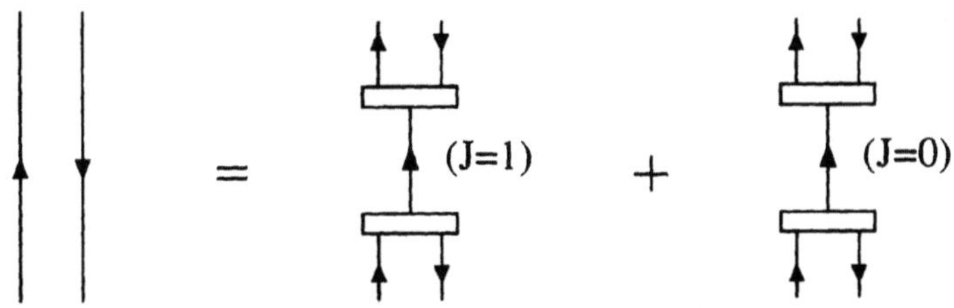

Figure 13.19.

An example of a tangle decomposition in terms of projectors is shown in Fig. 13.20. The relation shown in Fig. 13.20 can easily be understood. The string configuration shown in Fig. 13.20 is compatible with the existence of a meridinal disc (of the solid torus) which is not punctured; therefore, only the trivial $J = 0$ representation must appear on the right-hand-side of the relation; i.e. the coefficient a must vanish.

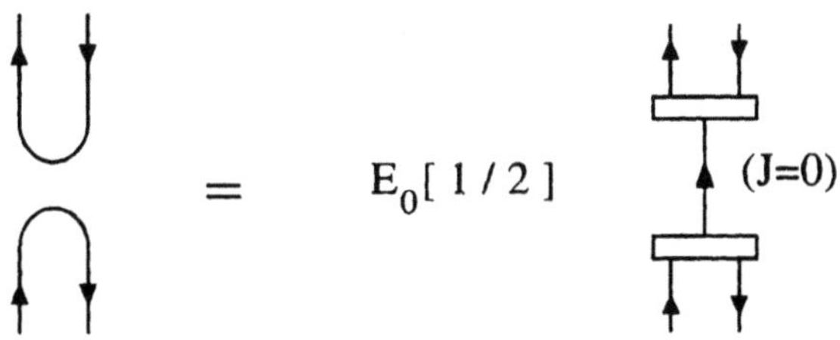

Figure 13.20.

In order to verify that the correct value of b is given by $E_0[1/2]$, as shown in Fig. 13.20, we only need to consider the closure of the pictures on both sides of the relation. The end-points of the strings on the top of the diagram are connected (in the usual ordering-preserving way) with the end-points on the bottom. On the left-hand-side we get one simple circle with $J = 1/2$; the associated value of the extended Jones polynomial is $E_0[1/2]$. On the right-hand-side, we get the value of the unknot, with $J = 0$, multiplied by b. This equation is satisfied if and only if the coefficient b is given by $b = E_0[1/2]$.

As a final check, consider the composition of several consecutive tangles of the type shown in Fig. 13.20; the resulting coefficients have the correct values, as they should.

Let us now reconsider the link B_R shown in Fig. 13.17. The interesting link, contained inside the solid torus which is the complement of one of its components in S^3, can be understood as the composition (obtained with the consecutive insertion) of the two tangles shown in Fig. 13.21.

Figure 13.21.

Let us compute the coefficients c and d entering the relation shown in Fig. 13.22.

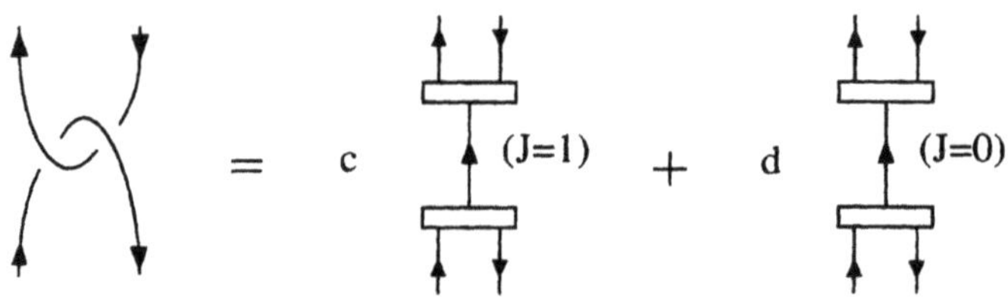

Figure 13.22.

The closure of the pictures appearing on both sides of this equation gives

$$q^{-3/2}\, E_0[1/2] \;=\; c\, E_0[1] \,+\, d\, E_0[0]\,. \tag{13.42}$$

The composition of the tangle appearing in Fig. 13.22 with the tangle of Fig. 13.20 is shown in Fig. 13.23; the closure of both sides of the resulting relation gives

$$E(\text{Hopf link};\, 1/2, 1/2) \;=\; d\, E_0[1/2]\, E_0[0]\,. \tag{13.43}$$

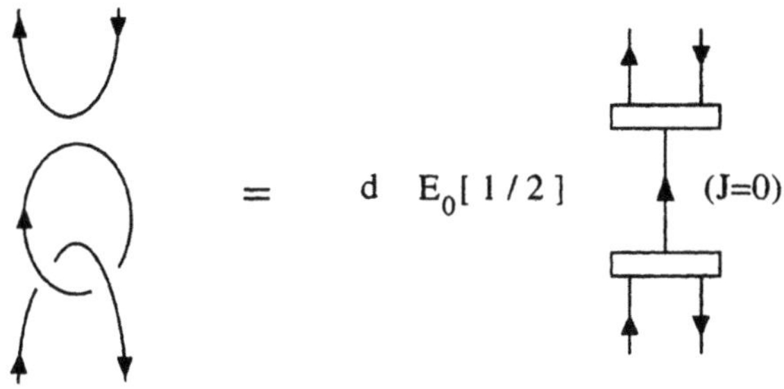

Figure 13.23.

From equations (13.42) and (13.43) it follows that

$$c = q^{-1} - 1\,, \quad d = q + q^{-1}\,. \tag{13.44}$$

It should be noted that in our computation of the coefficients c and d, appearing in the relation shown in Fig. 13.22, the universality of these coefficients has been used; i.e. the values of c and d are fixed and do not depend on the structure of the companions.

The second tangle shown in Fig. 13.21 can also be written as a suitable combination of the two projectors with $J = 1$ and $J = 0$. Let the second tangle be characterized by the coefficients (c', d') which are the analogues of (c, d) respectively. The new coefficients (c', d') can be computed by following the same method illustrated above. Equivalently, one can use the following argument: the second tangle of Fig. 13.21 is precisely the mirror image of the first tangle. Therefore the coefficients (c', d') can be obtained from (c, d) simply by replacing q with q^{-1}. In other words,

$$c' = q - 1\,, \quad d' = q + q^{-1}\,. \tag{13.45}$$

At this point, the projection decompositions of both the tangles shown in Fig. 13.21 are known. Consequently, any combination of these two tangles now gives a known result. In particular, the consecutive insertion of the tangles inside a solid torus gives the relation shown in Fig. 13.24, where the coefficients $\xi(1)$ and $\xi(0)$ are simply given by

$$\xi(1) = c\,c' = 2 - q - q^{-1}\,, \tag{13.46}$$

and

$$\xi(0) = d\,d' = 2 + q^2 + q^{-2}\,. \tag{13.47}$$

Finally, the value of the extended Jones polynomial $E(B_R)$ for the Borromean Rings (with $(J = 1/2)$) can be expressed in terms of the values $E(H; J_1, J_2)$ of the Hopf link as

$$E(B_R) = (2 - q - q^{-1})\, E(H; 1/2, 1) + (2 + q^2 + q^{-2})\, E(H; 1/2, 0)\,. \quad (13.48)$$

The value $E(H; J, J')$ of the extended Jones polynomial for the Hopf link is shown in equation (12.23); so, our computation is complete.

Exercise. Compute $E(B_R)$ by using the skein relation which is shown in equation (12.58) and verify that the result is in agreement with the expression (13.48).

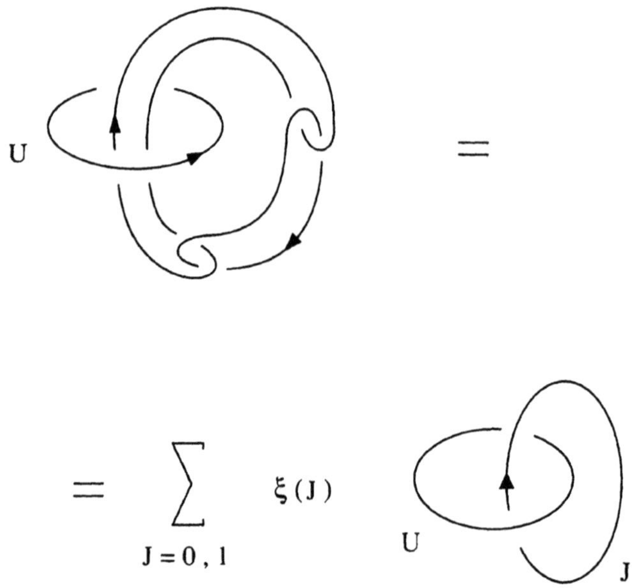

Figure 13.24.

It should be noted that our way of computing of $E(B_R)$ has produced much more information than just the value of the Borromean Rings. For example, we have found the explicit form of the projection decomposition of the tangles shown in Fig. 13.21. Since the derived coefficients are universal, they can be used in the computation of any link which contains the tangles we have mentioned. Similarly, any satellite constructed with the pattern link shown in Fig. 13.24 can immediately be related to its companion by using the coefficients shown in equations (13.46) and (13.47).

In the example we have considered, the direct computation of $E(B_R)$ by means of the skein relation (12.58) is quite simple. However, when the representations associated with the components of B_R have higher values of J, the naive use of the skein relation may involve a large number of operations. In these cases, the method illustrated in this section is more convenient.

13.8 Connected sums

A certain number of known properties of the skein invariants admit a simple derivation in the CS theory. In this section, we shall consider the relations satisfied by the connected sums of links.

Given two link diagrams L_1 and L_2, the link diagram $L_1\#L_2[\rho]$, shown in Fig. 13.25, represents the connected sum of L_1 and L_2 obtained by acting on two strings in the irreducible representation ρ.

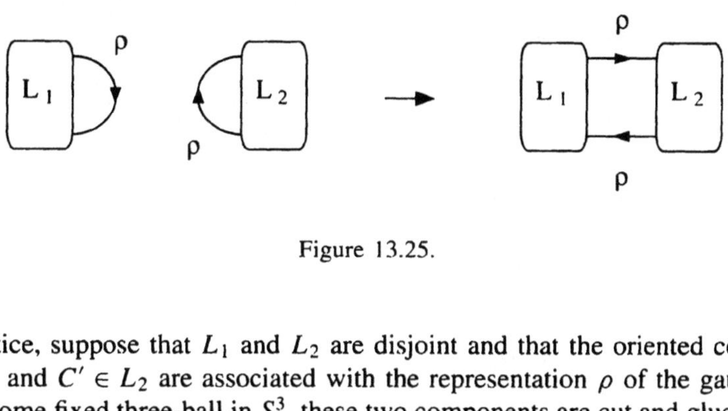

Figure 13.25.

In practice, suppose that L_1 and L_2 are disjoint and that the oriented components $C \in L_1$ and $C' \in L_2$ are associated with the representation ρ of the gauge group. Inside some fixed three-ball in S^3, these two components are cut and glued together as shown in Fig. 13.25. The resulting new link is the connected sum $L_1\#L_2[\rho]$. From the results of Section 13.3 and Section 13.4 it follows that

Property G5. *The universal link polynomial satisfies*

$$E(L_1\#L_2[\rho]) = \frac{E(L_1)\,E(L_2)}{E_0[\rho]} . \tag{13.49}$$

Proof. As shown in Fig. 13.26, the connected sum $L_1\#L_2[\rho]$ is the composition of two tangles which are obtained from the links L_1 and L_2 respectively by cutting

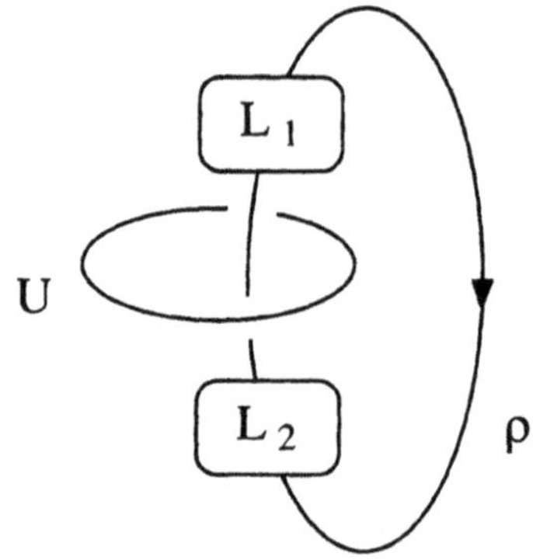

Figure 13.26.

the components $C \in L_1$ and $C' \in L_2$. Note that there is not a Wilson line operator associated with the circle U. The unknot U has been introduced simply to show that the tangles associated with L_1 and L_2 belong to the interior of a solid torus, which coincides with the complement of U in S^3.

Each tangle satisfies the relation shown in Fig. 13.27 and, according to equation (13.21), the coefficients d_1 and d_2 are given by

$$d_1 = \frac{E(L_1)}{E_0[\rho]}; \quad d_2 = \frac{E(L_2)}{E_0[\rho]}. \tag{13.50}$$

Therefore, the connected sum satisfies

$$E(L_1\#L_2[\rho]) = d_1\, d_2\, E_0[\rho] = \frac{E(L_1)\, E(L_2)}{E_0[\rho]}, \tag{13.51}$$

which is precisely the result reported in equation (13.49). □

Property **G5** represents a generalization of the known formula which holds for the ordinary HOMFLY polynomial. The generalization consists in the fact that equation (13.49) is valid not only in the case of the defining (fundamental) representation but also in the case of arbitrary irreducible representations of the gauge group.

Let us now consider tangles with two "incoming" and "outcoming" strings. In this case the irreducible representations $\{\rho(t)\}$, passing through a meridinal disc of the solid torus, are those entering the tensor product decomposition of the representations ρ_1 and ρ_2 associated with the two (incoming or outcoming) strings.

Every tangle can be expressed as a sum (with appropriate coefficients) of the corresponding projectors, as shown in Fig. 13.28. When these representations are

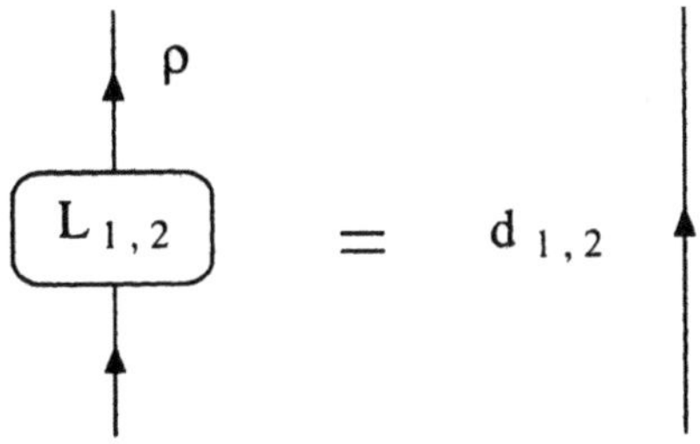

Figure 13.27.

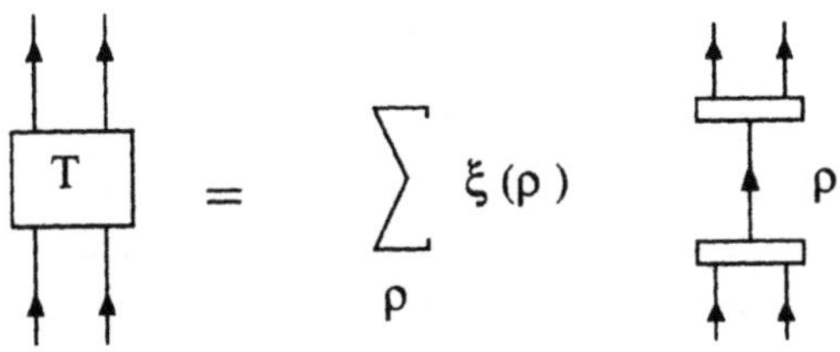

Figure 13.28.

Figure 13.29.

inequivalent, i.e. $\rho(t) \not\simeq \rho(t')$ for $t \neq t'$, each irreducible representation represents an independent "channel". The composition of several consecutive tangles is obtained by multiplying the coefficients which refer to the same "channel".

Consider for example the tangle shown in Fig. 13.22 in which each string is associated with the $(J = 1/2)$ representation of $SU(2)$. The projection decomposition of this tangle is shown in Fig. 13.22 and the coefficients c and d are given in equation (13.44). The link 6^3_1, shown in Fig. 13.29, is the closure of the composition of three consecutive tangles of this type. Therefore, the associated value of the extended Jones polynomial is

$$\begin{aligned} E(6_1^3; 1/2, 1/2, 1/2) &= c^3\, E_0[1] + d^3\, E_0[0] \\ &= (q^{-1} - 1)^3\,(q + 1 + q^{-1}) + (q + q^{-1})^3 \,. \end{aligned} \tag{13.52}$$

13.9 Mutations

Let us consider a given (framed and oriented) link L in S^3 and let us concentrate on a three-ball $\mathcal{B} \subset S^3$. The part of the link L which is contained in $\mathcal{B}$ is called the ***inhabitant*** T of $\mathcal{B}$. The boundary $\partial\mathcal{B}$ of the three-ball is a two-sphere which may contain punctures corresponding to the intersection points of $\partial\mathcal{B}$ with the components of L. Clearly, the number of punctures in $\partial\mathcal{B}$ is always even (or equal to zero) and the set of the irreducible representations associated with the punctures consists of couples of equivalent representations.

When the number of punctures is equal to two, the inhabitant T represents a tangle which admits a decomposition of the type discussed in Section 13.3.

Let us now consider the case in which there are four punctures on $\partial\mathcal{B}$ and, therefore, there are four strings intersecting $\partial\mathcal{B}$. In these conditions, the three-ball in S^3 is called Q and, by definition, ∂Q is a two-sphere with four punctures on it.

The inhabitant T in Q defines a tangle and, in order to introduce the projection decomposition shown in Fig. 13.28, we must specify how the set of the four strings is divided into the two subsets of the "incoming" and "outcoming" strings. In other words, the projectors always refer to a given solid torus $\mathcal{V}$ in S^3; indeed, each projector selects the irreducible representations of the group passing through a meridinal disc of the torus $\mathcal{V}$. Imagine now that a simple circle U, whose complement in S^3 defines $\mathcal{V}$, is introduced. For any fixed inhabitant T, there are three inequivalent ways of inserting U so that two strings intersect a meridinal disc of $\mathcal{V}$. These three possibilities correspond to the inequivalent partitions of the set of four strings into two couples of strings. Note that the circle U defines a plane (to which it belongs) and there are only three orthogonal inequivalent planes in three dimensions. It is always possible to find a correspondence between the three orthogonal planes and the possible couples of strings.

In conclusion, the projection decomposition of a tangle, as indicated in Fig. 13.28, has to be understood with respect to some given choice of $\mathcal{V}$; this choice of the solid torus $\mathcal{V}$ is not necessarily related to the orientations of the strings.

It is known [17] that the HOMFLY polynomial does not distinguish the mutations of a generic link. We now want to prove that the same property is satisfied by the universal link polynomial $E(L)$.

Let L_1 and L_2 be two oriented, framed and coloured links in S^3; then L_2 is a ***mutation*** of L_1 if L_2 can be obtained from L_1 by means of the following procedure:

(1) remove from L_1 an inhabitant T of a three-ball Q;
(2) rotate T of a π angle around a (arbitrary) central axis and, if necessary, modify the orientation of all the strings of T;
(3) place this new inhabitant in Q to obtain L_2.

The inhabitant T defines a tangle and, with an appropriate choice of the solid torus $\mathcal{V}$ in S^3, the relevant projection decomposition which has to be considered is shown in Fig. 13.30.

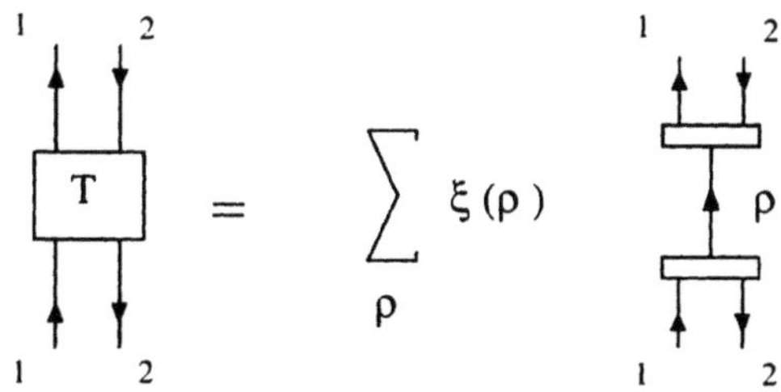

Figure 13.30.

There are four irreducible representations associated with the external strings; two of them are equivalent to the representation $\rho(1)$ and the remaining two representations are equivalent to $\rho(2)$. In Fig. 13.30, these representations have been denoted simply by 1 and 2.

Let us first consider the case in which $\rho(1)$ is not equivalent to $\rho(2)$. The orientation of the strings, associated with $\rho(1)$, is independent of the orientation of the strings associated with $\rho(2)$. On the other hand, for each couple of strings associated with equivalent representations, the orientations are specified by two arrows which always have opposite directions with respect to the interior (or centre) of the tangle. This being the case, the picture of Fig. 13.30 represents one of the four possible dispositions of the external arrows; the other three configurations can be obtained by modifying the directions of the arrows associated with $\rho(1)$ and/or $\rho(2)$. For each of these four possible arrangements of the external arrows, all the arguments we are going to present are valid. Therefore, we only need to consider the particular case shown in Fig. 13.30.

We shall now consider the effects of two consecutive modifications of the inhabitant T:

$$T \longrightarrow X \longrightarrow Y . \tag{13.53}$$

The tangle X is the image of T under a π rotation around a given central axis (specified below) and Y is obtained from X by a change of the orientation of all the strings of X.

Let us consider the tangle T shown in Fig. 13.30; the corresponding tangle X is shown in Fig. 13.31 and the associated tangle Y is shown in Fig. 13.32.

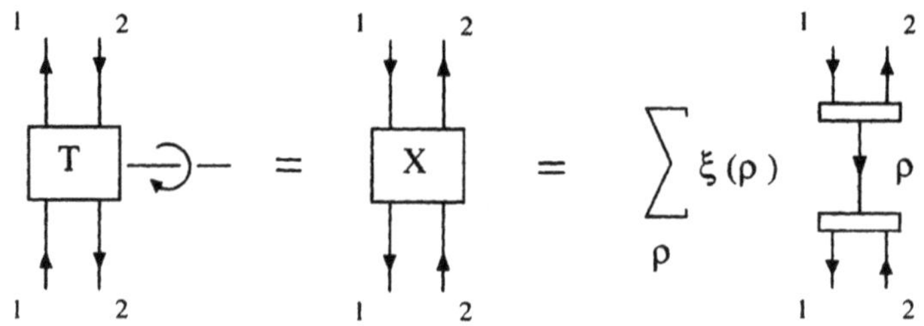

Figure 13.31.

By comparing the two relations shown in Fig. 13.30 and Fig. 13.32, we conclude that the tangles T and Y have the same projection decomposition; consequently, if in a generic link L the tangle T is removed and is substituted by Y, the value of the associated universal link polynomial is not modified.

When the representations $\rho(1)$ and $\rho(2)$ are inequivalent, any mutation of the link L is obtained by means of the transformations considered in Figs.13.31 and 13.32. Therefore, we have shown that, in this case, any link L and its mutations are associated with the same value $E(L)$ of the universal link polynomial.

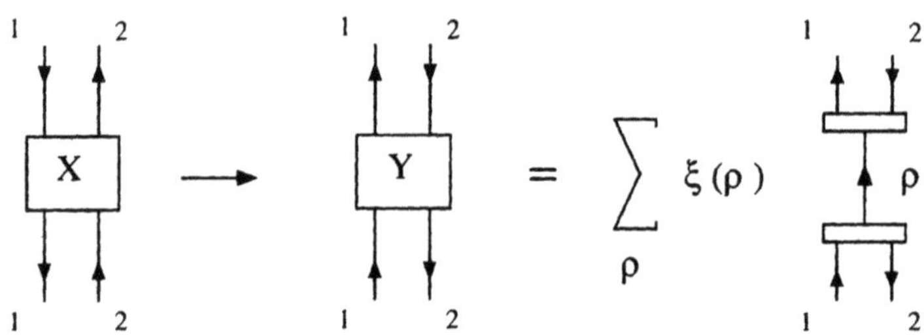

Figure 13.32.

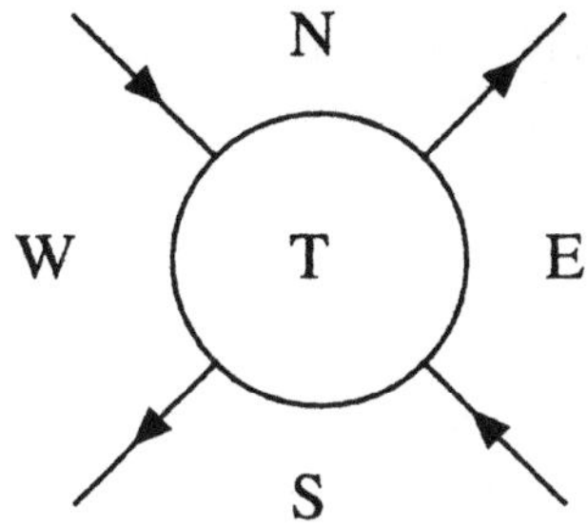

Figure 13.33.

We shall now analyse the case in which the representations $\rho(1)$ and $\rho(2)$ are equivalent; the inhabitant $T \subset Q$ is shown in Fig. 13.33.

Starting from a given link L and the associated inhabitant T, three (in general) different mutations L_1, L_2 and L_3 can be constructed. These three mutation links correspond to the three possible transformations t_1, t_2 and t_3 according to which the inhabitant T can be modified.

The transformation t_1 is the composition of a π-rotation around the N-S axis and a modification of the orientation of all the strings. The transformation t_2 is a π-rotation around the E-W axis followed by a change of the orientation of all the strings. Finally, the transformation t_3 is simply a π-rotation around the central axis which is perpendicular to the plane of the diagram (in this case there is no modification of the orientations).

For the mutation links L_1 and L_2, corresponding to the transformations t_1 and t_2 respectively, we can use the same argument presented above (when ρ_1 and ρ_2 were inequivalent) to show that $E(L_1) = E(L) = E(L_2)$.

Only the mutation link L_3 remains to be considered. One should note now that the transformation t_3 is the composition of t_1 and t_2. Indeed, the composition of two π-rotations around two orthogonal axes is equal to a π-rotation around the third axis; furthermore, two consecutive changes of the orientations of all the strings is equivalent to no modification at all. Since the universal link polynomial is invariant under the transformations t_1 and t_2, it is invariant also under the composition of the two transformations; therefore $E(L_3) = E(L)$.

To sum up, the universal link polynomial is invariant under the transformations which are relevant for the construction of mutations of links; therefore, it follows that

Property G6. *If the oriented, framed and coloured link L' in S^3 is a mutation of the link L, then*

$$E(L') = E(L). \tag{13.54}$$

As we have seen in the case of the connected sum formulae, property **G6** represents a generalization of the property which holds for the HOMFLY polynomial.

Property **G6** is a consequence of the symmetries of the CS theory. On the one hand, the non-Abelian electric flux associated with a meridinal disc of a solid torus (which contains an arbitrary link) is always conserved. On the other hand, a π-rotation which modifies the sign of the flux can be compensated by a "charge conjugation" transformation (modification of the orientations of all the strings). It is interesting to find that the (apparently mysterious) law concerning the mutations has really such a simple physical explanation.

Chapter 14

Unitary groups

In this chapter we shall consider the case in which the gauge group is $G = SU(N)$ with $N \geq 2$. The reconstruction theorems and the results obtained in Chapter 12 will be generalized to the case of arbitrary unitary groups.

Let $\{ \alpha^{(i)} \}$ be the $(N-1)$ simple roots of the Lie algebra of $SU(N)$; the fundamental weights are denoted by $\{ \mu^{(i)} \}$ and satisfy

$$2 \frac{\alpha^{(i)} \cdot \mu^{(j)}}{\alpha^{(i)} \cdot \alpha^{(i)}} = \delta^{ij} . \tag{14.1}$$

Each irreducible representation is characterized by the highest weight μ which admits an expansion

$$\mu = \sum_i n_i \, \mu^{(i)} , \tag{14.2}$$

where $(n_1, n_2, \ldots, n_{N-1})$ is a set of non-negative integers. Each irreducible representation can be labelled by the ordered set $(n_1, n_2, \ldots, n_{N-1})$. In these notations (Dynkin labels), the N-dimensional defining representation, with highest weight $\mu^{(1)}$, is denoted by $(1, 0, \ldots, 0)$.

The irreducible $SU(N)$ representations can also be described by means of the standard Young-tableau notation. Each Young tableau is characterized by a series of non-increasing integers $\{m_1, m_2, \ldots\}$ which represent the number of boxes in the columns of the tableau: the columns are ordered starting from the left. In this new notation, the defining N-dimensional representation is denoted by $\{1\}$. We will label the irreducible representations of $G = SU(N)$ by means of the corresponding Young tableaux. The basic property we shall use is that each irreducible representation of $SU(N)$ is contained in the tensor product $\{1\} \otimes \cdots \otimes \{1\}$ of an appropriate number of defining representations.

14.1 Fundamental skein relation

Let us consider the case in which all the components of the oriented framed links in S^3 are associated with the defining $\{1\}$ representation of $SU(N)$. In the decomposition of the tensor product

$$\{1\} \otimes \{1\} = \{2\} \oplus \{1,1\}, \tag{14.3}$$

the representation $\{2\}$ is antisymmetric (under a permutation of the two defining representations) and has dimension $N(N-1)/2$; whereas the representation $\{1,1\}$ is symmetric and has dimension $N(N+1)/2$. The associated values of the quadratic Casimir operator are

$$Q(\{1\}) = \frac{N^2-1}{2N}, \tag{14.4}$$

$$Q(\{2\}) = \frac{N^2-N-2}{N}, \tag{14.5}$$

$$Q(\{1,1\}) = \frac{N^2+N-2}{N}. \tag{14.6}$$

According to equation (14.3), the projection decomposition **(1)** contains only two projectors: one of them selects the representation $\{2\}$ and the other one selects the representation $\{1,1\}$. These two projectors also enter the crossing relations **(2)**. Consider now the three configurations L_+, L_- and L_0, corresponding to one overcrossing, one undercrossing and a no-crossing between two strings; L_+, L_- and L_0 admit a decomposition in terms of two projectors only. Consequently, a linear relation between them exists and this is simply the familiar skein relation. Thus, equation (14.3) implies that, when all the components of the links are associated with the representation $\{1\}$, the value $E(L)$ can be computed by using the skein relation.

Since $E(L)$ is a regular isotopy invariant, we must specify the value of the twist variable $\alpha(\{1\})$. According to rule **(3)**, one has

$$\alpha(\{1\}) = q^{Q(\{1\})} = q^{(N^2-1)/2N}. \tag{14.7}$$

We must determine then the value $E_0[\{1\}]$ of the unknot U_0 (with zero writhe). As we already know, this value cannot be arbitrarily chosen because $E_0[\{1\}]$ is fixed, for example, by the satellite relations. As in the case of $G = SU(2)$, $E_0[\{1\}]$ will be determined by means of three equations.

The first equation is an immediate consequence of equation (14.3)

$$E_0[\{1\}]\, E_0[\{1\}] = E_0[\{2\}] + E_0[\{1,1\}]. \tag{14.8}$$

The second equation is the analogue of equation (12.5) (which follows from the relation shown in Fig. 12.7); by taking into account the crossing rule **(2)**, one obtains

$$\begin{aligned} q^{Q(\{1\})}\, E_0[\{1\}] &= q^{-Q(\{1\})+Q(\{1,1\})/2}\, E_0[\{1,1\}] \\ &\quad - q^{-Q(\{1\})+Q(\{2\})/2}\, E_0[\{2\}]\,. \end{aligned} \tag{14.9}$$

Finally, the closure of one undercrossing between two strings gives the third equation

$$\begin{aligned} q^{-Q(\{1\})}\, E_0[\{1\}] &= q^{Q(\{1\})-Q(\{1,1\})/2}\, E_0[\{1,1\}] \\ &\quad - q^{Q(\{1\})-Q(\{2\})/2}\, E_0[\{2\}]\,. \end{aligned} \tag{14.10}$$

The system of equations (14.8)-(14.10) admits two solutions. The first solution is that all the values of the unknots are equal to zero; this solution must be discarded in order to be consistent with the field theory results. The second solution, which is the admissible one, is

$$E_0[\{1\}] = \frac{q^{N/2} - q^{-N/2}}{q^{1/2} - q^{-1/2}}\,, \tag{14.11}$$

$$E_0[\{2\}] = \frac{q^{(N-1)/2} - q^{-(N-1)/2}}{q - q^{-1}}\, E_0[\{1\}]\,, \tag{14.12}$$

$$E_0[\{1,1\}] = \frac{q^{(N+1)/2} - q^{-(N+1)/2}}{q - q^{-1}}\, E_0[\{1\}]\,. \tag{14.13}$$

The numerical coefficients, entering the (exchange) skein relation, can easily be derived from the values (14.4)-(14.6) of the Casimir operator and from the structure of the crossing relations (2). Finally, the results of this section are summarized by the following

Theorem R3. *For any oriented link diagram L, whose components $\{C_i\}$ are labelled by the defining irreducible representation $\{1\}$ of $SU(N)$, the associated non-trivial value*

$$E(L) = E(C_1, C_2, \ldots, C_m; \{1\}, \{1\}, \ldots, \{1\})\,, \tag{14.14}$$

is uniquely determined by

(i) *regular isotopy invariance,*

(ii) $E_0[\{1\}] = \left(q^{N/2} - q^{-N/2}\right) \Big/ \left(q^{1/2} - q^{-1/2}\right)$,

(iii) $E(L^{(+)}) = q^{(N^2-1)/2N}\, E(L^{(0)})$,
$E(L^{(-)}) = q^{-(N^2-1)/2N}\, E(L^{(0)})$,

(iv) $q^{1/2N}\, E(L_+) - q^{-1/2N}\, E(L_-) = (q^{1/2} - q^{-1/2})\, E(L_0)\,.$ (14.15)

The configurations L_+, L_- *and* L_0 *are shown in* Fig. 2.5 *and* $L^{(+)}$, $L^{(-)}$ *and* $L^{(0)}$ *appear in* Fig. 2.6. $E(L)$ *can easily be computed by using the skein relation recursively. Consequently,* $E(L)$ *is a regular isotopy invariant which takes the form of a (finite) Laurent polynomial in the variable* $q^{1/2N}$ *with integer coefficients.*

Proof. Properties (i)-(iv) have been derived from the defining relations **(1)-(5)**. Furthermore, $E(L)$ (shown in equation (14.14)) is a polynomial in $q^{1/2N}$ simply because all the numerical coefficients, entering (i)-(iv), are finite Laurent polynomials in $q^{1/2N}$ with integer coefficients. □

14.2 Casimir operator

In this section, we shall compute the values $Q(\rho)$ of the quadratic Casimir operator for the different irreducible representations $\{\rho\}$ of the group $SU(N)$.

Let us start by considering the N-dimensional defining representation $\{1\}$. We shall introduce a standard basis for the linear space of the $\{1\}$ representation. The orthonormal vectors of this basis will be denoted by $\{|i\rangle\}$ with $i = 1, 2, \ldots, N$. The highest-weight vector $|\mu^{(1)}\rangle$ is chosen to be

$$|\mu^{(1)}\rangle = |1\rangle . \tag{14.16}$$

The remaining vectors $\{|i\rangle\}$, with $i = 2, 3, \ldots, N$, are ordered in such a way that the vector $|i\rangle$ has level $(i-1)$.

Let us denote by $\{T^a\}$, with $a = 1, 2, \ldots, N^2 - 1$, the $SU(N)$ generators of the representation $\{1\}$ with normalization

$$\mathrm{Tr}\left(T^a T^b\right) = \frac{1}{2}\,\delta^{ab} . \tag{14.17}$$

The matrices $\{T^a\}$ satisfy the Fierz identity

$$\sum_a T^a_{ij}\, T^a_{k\ell} = \frac{1}{2}\,\delta_{i\ell}\,\delta_{jk} - \frac{1}{2N}\,\delta_{ij}\,\delta_{k\ell} . \tag{14.18}$$

By contracting the indices j and k in equation (14.18), one gets

$$\sum_a \left(T^a\, T^a\right)_{i\ell} = \frac{N^2-1}{2N}\,\delta_{i\ell} ; \tag{14.19}$$

therefore,

$$Q(\{1\}) = \frac{N^2-1}{2N} . \tag{14.20}$$

The one-column Young tableaux correspond to the $SU(N)$ fundamental representations $\{m\}$ with $m = 1, 2, \ldots, N-1$. These representations are also called the basic representations. The representation $\{m\}$ is the completely antisymmetric irreducible component of the tensor product $\{1\} \otimes \{1\} \otimes \cdots \otimes \{1\}$ of m representations $\{1\}$. The corresponding highest-weight vector $|\,\mu^{(m)}\rangle$ is

$$|\,\mu^{(m)}\rangle = \frac{1}{\sqrt{m!}} \left(|\,1\rangle |\,2\rangle \cdots |\,m\rangle - \text{permutations} \right) , \tag{14.21}$$

and will be denoted by

$$|\,\mu^{(m)}\rangle = |\,[1, 2, \ldots, m]\rangle . \tag{14.22}$$

In order to determine the value $Q(\{m\})$, we shall compute the eigenvalue of the quadratic Casimir operator on the highest-weight vector $|\,\mu^{(m)}\rangle$. In the tensor product of m representations $\{1\}$, the $SU(N)$ generators $\{ R^a \}$ are given by

$$R^a = T_1^a + T_2^a + \cdots + T_m^a , \tag{14.23}$$

and then

$$\sum_a R^a R^a = \sum_{r=1}^{m} \sum_a T_r^a T_r^a + 2 \sum_{r<s} \sum_a T_r^a T_s^a \tag{14.24}$$

From equation (14.19) we know that

$$\sum_a T_r^a T_r^a = \frac{N^2 - 1}{2N} \; \mathbb{1} . \tag{14.25}$$

On the other hand, from equation (14.18) it follows that

$$\sum_a T_r^a T_s^a = \frac{1}{2} \Pi_{r,s} - \frac{1}{2N} \; \mathbb{1} , \tag{14.26}$$

where $\Pi_{r,s}$ is the permutation operator of the two factor spaces labelled by r and s respectively. Therefore, equation (14.24) takes the form

$$\sum_a R^a R^a = \frac{m N^2 - m^2}{2N} \; \mathbb{1} + \sum_{r<s} \Pi_{r,s} . \tag{14.27}$$

Since $|\,\mu^{(m)}\rangle$ is completely antisymmetric, one has

$$\sum_{r<s} \Pi_{r,s} \,|\,\mu^{(m)}\rangle = -\frac{m\,(m-1)}{2} \,|\,\mu^{(m)}\rangle . \tag{14.28}$$

Thus

$$\sum_a R^a R^a \mid \mu^{(m)} \rangle = Q(\{m\}) \mid \mu^{(m)} \rangle , \tag{14.29}$$

with

$$Q(\{m\}) = \frac{N+1}{2N} (N-m)\, m . \tag{14.30}$$

The Young tableaux labelled by $\{0\}$ or by $\{N\}$ correspond to the trivial representation of $SU(N)$. The expression (14.30) has been obtained for $1 \leq m \leq N-1$ and remains valid also for $0 \leq m \leq N$.

Let us now consider the two-column Young tableaux which describe the irreducible representations $\{m_1, m_2\}$ with

$$N-1 \geq m_1 \geq m_2 \geq 1 . \tag{14.31}$$

The corresponding highest-weight vector $\mid \mu \rangle$ is given by

$$\mid \mu \rangle = \mid \mu^{(m_1)} \rangle \otimes \mid \mu^{(m_2)} \rangle . \tag{14.32}$$

Equation (14.32) together with equation (14.22) determine the symmetry of the irreducible component $\{m_1, m_2\}$ contained in the tensor product of $m_1 + m_2$ representations $\{1\}$. In this tensor product, the group generators $\{ R^a \}$ are

$$R^a = T_1^a + \cdots + T_{m_1}^a + T_{m_1+1}^a + \cdots + T_{m_1+m_2}^a , \tag{14.33}$$

and therefore

$$\begin{aligned} R^a R^a = & \left(T_1^a + \cdots + T_{m_1}^a\right)^2 + \left(T_{m_1+1}^a + \cdots + T_{m_1+m_2}^a\right)^2 \\ & + 2 \sum_{r=1}^{m_1} \sum_{s=1}^{m_2} T_r^a T_{m_1+s}^a . \end{aligned} \tag{14.34}$$

In order to determine $Q(\{m_1, m_2\})$, we shall compute the eigenvalue of the operator (14.34) on the highest-weight vector (14.32). The operator (14.34) is the sum of three terms. The first term gives

$$\begin{aligned} & \left(T_1^a + \cdots + T_{m_1}^a\right)^2 \mid \mu^{(m_1)} \rangle \otimes \mid \mu^{(m_2)} \rangle = \\ & \qquad = Q(\{m_1\}) \mid \mu^{(m_1)} \rangle \otimes \mid \mu^{(m_2)} \rangle ; \end{aligned} \tag{14.35}$$

and the second term gives

$$\begin{aligned} & \left(T_{m_1+1}^a + \cdots + T_{m_1+m_2}^a\right)^2 \mid \mu^{(m_1)} \rangle \otimes \mid \mu^{(m_2)} \rangle = \\ & \qquad = Q(\{m_2\}) \mid \mu^{(m_1)} \rangle \otimes \mid \mu^{(m_2)} \rangle . \end{aligned} \tag{14.36}$$

By means of equation (14.26), the third term can be written as

$$2\sum_{r=1}^{m_1}\sum_{s=1}^{m_2} T_r^a T_{m_1+s}^a = \sum_{r=1}^{m_1}\sum_{s=1}^{m_2}\left(\Pi_{r,m_1+s} - \frac{1}{N}\,\mathbb{1}\right). \tag{14.37}$$

Clearly, one has

$$\begin{aligned}\sum_{r=1}^{m_1}\sum_{s=1}^{m_2}\left(-\frac{1}{N}\,\mathbb{1}\right)|\,\mu^{(m_1)}\rangle \otimes |\,\mu^{(m_2)}\rangle &= \\ &= -\frac{m_1 m_2}{N}\,|\,\mu^{(m_1)}\rangle \otimes |\,\mu^{(m_2)}\rangle\,.\end{aligned} \tag{14.38}$$

On the other hand, since $m_2 \leq m_1$, it is easy to verify that

$$\sum_{r=1}^{m_1}\sum_{s=1}^{m_2}\left(\Pi_{r,m_1+s}\right)|\,\mu^{(m_1)}\rangle \otimes |\,\mu^{(m_2)}\rangle = m_2\,|\,\mu^{(m_1)}\rangle \otimes |\,\mu^{(m_2)}\rangle\,. \tag{14.39}$$

Finally, from equations (14.35), (14.36), (14.38) and (14.39), it follows that

$$Q(\{m_1, m_2\}) = Q(\{m_1\}) + Q(\{m_2\}) + \frac{N - m_1}{N}\,m_2\,. \tag{14.40}$$

This expression, which has been derived for the values of m_1 and m_2 shown in equation (14.31), is also valid for

$$N \geq m_1 \geq m_2 \geq 0\,. \tag{14.41}$$

Let us now consider a generic representation $\{m_1, m_2, \ldots, m_t\}$, with

$$N - 1 \geq m_1 \geq m_2 \geq \cdots \geq m_t \geq 1\,, \tag{14.42}$$

contained in the tensor product of $m_1 + m_2 + \cdots + m_t$ representations $\{1\}$ and whose corresponding highest-weight vector $|\,\mu\,\rangle$ is

$$|\,\mu\,\rangle = |\,\mu^{(m_1)}\rangle \otimes |\,\mu^{(m_2)}\rangle \otimes \cdots \otimes |\,\mu^{(m_t)}\rangle\,. \tag{14.43}$$

A straightforward extension of the arguments presented above gives

$$Q(\{m_1, m_2, \ldots, m_t\}) = \sum_{r=1}^{t} Q(\{m_r\}) + \sum_{r=2}^{t}\sum_{s=1}^{r-1}\frac{N - m_s}{N}\,m_r \tag{14.44}$$

and, as usual, this expression is valid also when $m_1 = N$ and/or $m_t = 0$.

We shall conclude this section by proving a lemma which will be used later.

Lemma 1. *Consider the tensor product of the irreducible representations* $\{m_1, m_2, \cdots, m_t\}$ *and* $\{1\}$. *Any two irreducible representations contained in the decomposition of* $\{m_1, m_2, \cdots, m_t\} \otimes \{1\}$ *are inequivalent and the associated values of the quadratic Casimir operator are different.*

Proof. Let us recall that the integers $\{m_1, m_2, \cdots, m_t\}$ satisfy the relation (14.42). Each irreducible representation, contained in the tensor product $\{m_1, m_2, \cdots, m_t\} \otimes \{1\}$, is described by a Young tableau belonging to a finite set S of tableaux. Each element of this set is labelled by an integer number p. The Young tableau associated with a fixed value of p is described by

$$\{m_1, m_2, \cdots, m_p + 1, \cdots, m_t, m_{t+1} = 0\} . \tag{14.45}$$

In this expression, the extra entry $m_{t+1} = 0$ has only been introduced for convenience. The permitted values of p must satisfy $1 \leq p \leq t + 1$ and are those for which $m_{p-1} > m_p$; the values $p = 1$ and $p = t + 1$ are always permitted. All the tableaux of the set S are inequivalent; therefore, the associated irreducible representations are inequivalent.

Let us now consider the values of the quadratic Casimir operator. From equation (14.44), it follows that

$$\begin{aligned} Q(\{m_1, \cdots, m_p + 1, \cdots, m_t, m_{t+1} = 0\}) &= Q(\{m_1, \cdots, m_t\}) + \\ + \frac{N^2 - 1}{2N} - \frac{1}{N}\left(\sum_{r=1}^{t} m_r\right) &+ (p - m_p) - 1 . \end{aligned} \tag{14.46}$$

This equation shows that the value of the Casimir operator depends on p through the additive term

$$A(p) = (p - m_p) , \tag{14.47}$$

appearing in the expression (14.46). By taking into account the relation (14.42), one finds that $A(p + 1) > A(p)$. Therefore, any two irreducible representations, appearing in the decomposition of the tensor product $\{m_1, m_2, \cdots, m_t\} \otimes \{1\}$, are associated with different values of the quadratic Casimir operator. □

14.3 Composite states

We are now interested in the computation of the universal link polynomial in the case in which the different components of the links are associated with arbitrary irreducible representations of $SU(N)$. In order to solve this problem, we shall make use of the general satellite relations described in Section 13.4.

Before considering the general solution, let us make a few remarks and present a simple example. Obviously, any link component associated with the trivial representation $\{0\}$ can be eliminated. The complex conjugate of the representation $\{m_1, m_2, \ldots, m_t\}$ is given by

$$\{m_1, m_2, \ldots, m_t\}^* = \{N - m_t, \ldots, N - m_2, N - m_1\}. \tag{14.48}$$

Let us recall that if C is an oriented (framed) component of a link, C^{-1} denotes the component obtained from C by a modification of its orientation. Clearly, any oriented component C associated with the representation $\{m_1, m_2, \ldots, m_t\}$ is equivalent to the component C^{-1} associated with the representation $\{m_1, m_2, \ldots, m_t\}^*$.

Let L be an oriented framed link in S^3 with components $\{C_1, C_2, \ldots\}$. Assume that all the components of L are associated with the representation $\{1\}$ except for C_1 which is associated with the representation $\{2\}$. How can we compute the corresponding value $E(C_1, C_2, \ldots; \{2\}, \{1\}, \ldots)$ of the universal link polynomial?

The idea is to use satellites obtained from L by replacing the component C_1 with the image of certain pattern links.

Consider the case in which the two oriented components of the pattern link B, shown in Fig. 6.6, are associated with the representation $\{1\}$. Let L_1 be the satellite of L with components $\{h^\circ(B), C_2, \ldots\}$ where $h^\circ(B)$ is the image under the homeomorphism h° (specified in Chapter 2) of B. Both components of $h^\circ(B)$ are oriented in the same direction as the original component C_1.

According to the satellite formula (6.35) and the decomposition (14.3), one has

$$\begin{aligned} E(h^\circ(B), C_2, \ldots; \{1\}, \{1\}, \{1\}, \ldots) &= E(C_1, C_2, \ldots; \{2\}, \{1\}, \ldots) \\ &\quad + E(C_1, C_2, \ldots; \{1,1\}, \{1\}, \ldots). \end{aligned} \tag{14.49}$$

The term on the left-hand-side of equation (14.49) is known because all the different components of the link are associated with the fundamental representation $\{1\}$. The values of the universal link polynomial appearing on the right-hand-side of equation (14.49) are unknown; one of them, $E(C_1, C_2, \ldots; \{2\}, \{1\}, \ldots)$, is actually the value in which we are interested.

Clearly, a second equation is needed. So, let us consider the new pattern link $B^{(1)}$, shown in Fig. 14.1. As we have already mentioned, $B^{(1)}$ is simply the image of the link B under a right-handed twist of the solid torus which contains B.

Consider now the satellite L_2 of L with components $\{h^\circ(B^{(1)}), C_2, \ldots\}$; in this case, the satellite formula reads

Figure 14.1.

$$\begin{aligned} E(h^\circ(B^{(1)}), C_2, \ldots; \{1\}, \{1\}, \{1\}, \ldots) &= \\ &= q^{Q(\{2\})}\, E(C_1, C_2, \ldots; \{2\}, \{1\}, \ldots) \\ &\quad + q^{Q(\{1,1\})}\, E(C_1, C_2, \ldots; \{1,1\}, \{1\}, \ldots)\,. \end{aligned} \tag{14.50}$$

Again, the term on the left-hand-side of equation (14.50) is known (all the components of the link are associated with the representation $\{1\}$). By combining equations (14.49) and (14.50), we get the desired result

$$\begin{aligned} E(C_1, C_2, \ldots; \{2\}, \{1\}, \ldots) &= \\ &= x_1\, E(h^\circ(B), C_2, \ldots; \{1\}, \{1\}, \{1\}, \ldots) \\ &\quad + x_2\, E(h^\circ(B^{(1)}), C_2, \ldots; \{1\}, \{1\}, \{1\}, \ldots)\,. \end{aligned} \tag{14.51}$$

where

$$x_1 = -q^{Q(\{1,1\})} \left[q^{Q(\{2\})} - q^{Q(\{1,1\})} \right]^{-1}, \tag{14.52}$$

and

$$x_2 = \left[q^{Q(\{2\})} - q^{Q(\{1,1\})} \right]^{-1}. \tag{14.53}$$

Since $Q(\{2\}) \neq Q(\{1,1\})$, for generic values of the deformation parameter q the coefficients x_1 and x_2 are well defined. The roots of unity will be considered in Chapter 15.

In conclusion, the link component associated with the representation $\{2\}$ can be substituted by a suitable combination of $h^\circ(B)$ and $h^\circ(B^{(1)})$ which contain two components labelled by $\{1\}$. But when all the components are characterized by the representation $\{1\}$, we know how to compute $E(L)$ by means of the skein relation of theorem **R3**. Therefore, our problem is solved.

The structure of equation (14.51) should be noted. The original link component C_1 determines the homeomorphism $h^\diamond$ which is used in the construction of the satellites; whereas the particular representation of the gauge group associated with C_1 has determined the values of the coefficients x_1 and x_2 entering the linear combination (14.51) of the two satellites. In order to organize the whole information connected with the representations of the group, let us denote by $\widetilde{P}[\{2\}]$ the linear combination of pattern links associated with the representation $\{2\}$:

$$\widetilde{P}[\{2\}] = x_1 \, B(\{1\},\{1\}) + x_2 \, B^{(1)}(\{1\},\{1\}) , \tag{14.54}$$

where the coefficients x_1 and x_2 are shown in equations (14.52) and (14.53). Then, equation (14.51) can be rewritten as

$$E(C_1, C_2, \ldots; \{2\}, \{1\}, \ldots) = E(h^\diamond(\widetilde{P}[\{2\}]), C_2, \ldots; \{1\}, \{1\}, \{1\}, \ldots) . \tag{14.55}$$

In this equation it is understood that, by linearity, the sum (14.54) entering the definition of $\widetilde{P}[\{2\}]$ results in the sum shown in equation (14.51).

Suppose now that the link component C_1 is associated with the representation $\{1, 1\}$. How can we compute, in this case, the corresponding value $E(C_1, C_2, \ldots; \{1, 1\}, \{1\}, \ldots)$ of the universal link polynomial ?
By comparing equation (14.49) with equation (14.50), the answer is immediately obtained; $E(C_1, C_2, \ldots; \{1, 1\}, \{1\}, \ldots)$ also can be written as a linear combination of the form shown in equation (14.51) but with new coefficients.
According to the notations introduced above, the representation $\{1, 1\}$ is associated with the pattern link $\widetilde{P}[\{1, 1\}]$ given by

$$\widetilde{P}[\{1,1\}] = y_1 \, B(\{1\},\{1\}) + y_2 \, B^{(1)}(\{1\},\{1\}) , \tag{14.56}$$

where

$$y_1 = -q^{Q(\{2\})} \left[q^{Q(\{1,1\})} - q^{Q(\{2\})} \right]^{-1} , \tag{14.57}$$

and

$$y_2 = \left[q^{Q(\{1,1\})} - q^{Q(\{2\})} \right]^{-1} . \tag{14.58}$$

The analogue of equation (14.55) is simply

$$\begin{aligned} E(C_1, C_2, \ldots; \{1, 1\}, \{1\}, \ldots) &= \\ &= E(h^\diamond(\widetilde{P}[\{1, 1\}]), C_2, \ldots; \{1\}, \{1\}, \{1\}, \ldots) . \end{aligned} \tag{14.59}$$

At this stage, let us consider a generic link L in which any component is associated with an irreducible representation which is equivalent to one of the following three representations: $\{1\}$, $\{2\}$ and $\{1, 1\}$. By using the results of this section, we can now compute the corresponding value of the universal link polynomial. Indeed, we need only to replace each link component, labelled by {2} or by {1, 1}, with the corresponding image of the appropriate pattern link $\widetilde{P}[\{2\}]$ or $\widetilde{P}[\{1,1\}]$. In this way, we obtain a linear combination (with known coefficients) of new links in which all the components are associated with the defining {1} representation.

The construction illustrated above, and based on the use of the satellite relations, will be generalized to the case of arbitrary representations. Unlike the case in which $G = SU(2)$, the method of standard cabling is not practical for $SU(N)$ with $N > 3$; yet, a slightly modified version of cabling (as illustrated in this section) works for arbitrary unitary groups. This method is based on a simple idea which is the analogue of the usual picture in which baryons (or mesons) are "made" of quarks. The strings associated with the defining representation {1} play the role of quarks, and the link components associated with the higher dimensional representations are the analogue of the "composite" states. Really, in our case, there is a perfect "democracy" between the different representations of the group and we can construct satellites with arbitrary representations. The reason why the representation {1} has a special role is that the decomposition (14.3) contains only two terms: this implies that an ordinary skein relation is satisfied. Moreover, each irreducible representation of the group $SU(N)$ is contained in the tensor product of an appropriate number of defining representations. In this picture, the expressions (14.54) and (14.56) can be interpreted as two "wave functions" showing how the strings, labelled by {2} and {1, 1}, are made of two "elementary" {1}-strings.

There is a final point to be considered. The coefficients x_1 and x_2 (or y_1 and y_2) are not finite Laurent polynomials in the variable $q^{1/2N}$. Consequently, it is not evident that the expressions we obtain for $E(L)$ (when some link components are associated with higher dimensional representations) are polynomials. Still, all the direct computations performed on explicit examples show that $E(L)$ is a polynomial. This is not a coincidence: for arbitrary irreducible representations of the gauge group $SU(N)$, the universal link polynomial $E(L)$ is in fact a finite Laurent polynomial in $q^{1/2N}$ with integer coefficients. The proof will be presented in Section 14.6.

14.4. Pattern links

In order to generalize the construction presented in Section 14.3, we need to consider the satellite relations discussed in Chapter 13.

Let us consider an oriented framed link L in $\mathbb{R}^3$ or in S^3. Let C_1 be a component of L. We shall denote by ρ_1 the representation associated with C_1, and by N_1 a tubular neighbourhood of C_1.

Let $\mathcal{V}$ be a solid torus standardly embedded in S^3. An orientation for $\mathcal{V}$ is introduced; for example, the core C of $\mathcal{V}$ is a circle with some given orientation.

As we have already mentioned in Chapter 2 and Chapter 13, the orientation and the framing of C_1 determine a unique (up to ambient isotopy) homeomorphism $h^\diamond : \mathcal{V} \to N_1$. The image of C under $h^\diamond$ is precisely the component C_1.

We shall now construct a satellite L' of L. L' is obtained from L by substituting $h^\diamond(P)$ for the component C_1, $P \subset \mathcal{V}$ being the pattern link. According to the properties **G3** and **G4**, the satellite L' is related to L by the ξ-coefficients $\{\xi(\rho)\}$ entering the expansion (13.18)

$$W(P) = \sum_\rho \xi(\rho)\, W(C;\rho) \tag{14.60}$$

of the Wilson operator associated with P inside $\mathcal{V}$.

We shall now consider a linear combination of several satellites defined by means of different pattern links. Because of the linearity of the satellite relations, this is equivalent to considering a linear combination $\widetilde{P}$ of several pattern links. The resulting ξ-coefficients are the linear combinations of the ξ-coefficients of the single pattern links. Suppose now that, for some given $\widetilde{P}[\rho_1]$, the resulting coefficients $\{\xi(\rho)\}$ are vanishing except for $\xi(\rho_1)$ which is actually equal to unity. In other words,

$$W(\widetilde{P}[\rho_1]) = W(C,\rho_1)\,. \tag{14.61}$$

In this particular case, according to equation (13.34) one finds

$$E(L') = E(L)\,, \tag{14.62}$$

where L' now stands for the linear combination of the satellites corresponding to $\widetilde{P}[\rho_1]$. In conclusion, the value $E(L)$ of the universal link polynomial can be computed by analysing L' instead of L.

The same argument can be repeated for all the link components which are associated with the higher dimensional representations of the gauge group. Consequently, we can compute $E(L)$ by considering a certain set of satellite links instead of the original companion L. But here we find the crucial point; perhaps the satellites can be chosen in such a way that all their components are associated with the defining representation $\{1\}$. If this is the case, we can compute $E(L)$ for arbitrary representations by means of theorem **R3**.

Therefore, our problem consists of finding a convenient set of pattern links.

In order to illustrate the method we shall follow, let us reconsider the example illustrated in Section 14.3. In this case, we are interested in the representations $\{2\}$ and $\{1,1\}$. Consider the pattern link B in $\mathcal{V}$ when both the components of B are associated with the representation $\{1\}$. The corresponding decomposition (14.60) takes the form

$$W(B;\{1\},\{1\}) = W(C;\{2\}) + W(C;\{1,1\})\,. \tag{14.63}$$

On the other hand, the decomposition associated with the pattern link $B^{(1)}$ (with both components labelled by $\{1\}$) in $\mathcal{V}$ reads

$$W(B^{(1)}; \{1\}, \{1\}) = q^{Q(\{2\})}\, W(C; \{2\}) + q^{Q(\{1,1\})}\, W(C; \{1, 1\}) . \qquad (14.64)$$

Since $Q(\{2\}) \neq Q(\{1, 1\})$, the system of equations (14.63) and (14.64) is soluble for $W(C; \{2\})$ and $W(C; \{1, 1\})$ and we obtain

$$x_1\, W(B; \{1\}, \{1\}) + x_2\, W(B^{(1)}; \{1\}, \{1\}) = W(C; \{2\}) , \qquad (14.65)$$

and

$$y_1\, W(B; \{1\}, \{1\}) + y_2\, W(B^{(1)}; \{1\}, \{1\}) = W(C; \{1, 1\}) , \qquad (14.66)$$

where the coefficients x_1, x_2, y_1 and y_2 are given in equations (14.52), (14.53), (14.57) and (14.58). Equations (14.65) and (14.66) represent two explicit solutions of equation (14.61) and the corresponding pattern links $\widetilde{P}[\{2\}]$ and $\widetilde{P}[\{1, 1\}]$ are shown in equations (14.54) and (14.56).

Let us now consider the next step. Suppose we are interested in the representation $\{3\}$. We shall analyse the tensor product $\{2\} \otimes \{1\}$

$$\{2\} \otimes \{1\} = \{3\} \oplus \{2, 1\} . \qquad (14.67)$$

Consider then the pattern link $B \subset \mathcal{V}$ in which one component is associated with the representation $\{2\}$ and the second component is labelled by $\{1\}$. From equation (14.67) it follows that, in this case, the decomposition (14.60) takes the form

$$W(B; \{2\}, \{1\}) = W(C; \{3\}) + W(C; \{2, 1\}) . \qquad (14.68)$$

Under a right-handed twist of $\mathcal{V}$, the link B is transformed in $B^{(1)}$ and the analogue of equation (14.64) reads

$$W(B^{(1)}; \{2\}, \{1\}) = q^{Q(\{3\})}\, W(C; \{3\}) + q^{Q(\{2,1\})}\, W(C; \{2, 1\}) . \qquad (14.69)$$

The values of the quadratic Casimir operator are

$$Q(\{3\}) = \frac{3N^2 - 6N - 9}{2N} , \qquad (14.70)$$

and

$$Q(\{2, 1\}) = \frac{3N^2 - 9}{2N} . \qquad (14.71)$$

Since $Q(\{3\}) \neq Q(\{2, 1\})$, the system of equations (14.68) and (14.69) is soluble for $W(C; \{3\})$ and $W(C; \{2, 1\})$. One finds, in particular,

$$z_1 \, W(B; \{2\}, \{1\}) + z_2 \, W(B^{(1)}; \{2\}, \{1\}) \;=\; W(C; \{3\}) \,, \tag{14.72}$$

where

$$z_1 \;=\; -q^{Q(\{2,1\})} \left[q^{Q(\{3\})} - q^{Q(\{2,1\})} \right]^{-1} \,, \tag{14.73}$$

and

$$z_2 \;=\; \left[q^{Q(\{3\})} - q^{Q(\{2,1\})} \right]^{-1} \,. \tag{14.74}$$

From equation (14.72) it follows that any link component associated with the representation $\{3\}$ can be substituted by the linear combination (14.72) of two strings; one of them is labelled by $\{2\}$ and the remaining string is labelled by $\{1\}$. To be more precise, let us denote by B_1 and B_2 the two components of the the link $B \subset \mathcal{V}$; similarly, $B_1^{(1)}$ and $B_2^{(1)}$ are the two components of $B^{(1)} \subset \mathcal{V}$. Equation (14.72) means that the link component C_1, associated with $\{3\}$, can be substituted by $h^\circ(P[\{3\}])$, where

$$P[\{3\}] \;=\; z_1 \, B(B_1, B_2; \{2\}, \{1\}) + z_2 \, B^{(1)}(B_1^{(1)}, B_2^{(1)}; \{2\}, \{1\}) \,. \tag{14.75}$$

On the other hand, we already know how to deal with strings characterized by the representation $\{2\}$. By using equation (14.54), we finally obtain

$$\begin{aligned} \widetilde{P}[\{3\}] &= z_1 \, B(h_1^\circ(\widetilde{P}[\{2\}]), B_2; \{1\}, \{1\}, \{1\}) \\ &\quad + z_2 \, B^{(1)}(\widehat{h}_1^\circ(\widetilde{P}[\{2\}]), B_2^{(1)}; \{1\}, \{1\}, \{1\}) \,, \end{aligned} \tag{14.76}$$

where the homeomorphism h_1° is determined by the first component B_1 of B and $\widehat{h}_1^\circ$ is determined by the first component $B_1^{(1)}$ of $B^{(1)}$.

In conclusion, if the link component C_1 is associated with the representation $\{3\}$, C_1 can be interpreted as a "composite state" made of two strings: the first string has colouring $\{2\}$ and the second string has colouring $\{1\}$. The first string, in turn, is a composite state made of two strings, both associated with $\{1\}$. So, in the end, the component C_1 can be understood as a composite state made of three elementary $\{1\}$-strings. The corresponding "wave function" (pattern link $\widetilde{P}[\{3\}]$) is given precisely in equation (14.76).

Summarizing, we have shown in a few particular examples that, by means of a recursive use of the satellite relations, we end up with links in which all the components are characterized by the defining representation $\{1\}$. This is the desired result because, at this point, we can use the skein relation of theorem **R3**.

14.5 Higher dimensional representations

Let us now consider the general construction for arbitrary representations of $SU(N)$. We shall be interested in oriented framed links in $\mathbb{R}^3$ (or in S^3) in which each component is associated with an irreducible representation of the gauge group $SU(N)$. We shall now prove the reconstruction theorem:

Theorem R4. *For any oriented framed link $L \subset \mathbb{R}^3$, whose components are labelled by arbitrary irreducible representations of $SU(N)$, the associated value $E(L)$ can be computed by means of a modified cabling method. The method consists of replacing each link component, associated with a higher dimensional representation ρ of the group, with the image of an appropriate pattern link $\widetilde{P}[\rho]$. All the components of the pattern link $\widetilde{P}[\rho]$ are associated with the defining representation $\{1\}$ of $SU(N)$. Consequently, we obtain a set of satellites in which all the components are associated with the representation $\{1\}$ and theorem* **R3** *can be applied.*

Proof. The proof consists of determining $\widetilde{P}[\rho]$ for each irreducible representation ρ of $SU(N)$. Of course, several solutions can a priori be found. Different (combinations of) pattern links may give rise to identical results. For example, the different mutations of the same link will always produce the same results. For our purposes it is sufficient to present, for each irreducible representation ρ of the group $SU(N)$, just one example of pattern link $\widetilde{P}[\rho]$.

As we have already mentioned, each irreducible representation of $SU(N)$ is characterized by the set $\{m_1, m_2, \ldots, m_t\}$ of nonvanishing integers representing the number of boxes in the columns of the Young tableau. Let I_n be the set of the irreducible representations described by the Young tableaux in which

$$m_1 + m_2 \cdots + m_t = n . \tag{14.77}$$

Assume by induction that the pattern links $\{ \widetilde{P}[\rho'] \}$, associated with the representations $\rho' \in I_{n-1}$, are known. We shall now construct the new pattern links $\{ \widetilde{P}[\rho] \}$ corresponding to the representations $\rho \in I_n$.

Each representation $\rho \in I_n$ appears in decomposition of the tensor product of a certain representation $\rho' \in I_{n-1}$ and $\{1\}$. Indeed, if the representation ρ is described by the Young tableau $\{m_1, m_2, \ldots, m_t\}$, then

$$\{m_1, m_2, \ldots, m_t\} \subset \{m_1, m_2, \ldots, m_t - 1\} \otimes \{1\} . \tag{14.78}$$

Consider now the pattern link $B \subset V$, shown in Fig. 14.2, with components B_1 and B_2. Let B_1 be associated with the representation

$$\rho' = \{m_1, m_2, \ldots, m_t - 1\} , \tag{14.79}$$

and B_2 be labelled by $\{1\}$.

Let us denote by $\mathcal{S}$ the set of the irreducible representations appearing in the tensor product decomposition

$$\{m_1, m_2, \ldots, m_t - 1\} \otimes \{1\} = \oplus_{M \in \mathcal{S}} \{M\}, \tag{14.80}$$

where $\{M\}$ stands for the appropriate sequence of indices. According to Lemma 1, any two representations of the set $\mathcal{S}$ are inequivalent. Moreover, $\rho \in \mathcal{S}$ by construction.

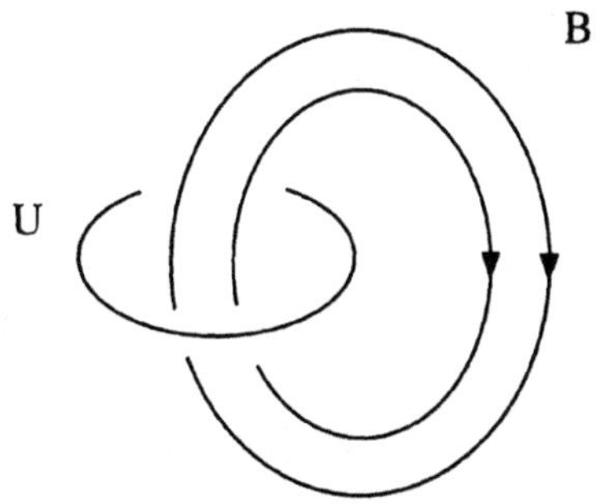

Figure 14.2.

The Wilson operator $W(B)$, associated with B inside $\mathcal{V}$, admits the decomposition

$$W(B) = \sum_{M \in \mathcal{S}} W(C; \{M\}), \tag{14.81}$$

where C is the core of $\mathcal{V}$. Let $B^{(p)}$ be the image of B under the action of p consecutive right-handed twist homeomorphisms of $\mathcal{V}$. Then, one has

$$W(B^{(p)}) = \sum_{M \in S} \alpha^p(\{M\}) \; W(C; \{M\}), \tag{14.82}$$

where

$$\alpha(\{M\}) = q^{Q(\{M\})}. \tag{14.83}$$

Suppose that $\mathcal{S}$ contains z elements; then, by varying p from 0 to $z - 1$ in equation (14.82), we obtain z equations for the z unknown quantities $W(C; \{M\})$ with $\{M\} \in \mathcal{S}$. This system of equations in invertible and each $W(C; \{M\})$ can be written as a linear combination of $\{ W(B^{(p)}) \}$. Indeed, according to Lemma 1, for each couple of representations $\{M\}$ and $\{M'\}$ belonging to $\mathcal{S}$ one has $Q(\{M\}) \neq$

$Q(\{M'\})$ and thus

$$\alpha(\{M\}) \neq \alpha(\{M'\}) . \tag{14.84}$$

Consequently, each operator $W(C\,;\{M\})$, can be written as

$$W(C\,;\{M\}) = \sum_{p=0}^{z-1} x_p(\{M\})\; W(B^{(p)}) , \tag{14.85}$$

for certain coefficients $x_p(\{M\})$.

Equation (14.85) implies that each representation $\{M\} \in \mathcal{S}$ is associated to the pattern link

$$\{M\} \Longrightarrow P[\{M\}] = \sum_{p=0}^{z-1} x_p(\{M\})\; B^{(p)} , \tag{14.86}$$

where the two link components of $B^{(p)}$ are labelled by the representations $\{m_1, m_2, \ldots, m_t - 1\}$ and $\{1\}$ respectively.

Since $\rho \in \mathcal{S}$, we obtain

$$P[\rho] = \sum_{p=0}^{z-1} x_p(\rho)\; B^{(p)}(B_1, B_2; \rho', \{1\}) . \tag{14.87}$$

The links, which appear on the right-hand-side of this equation have two components B_1 and B_2. The component B_1 is associated with the representation $\rho' \in I_{n-1}$ and the component B_2 is labelled by $\{1\}$. At this point, the component B_1 can be substituted by the image of the pattern link $\widetilde{P}[\rho']$. Indeed, because of the induction assumption, $\widetilde{P}[\rho']$ is known. Therefore,

$$\widetilde{P}[\rho] = \sum_{p=0}^{z-1} x_p(\rho)\; B^{(p)}(h_1^{\circ}(\widetilde{P}[\rho']), B_2; \{1\}, \ldots, \{1\}) \tag{14.88}$$

represents an explicit solution for $\widetilde{P}[\rho]$.

In conclusion, if the pattern links $\widetilde{P}[\rho]$ are known for $\rho \in I_{n-1}$, we can construct $\widetilde{P}[\rho]$ for $\rho \in I_n$. On the other hand, the starting set I_1 of irreducible representations contains only one element: $\{1\}$. In this case, $\widetilde{P}[\{1\}]$ is precisely the core C of $\mathcal{V}$ associated with the defining representation $\{1\}$. □

14.6 Polynomial structure

The constructive proof of the existence of the universal link polynomial for unitary groups is completed by the following

Theorem R5. *Consider a generic oriented framed link $L \subset \mathbb{R}^3$ whose components are labelled by arbitrary irreducible representations of $SU(N)$. The associated value $E(L)$, defined by theorems* **R3** *and* **R4**, *is a finite Laurent polynomial in $q^{1/2N}$ with integer coefficients.*

Proof. The simplest proof consists of showing that $E(L)$ admits an explicit presentation as a Markov trace of the type considered in Chapter 9. If both the enhancement and the braid generators are represented by finite dimensional matrices in which each matrix element belongs to $Z[q^{\pm 1/2N}]$, then $E(L)$ clearly belongs to $Z[q^{\pm 1/2N}]$. Such a presentation of $E(L)$ can easily be obtained by means of the quantum group construction mentioned in Chapter 9. Indeed, all the work has already been done in the articles on this subject quoted in this chapter. With the deformation parameter q playing the role of a free indeterminate, the representation rings of "quantum groups" coincide with the representation rings of ordinary simple Lie algebras. Therefore, all the satellite relations are satisfied. We only need to fix the normalization of the braid generators and the normalization of the enhancement. (Remember that one can always introduce an abelian factor multiplying the R-matrix; moreover, the enhancement can be modified according to the inequivalent representations of a normal subgroup of the gauge group). These normalizations are completely fixed by conditions (i)-(iv) of theorem **R3**. Indeed, conditions (i)-(iv) combined with the satellite relations uniquely determine $E(L)$. □

Exercise. Produce an alternative proof of theorem **R5**. This proof should be based on properties (i)-(iv) of theorem **R3** and on the satellite relations.

As we have seen in Section 14.2, the values $Q(\rho)$ of the quadratic Casimir operator for $G = SU(N)$ are of the type $Q(\rho) = n/2N$, where n is an integer. Roughly speaking, the variable of the polynomials is $q^{\pm 1/2N}$ simply because, in the relevant formulae, the deformation parameter q appears in the combination $q^{Q(\rho)}$.

In Chapter 12 we already proved that, for the group $G = SU(2)$, $E(L) \in Z[q^{\pm 1/4}]$; the proof was based on the recursive use of the standard cabling method. In certain particular cases, standard cabling can also be applied when $G = SU(N)$ with $N > 2$. For example, consider the tensor product decomposition

$$\{1\} \otimes \{N-1\} = \{0\} \oplus \{N-1, 1\} . \tag{14.89}$$

Suppose that the link component C_1 is associated with the adjoint representation $\{N-1, 1\}$ and the remaining components of a given link are labelled by $\{1\}$. Then, equation (14.89) implies that

$$\begin{aligned} E(C_1, \ldots; \{N-1, 1\}, \ldots) &= E(h_1^\circ(B), \ldots; \{1\}, \{N-1\}, \ldots) \\ &- E(C_1, \ldots; \{0\}, \ldots) . \end{aligned} \tag{14.90}$$

The first of the two oriented components of $h_1^\circ(B)$ is labelled by $\{1\}$ and the second is labelled by $\{N-1\}$. Since $\{N-1\}$ is the complex conjugate of $\{1\}$, with a modification of the orientation of the second component we get a link in which all the components are associated with $\{1\}$. On the other hand, when C_1 is labelled by the trivial representation $\{0\}$, C_1 can simply be removed. In conclusion, equation (14.90) shows that $E(C_1, \ldots; \{N-1, 1\}, \ldots)$ can be computed by analysing two links in which all the components are labelled by the defining representation $\{1\}$. According to theorem **R3**, the associated values of $E(L)$ are polynomials; therefore, their difference also is a polynomial. Clearly, the same argument can be used in the case in which an arbitrary number of link components are associated with the adjoint $\{N-1, 1\}$ representation of the group $SU(N)$.

Let us now consider the consequences of theorem **R5**. As we have already mentioned, the coefficients $\{x_p(\rho)\}$, entering the expression (14.87) of the standard pattern links, do not belong to $Z[q^{\pm 1/2N}]$. Indeed, it is easy to see that each $x_p(\rho)$ is in general the ratio of two finite Laurent polynomials in $q^{1/2N}$ with integer coefficients. Suppose now that the framed link component C_1 is associated with the representation ρ; equation (14.87) implies that

$$E(C_1, \ldots; \rho, \ldots) = \sum_{p=0}^{z-1} x_p(\rho)\, E(h_1^\circ(B^{(p)}), \ldots; \rho', \{1\}, \ldots) , \tag{14.91}$$

where h_1° is the homeomorphism defined by the framing of C_1. From this equation it is not evident that $E(C_1, \ldots; \rho, \ldots) \in Z[q^{\pm 1/2N}]$. However, theorem **R5** guarantees that this is always the case. In other words, the single terms of the sum (14.91) always combine in such a way that the final result is a polynomial.

The implications of theorem **R5** in quantum field theory are quite remarkable. Since the deformation parameter is given by

$$q = \exp(-2\pi i/k) , \tag{14.92}$$

theorem **R5** implies that $E(L)$ is well defined for arbitrary nonvanishing values of the coupling constant k. Let us recall that the standard perturbative series, defined by the Feynman diagrams of the quantum field theory in $\mathbb{R}^3$, corresponds to the Taylor expansion of $E(L)$ in powers of $\lambda = (2\pi/k)$ around $\lambda = 0$, i.e. around

$q = 1$. Thus, this series is summable and the result presents no singularities on the complex λ plane.

14.7 SU(3) examples

In this section, we shall consider the case in which the gauge group is $G = SU(3)$. A few examples, corresponding to a certain number of simple links, will be analysed. We shall produce the explicit expressions of the associated polynomials.

Let us start by giving the values $E_0[\rho]$ of the unknot (with zero writhe) associated with the irreducible representation ρ of $SU(3)$. By using the fundamental skein relation and the satellite formulae, one finds

$$E_0[\{1\}] = E_0[\{2\}] = \frac{(1-q^3)}{(1-q)}\, q^{-1}, \tag{14.93}$$

$$E_0[\{1,1\}] = E_0[\{2,2\}] = (1+q^{-2})\,\frac{(1-q^3)}{(1-q)}, \tag{14.94}$$

$$E_0[\{2,1\}] = (1+q)^2\,(1+q^{-2}), \tag{14.95}$$

$$E_0[\{1,1,1\}] = E_0[\{2,2,2\}] = q^{-3}\,(1+q^2)\,\frac{(1-q^5)}{(1-q)}, \tag{14.96}$$

$$E_0[\{2,1,1\}] = E_0[\{2,2,1\}] = \frac{(q^{-3}-1)\,(1-q^5)}{(1-q)^2}, \tag{14.97}$$

$$E_0[\{1,1,1,1\}] = E_0[\{2,2,2,2\}] = q^{-4}\,\frac{(1-q^6)\,(1-q^5)}{(1+q)\,(1-q)^2}, \tag{14.98}$$

$$E_0[\{2,2,1,1\}] = q^{-4}\,\frac{(1-q^3)^2\,(1-q^6)}{(1-q)^3\,(1+q)}, \tag{14.99}$$

$$\begin{aligned} E_0[\{2,1,1,1\}] &= E_0[\{2,2,2,1\}] = \\ &= q^{-4}\,\frac{(1-q^3)}{(1-q)}\,(1+q)\,(1+q^2)\,(1+q^3), \end{aligned} \tag{14.100}$$

$$E_0[\{1,1,1,1,1\}] = E_0[\{2,2,2,2,2\}] = q^{-5}\,\frac{(1-q^6)\,(1-q^7)}{(1+q)\,(1-q)^2}, \tag{14.101}$$

$$E_0[\{2,1,1,1,1\}] = E_0[\{2,2,2,2,1\}] = \frac{(q^{-5}-1)\,(1-q^7)}{(1-q)^2}, \tag{14.102}$$

$$E_0[\{2,2,1,1,1\}] = E_0[\{2,2,2,1,1\}] = \\ = \frac{(1+q^{-2})(q^{-3}-1)(1-q^7)}{(1-q)^2}, \qquad (14.103)$$

$$E_0[\{1,1,1,1,1,1\}] = E_0[\{2,2,2,2,2,2\}] = \\ = \frac{(q^{-2}+1)(q^{-4}+1)(1-q^7)}{(1-q)}. \qquad (14.104)$$

We shall now consider the first three nontrivial knots and links of the Rolfsen classification table [12]. The right-handed trefoil knot 3_1 is shown in Fig. 14.3.

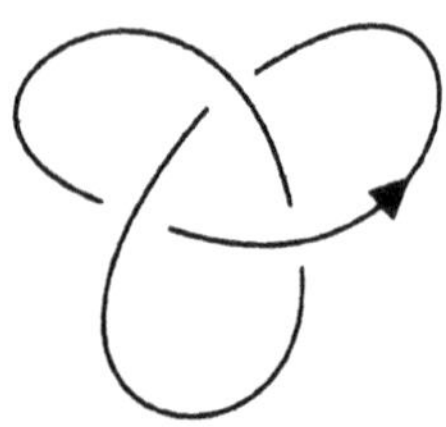

Figure 14.3. Right-handed trefoil knot 3_1

For the right-handed trefoil labelled by $\{1\}$, one has

$$E(3_1;\{1\}) = (q+1+q^{-1})(q^2+1-q^{-2}); \qquad (14.105)$$

whereas, for the six-dimensional representation $\{1, 1\}$, one gets

$$E(3_1;\{1,1\}) = \frac{(1+q^{-2})(1-q^3)}{(1-q)}[q^6+q^3+q^2-q-q^{-2}-q^{-3}+q^{-5}]. \qquad (14.106)$$

The figure-eight knot 4_1 is shown in Fig. 14.4. By means of the five defining relations of the universal link polynomial, one finds

$$E(4_1;\{1\}) = q^4+q^3-1+q^{-3}+q^{-4}, \qquad (14.107)$$

and

$$E(4_1;\{1,1\}) = q^{-10}(1+q^{18})\frac{(1-q^3)}{(1-q)} - q^{-7}(1+q^{11})\frac{(1-q^4)}{(1-q)} + q^{-2}(1+q^2)(1+q)^2 . \tag{14.108}$$

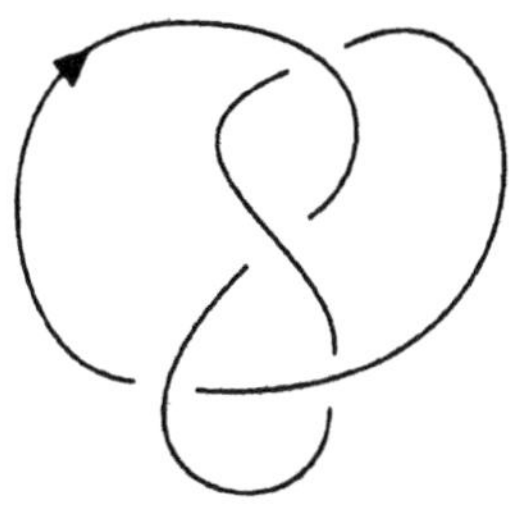

Figure 14.4. Figure-eight knot 4_1

Exercise. Compute $E(4_1;\{1,1\})$ by means of quantum group methods and verify that the result coincides with the expression (14.108).

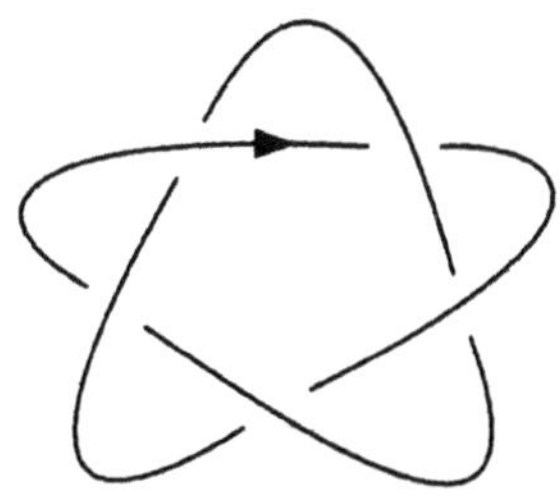

Figure 14.5. Right-handed knot 5_1

For the right-handed knot 5_1, shown in Fig. 14.5, we obtain

$$E(5_1,\{1\}) = \frac{(1-q^3)(1+q^2-q^{-4})}{(1-q)}\, q^{-1/3} , \tag{14.109}$$

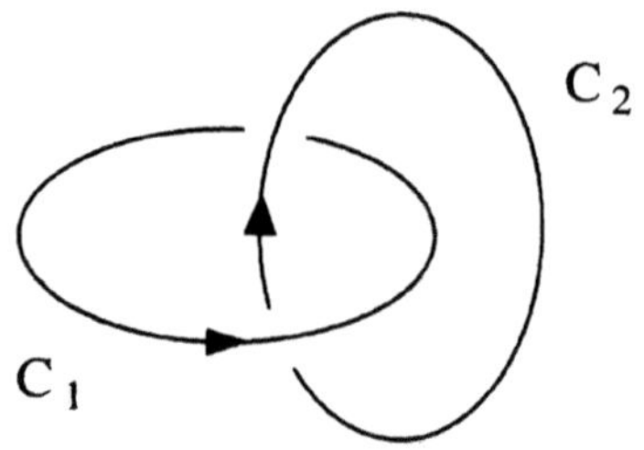

Figure 14.6. Oriented Hopf link 2_1^2

and

$$E(5_1;\{1,1\}) = \frac{(1-q^3)(1+q^2)}{(1-q)}\,q^{-31/3} + \frac{(1-q^3)(1-q^5)}{(1-q)^2}\left[\frac{(1+q^3)}{(1+q)} - q^{-9}\right] q^{8/3}. \qquad (14.110)$$

The oriented Hopf link 2_1^2, with writhe $w = +2$, is shown in Fig. 14.6. As we shall see, the values $E(2_1^2;\rho_1,\rho_2)$ of the universal link polynomial for the Hopf link are particularly interesting. One has

$$E(2_1^2;\{1\},\{1\}) = (1+q+q^2)(1+q+q^3)\,q^{-7/3}, \qquad (14.111)$$

$$E(2_1^2;\{1\},\{2\}) = (1+q+q^2)(1+q^2+q^3)\,q^{-8/3}, \qquad (14.112)$$

$$E(2_1^2;\{1\},\{1,1\}) = \frac{(1+q^2)(1-q^3)}{(1-q)}\left(1+q+q^4\right) q^{-11/3}, \qquad (14.113)$$

$$E(2_1^2;\{1\},\{2,1\}) = \frac{(1-q^4)(1-q^6)}{(1-q)^2}\,q^{-4}, \qquad (14.114)$$

$$E(2_1^2;\{1\},\{1,1,1\}) = \frac{(1-q^3)(1-q^5)}{(1-q)^2}\left[1+q^3-q^4+q^5\right] q^{-5}, \qquad (14.115)$$

$$E(2_1^2;\{1,1\},\{1,1\}) = \frac{(1-q^3)(1+q^2)}{(1-q)}\,q^{-16/3} + \frac{(1-q^3)(1-q^5)}{(1-q)^2}\left[1+q^3-q^4+q^5\right] q^{-13/3}, \qquad (14.116)$$

$$E(2_1^2;\{1,1\},\{2,1\}) = \frac{(1-q^6)(1-q^8)}{(1-q)^2}\,q^{-6}, \qquad (14.117)$$

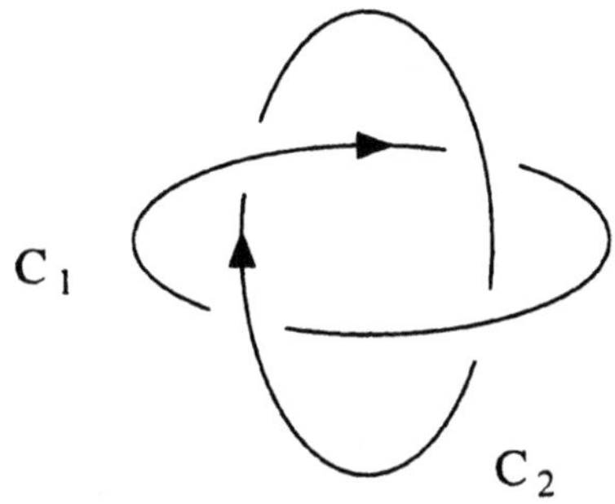

Figure 14.7. Oriented link 4_1^2

$$E(2_1^2;\{1,1\},\{1,1,1\}) = \frac{(1-q^3)(1-q^5)}{(1-q)^2}\,q^{-7} + \frac{(1-q^5)(1-q^6)}{(1-q)(1-q^2)}\left[1+q+q^6\right]q^{-5}. \tag{14.118}$$

For the oriented link 4_1^2, shown in Fig. 14.7, we obtain

$$E(4_1^2;\{1\},\{1\}) = (q+1+q^{-1})(1+q^3+q^5)\,q^{-8/3}, \tag{14.119}$$

$$E(4_1^2;\{1\},\{2\}) = (q+1+q^{-1})(1+q^2+q^5)\,q^{-7/3}, \tag{14.120}$$

$$E(4_1^2;\{1\},\{1,1\}) = \frac{(1+q^2)}{(1-q)}\left[1+q-q^2-q^3+q^5-q^{10}\right]q^{-16/3}, \tag{14.121}$$

$$E(4_1^2;\{1\},\{2,1\}) = \frac{(1-q^6)}{(1-q)}\left[1+q^5+q^6+q^7\right]q^{-7}. \tag{14.122}$$

$$E(4_1^2;\{1\},\{1,1,1\}) = \frac{(1-q^3)(1-q^5)}{(1-q)^2}\left[q^2-q+1+q^{-7}\right], \tag{14.123}$$

$$E(4_1^2;\{1,1\},\{1,1\}) = \frac{(1-q^3)(1+q^2)}{(1-q)}\,q^{-26/3} + \frac{(1-q^3)(1-q^5)}{(1-q)^2}\left[1-q+q^2+q^7\right]q^{4/3}. \tag{14.124}$$

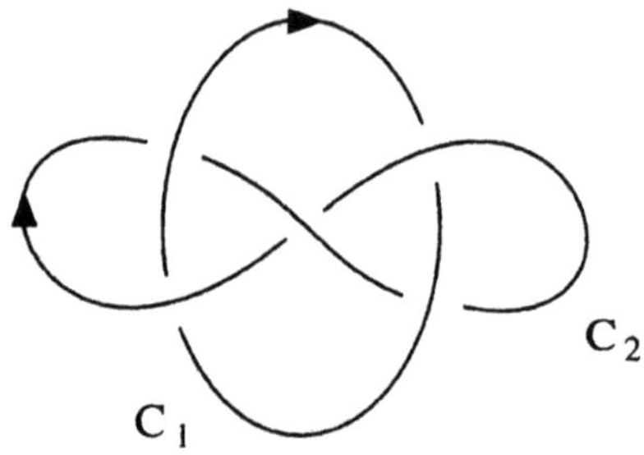

Figure 14.8. Oriented Whitehead's link 5_1^2

Finally, let us consider the Whitehead's link 5_1^2 shown in Fig. 14.8. By using the calculus rules of the universal link polynomial, we find

$$E(5_1^2; \{1\}, \{1\}) = \frac{(1-q^3)}{(1-q)} \left[q^{-3} + 2 - q + q^2 + q^3 - q^4 \right] q^{-4/3}, \quad (14.125)$$

and

$$\begin{aligned} E(5_1^2; \{1,1\}, \{1\}) &= \frac{(1+q^2)(1-q^3)}{(1-q)} q^{-1/3} \\ &+ \frac{(1-q^6)}{(1-q)} \left[1 - q^2 + 2q^4 - q^6 \right] (1+q^2) q^{-19/3}. \end{aligned} \quad (14.126)$$

Chapter 15

Reduced tensor algebra

In the previous chapters we presented the exact solution of the quantum CS field theory in $\mathbb{R}^3$. At this stage, producing the expression $E(L)$ of the universal link polynomial for different links in $\mathbb{R}^3$ is a simple matter of elementary algebra. Our main purpose now is to solve the topological CS field theory in any closed, connected and orientable three-manifold $\mathcal{M}$. In order to solve this problem, we must find a general strategy. Therefore, we shall analyse the possible methods which can, in principle, be used. We will see that, by means of surgery on three-manifolds, the problem can really be solved.

In this chapter, we shall first review the basic arguments which have been used to solve the theory in $\mathbb{R}^3$. Then we shall consider the theory in the three-sphere S^3 and we shall describe the remarkable properties of the model, with gauge group $SU(2)$, when the coupling constant k is an integer.

15.1 The restated solution

The quantum CS field theory is defined by the action

$$S_{CS} = \frac{k}{4\pi} \int d^3x \, \epsilon^{\mu\nu\rho} \, \mathrm{Tr}\left(A_\mu \partial_\nu A_\rho + i \, \tfrac{2}{3} \, A_\mu A_\nu A_\rho \right) . \tag{15.1}$$

All the properties of the model and the numerical results associated with the observables must be consequences of the particular structure of the action (15.1). Indeed, all the results that we obtained in the previous chapters were derived from the action (15.1). Let us briefly recall the main arguments which were used in our derivation.

The starting point was to consider the model in $\mathbb{R}^3$. The quantum field theory is known to be characterized by the renormalized vacuum expectation values of the fields. Because of local gauge invariance, a gauge fixing procedure was introduced to define the correlation functions. Clearly, the correlation functions are not gauge invariant. Nevertheless, the gauge-invariant content of the theory is included in the structure of the correlation functions. So, the first problem is to construct well defined correlation functions,

$$\text{action principle} \iff \text{correlation functions};$$

how to extract the physical gauge-invariant information from them is the second problem.

The correlation functions were defined by means of the standard method of the Feynman diagrams. For several reasons, our choice of $\mathbb{R}^3$ as the starting manifold was crucial. First of all, in $\mathbb{R}^3$ a well defined gauge-fixing procedure is known. (Of course, some suitable gauge-fixing procedure should in principle exist also in more complicated manifolds: for example, the Poincaré sphere. The point is that such a procedure is unknown and is therefore of no help in trying to solve the theory.) Secondly, in $\mathbb{R}^3$ the standard gauge-fixing procedure is consistent with the structure of the observables we are interested in (remember our observations at the end of Chapter 5). Finally, in $\mathbb{R}^3$ there are no constraints to be imposed on the value of the renormalized coupling constant. This fact, combined with the dynamical information that the model has a vanishing Hamiltonian, strongly supports the validity of perturbation theory.

We have seen that the perturbative expansion is well defined. The correlation functions of the CS theory can be computed by means of the Feynman diagrams:

$$\text{correlation functions} \iff \text{perturbation theory}.$$

At any given order of perturbation theory, the correlation functions satisfy an action principle based on the CS functional (15.1). This means that the renormalized Schwinger-Dyson equations are valid and there are no anomalies of any kind; all the symmetries of the system are preserved at the full quantum level and the associated Ward-identities are satisfied. As expected, the renormalized theory depends on a unique dimensionless parameter: the renormalized coupling constant k. Since the normalization of the fields is fixed by the form of the BRS transformations, the value of k is uniquely determined for example by the three-point proper vertex.

The renormalized correlation functions specify the model completely; all the physical effects and the associated numerical results are consequences of the structure of the renormalized correlation functions.

We showed that a complete set of gauge-invariant observables associated with links is given by the expectation values of Wilson line operators. In computing Wilson line operators, the correlation functions have to be integrated not with ordinary test functions but with rather "singular" functions which are associated with the path-ordered exponential of the fields defined along closed contours in $\mathbb{R}^3$. In other words, Wilson line operators are composite operators. Quite remarkably, the integrals of the correlation functions with such singular functions are finite. However, there are certain ambiguities which are due to finite terms. These ambiguities are completely removed if one requires general covariance. Indeed we have seen that, by introducing a framing procedure, the composite Wilson line operators are well defined and general covariance is maintained.

At this point, we computed the first three terms of the perturbative expansion of the observables of the CS theory: the result is a function of the Casimir operators of the gauge group and represents an ambient isostopy invariant of oriented framed links in $\mathbb{R}^3$. All this is in perfect agreement with the expected properties of the topological quantum field theory.

The next problem was to give the complete expression of the observables (to all orders in the coupling constant). In the Abelian case, we simply summed all the terms of the perturbative series. This direct method, however, cannot be used in the non-Abelian case. We found then an alternative method which is based on the peculiar properties of the topological field theory and of the Wilson line observables.

We made a list of all the general properties of the expectation values of the observables. We found that the structure of the composite Wilson line operators implies the existence of satellite relations between the link invariants. Then, we combined the general properties of the expectation values with the numerical information derived in the canonical approach defined by the action (15.1). At this point, we discovered that the expectation values are completely fixed. The values of the observables are uniquely determined by the satellite relations, by the factorization property of the path-integral and by the structure of the monodromy matrix.

We organized all this information, obtained in the field theory context, into a set of rules which permit us to compute the observables in closed form. The resulting expectation values $\{ E(L) \}$ can conveniently be expressed as functions of the deformation parameter

$$q = \exp\left(-i \frac{2\pi}{k} \right) , \tag{15.2}$$

where k is the coupling constant of the CS theory.

Let the oriented framed components $\{ C_1, C_2, \ldots, C_m \}$ of the link L be associated with the irreducible representations $\{ \rho_1 \rho_2, \ldots, \rho_m \}$ of the gauge group. The so-called semiclassical ($k \to \infty$) limit is well defined:

$$\lim_{k \to \infty} E(L) = \prod_j D(\rho_j) , \tag{15.3}$$

where $D(\rho_j)$ is the dimension of the representation ρ_j. On the other hand, the $k \to 0$ limit of $E(L)$ does not exist; this is in perfect agreement with the fact that, in the $k \to 0$ limit, the path-integral

$$E(L) = \langle W(L) \rangle = Z^{-1} \int [\mathcal{D}A_\mu] \, e^{ikS_0} \, W(L) \tag{15.4}$$

is meaningless.

The expressions $\{E(L)\}$ of the expectation values admit a Taylor expansion in powers of $\lambda = (2\pi/k)$ around $\lambda = 0$. The first three terms of this expansion coincide with the computations of perturbation theory based on ordinary Feynman diagrams.

In conclusion, all our results were derived exclusively from the action (15.1). Assuming that perturbation theory is meaningful, the solution of the quantum CS field theory in $\mathbb{R}^3$ is unique. This unique solution, which satisfies an action principle based on the CS functional (15.1), was presented in the previous chapters. At this stage, all the intrinsic and gauge-invariant content of the CS theory in $\mathbb{R}^3$ is expressed by the known values of the observables

$$\text{observables} \iff \text{link invariants}\,.$$

15.2 Outlook

Let us now consider the CS theory defined in a generic three-manifold $\mathcal{M}$ which is closed, connected and orientable. In this case, we cannot rely upon perturbation theory to compute the expectation values of the observables. One reason is that a general gauge-fixing procedure, which is consistent with the nature of the observables we are considering, is not known. Another reason is that, unlike the situation we had in $\mathbb{R}^3$, the coupling constant k can no longer be considered a free parameter. Only for certain integer values of k is the CS theory well defined in $\mathcal{M}$. Consequently, the expressions of the observables do not necessarily admit a Taylor expansion in powers of $\lambda = (2\pi/k)$ around $\lambda = 0$. Indeed, we shall consider explicit examples in which such a Taylor expansion does not exist.

In order to solve the CS theory in $\mathcal{M}$, we shall introduce a new method in which the topological nature of the theory plays an essential role. In practice, we will use

(a) the explicit results obtained in $\mathbb{R}^3$;
(b) "symmetry transformations" connecting different manifolds.

The basic idea is that once the solution of the topological field theory has been produced in some given manifold we only need to use appropriate "symmetry" arguments to solve the theory in any other three-manifold. The peculiar "symmetry" I am referring to is rather unusual in the context of ordinary field theories; the relevant symmetry transformations we will consider are twist homeomorphisms of solid tori.

Let me briefly explain the main points of the general strategy which will be used to solve the CS theory in a three-manifold $\mathcal{M}$. The first step consists of finding the solution of the CS theory in the three-sphere S^3. Clearly, from the knot theory point of view, the substitution

$$\mathbb{R}^3 \Longrightarrow S^3 \tag{15.5}$$

does not really change things. The field theory solution in S^3 coincides with the solution in $\mathbb{R}^3$ apart from the constraint which must be imposed on the values of the coupling constant k. As we have already mentioned in Chapter 4, gauge invariance requires that, in S^3, k must be an integer.

The crucial fact now is that any three-manifold $\mathcal{M}$ can be obtained by surgery on S^3. To be more precise, one can always find such a surgery presentation in terms of Dehn surgery on S^3. The second step consists of finding the connection between surgery operations and appropriate symmetry transformations

$$\text{surgery} \quad \Longleftrightarrow \quad \text{symmetry} \quad \text{transformations}\,. \tag{15.6}$$

Then, since the complete solution of the theory in S^3 is known, we shall consider the third step which consists of producing an explicit operator realization of surgery

$$\text{surgery} \quad \Longleftrightarrow \quad \text{operator} \quad \text{construction}\,. \tag{15.7}$$

The explicit construction of the field theory operator which represents Dehn surgery on S^3 is the only technical point of the whole construction. Here, the fact that k is an integer will play an important role. We shall concentrate on the algebraic simple case in which the gauge group is $G = SU(2)$. The generalization for arbitrary unitary groups is straightforward.

Finally, the last step consists of applying the operator surgery method to compute the values of the observables in any three-manifold $\mathcal{M}$.

To sum up, the solution of the quantum CS field theory in a generic three-manifold $\mathcal{M}$ will ultimately be determined by the Feynman path-integral, which satisfies an action principle based on the CS functional (15.1), and by symmetry arguments.

Instead of using the Feynman path-integral, some authors prefer to consider a set of "axioms" which should specify, with absolute mathematical rigour, what a "topological quantum field theory" should be. Within these "axioms", a given linear representation of standard cobordism is introduced and the solution of the "theory" is already contained, by assumption, in the starting "axioms". The great advantage of this approach is that, regardless of the existence or not of any real underlying field theory in three dimensions, one can consider specific braid group representations associated with certain quasi-tensor categories from an abstract point of view.

We shall not discuss this "axiomatic" method; our only purpose is to solve the three-dimensional quantum CS field theory defined by the action (15.1).

15.3 Representation ring

The expectation values $\{ E(L) \}$ form a complete set of observables; therefore, all the physical effects must be derived from $\{ E(L) \}$. Let us consider, in particular, the structure [59] of the quantum numbers which appear in the solution of the CS theory in $\mathbb{R}^3$.

Let C be an oriented framed knot in $\mathbb{R}^3$ labelled by the representation ρ of the gauge group G; the expectation value of the associated Wilson line operator is denoted by $E(C;\rho)$. Suppose that ρ is not irreducible but is equivalent to the direct sum

$$\rho = \rho_1 \oplus \rho_2 , \tag{15.8}$$

where ρ_1 and ρ_2 are irreducible representations. Since the Wilson operator is the trace of the quantum holonomy, it follows that

$$E(C;\rho) = E(C;\rho_1 \oplus \rho_2) = E(C;\rho_1) + E(C;\rho_2) . \tag{15.9}$$

Suppose now that ρ is equivalent to the tensor product

$$\rho = \rho_1 \otimes \rho_2 . \tag{15.10}$$

We can decompose this tensor product into a sum of irreducible components $\{ \rho(i) \}$

$$\rho = \rho_1 \otimes \rho_2 = \oplus_{i=1}^{n} \rho(i) . \tag{15.11}$$

Then, the analogue of equation (15.9) is simply

$$E(C;\rho) = E(C;\rho_1 \otimes \rho_2) = \sum_{i=1}^{n} E(C;\rho(i)) . \tag{15.12}$$

Equations (15.9) and (15.12) show that $E(C)$ is a linear function on the so-called representation ring of the gauge group G.

Let all the inequivalent irreducible representations $\{ \rho(i) \}$ of G be labelled by the index i. For each representation $\rho(i)$, we may introduce an element $\chi[\, i \,]$ of a ring R. The multiplication operation in R is defined according to the decomposition rule of the tensor product of the representations of the group G. In other words, let $t_{ij\ell}$ be the multiplicity of $\rho(\ell)$ in the decomposition of the tensor product of $\rho(i)$ and $\rho(j)$,

$$\rho(i) \otimes \rho(j) = \sum_{\ell} t_{ij\ell} \; \rho(\ell) . \tag{15.13}$$

Then $\{t_{ij\ell}\}$ are taken to be the structure constants of R:

$$\chi[i]\,\chi[j] = \sum_{\ell} t_{ij\ell}\;\; \chi[\ell]. \qquad (15.14)$$

The coefficients $\{t_{ij\ell}\}$ are symmetric in (ij) and take non-negative integer values. If, for some fixed ℓ, $\rho(\ell)$ does not appear in the decomposition of $\rho(i) \otimes \rho(j)$, then $t_{ij\ell} = 0$.

An addition operation for the elements of R is introduced with the standard properties: R is an Abelian group under the addition operation, R is a semigroup under multiplication and the multiplication is distributive over the addition. For a generic compact Lie group G, R is called the associated ***representation ring*** and is a commutative ring with identity.

In conclusion, $E(C)$ is a linear function on R. In general, for any given link L with m components, the associated expectation value $E(L)$ is a multilinear function on $R^{\otimes m}$.

As we have seen in Section 13.3, for our purposes it is both convenient and necessary to extend the representation ring R to a full algebra $\mathcal{T}$ over the complex numbers; this algebra is called the ***tensor algebra***. As a linear space, $\mathcal{T}$ is generated by $\{\chi[i]\}$ and its structure constants are shown in equation (15.14).

Any framed and oriented knot C can now be labelled by an element $\chi \in \mathcal{T}$

$$\chi = \sum_i x_i\; \chi[i], \qquad (15.15)$$

where $\{x_i\}$ are complex coefficients. Then, $E(C;\chi)$ is defined by

$$E(C;\chi) = \sum_i x_i\; E(C;\chi[i]) = \sum_i x_i\; E(C;\rho(i)). \qquad (15.16)$$

For any given knot C, $E(C;\chi)$ represents a linear function on $\mathcal{T}$. Let us now consider a two-component link $L = \{C_1 \cup C_2\}$; if the two components of L are labelled by the elements $\chi_1 = \sum_i x_i\, \chi[i]$ and $\chi_2 = \sum_j y_j\, \chi[j]$ of $\mathcal{T}$, the associated link invariant is

$$E(C_1, C_2; \chi_1, \chi_2) = \sum_i \sum_j x_i\, y_j\;\; E(C_1, C_2; \rho(i), \rho(j)). \qquad (15.17)$$

In general, we can associate the elements $\{\chi_1, \dots, \chi_m\}$ of $\mathcal{T}$ to the components $\{C_1, \dots, C_m\}$ of a given link L. In this case, the obvious extension of equation (15.17) shows that $E(L; \chi_1, \dots, \chi_m)$ is a multilinear function on $\mathcal{T}^{\otimes m}$.

To sum up, the group-theory labels, associated with the different link components, represent the gauge-invariant "quantum numbers" which characterize the observables of the CS theory. We have seen that an algebra structure is naturally associated with these quantum numbers; this algebra is precisely the tensor algebra $\mathcal{T}$ of the gauge (compact Lie) group G of the theory.

The algebra $\mathcal{T}$ is of infinite order. The elements $\{\chi[i]\}$ form the so-called standard basis of the linear space $\mathcal{T}$. This basis has a distinguished role because, as we have seen in the previous chapters, the elements $\{\chi[i]\}$ diagonalize twist transformations; i.e. the transformations corresponding to modifications of framing.

The structure constants (15.14) describe the rules which must be used in the construction of the elementary satellite relations for pattern links with two components.

As a linear space, $\mathcal{T}$ can be interpreted as a state space; the elements of $\mathcal{T}$ correspond to the physical (gauge-invariant) states of the system associated with the boundary of a solid torus. Indeed, as we have seen in Chapter 13, the linear structure of $\mathcal{T}$ enters the decomposition (13.18) of the Wilson line operators associated with links inside a solid torus. Each decomposition of a given Wilson operator inside a solid torus specifies a particular element χ of $\mathcal{T}$; this element χ is used in the construction of the generalized satellite relations (13.34).

Let us now consider the case in which the gauge group is $G = SU(2)$. The inequivalent irreducible representations of $SU(2)$ are labelled by J with $2J = 0, 1, 2, \ldots$ and so the elements of the standard basis of the tensor algebra $\mathcal{T}$ will be denoted by $\{\chi[J]\}$. The structure constants of $\mathcal{T}$ are given by

$$\chi[J_1]\,\chi[J_2] = \sum_{J=|J_1-J_2|}^{J_1+J_2} \chi[J]\,, \tag{15.18}$$

where, of course, one has to sum over the values of J which differ by unity.

Let us recall that, for the extended Jones polynomial, the values of the unknot U_0 (with zero writhe) are given by

$$E(U_0; \chi[J]) = E_0[J] = \frac{q^{(2J+1)/2} - q^{-(2J+1)/2}}{q^{1/2} - q^{-1/2}}\,. \tag{15.19}$$

As we have shown in Chapter 12, the extended Jones polynomial is a finite Laurent polynomial in $q^{1/4}$ with integer coefficients. We shall now prove a property [59] which will be used in Section 15.5.

Property G7. *For any given link L, in which one link component is labelled by* $\chi[J]$, *the associated extended Jones polynomial takes the form*

$$E(L) = E_0[J]\ \mathcal{F}_J\,, \tag{15.20}$$

with

$$\mathcal{F}_J \in Z[q^{\pm 1/4}]\,. \tag{15.21}$$

Proof. Consider the connected sum $L_1\#L_2$ in which the two links L_1 and L_2 are copies of the link L; in other words, L_1 and L_2 are separately ambient isotopic with the framed link L. Let the connected sum $L_1\#L_2$ be obtained from L_1 and L_2 by acting on the link component associated with $\chi[J]$. Then, from equation (13.49) it follows that

$$E(L_1\#L_2) = \frac{E(L)\,E(L)}{E_0[J]}. \tag{15.22}$$

Since $E(L) \in Z[q^{\pm 1/4}]$, one can factorize the maximal power of $q^{-1/4}$ in $E(L)$ and write

$$E(L) = \left(q^{-1/4}\right)^b \ \mathcal{P}(q^{1/4}), \tag{15.23}$$

where b is an integer and $\mathcal{P}(q^{1/4})$ is an ordinary finite polynomial in $q^{1/4}$ with integer coefficients. Similarly, the expression (15.19) can be written as

$$E_0[J] = \left(q^{-1/4}\right)^{4J} \ \mathcal{P}_0\,(q^{1/4}), \tag{15.24}$$

where $\mathcal{P}_0$ is the polynomial

$$\mathcal{P}_0 = 1 + q + q^2 + \cdots + q^{2J}. \tag{15.25}$$

By using equations (15.23) and (15.24), equation (15.22) takes the form

$$E(L_1\#L_2) = \left(q^{-1/4}\right)^{2b-4J} \ \frac{\mathcal{P}(q^{1/4})\,\mathcal{P}(q^{1/4})}{\mathcal{P}_0\,(q^{1/4})}. \tag{15.26}$$

Since $E(L_1\#L_2) \in Z[q^{\pm 1/4}]$, equation (15.26) implies that all the roots of the polynomial $\mathcal{P}_0$ must also be roots of the product $\mathcal{P}\mathcal{P}$. On the other hand, all the roots of $\mathcal{P}_0$ are inequivalent; i.e. each root of $\mathcal{P}_0$ has multiplicity equal to unity. Therefore, $\mathcal{P}_0$ also divides each single polynomial $\mathcal{P}$. We have then

$$\mathcal{P}(q^{1/4}) = \mathcal{P}_0\,(q^{1/4}) \ \mathcal{Q}(q^{1/4}), \tag{15.27}$$

where $\mathcal{Q}(q^{1/4})$ is a finite polynomial. Since $\mathcal{P}$ has integer coefficients, the particular structure (15.25) of $\mathcal{P}_0$ implies that $\mathcal{Q}(q^{1/4})$ also has integer coefficients. Therefore, property **G7** is proved. □

15.4 The three-sphere

Among the closed connected and orientable three-manifolds, the three-sphere S^3 plays a distinguished role. S^3 is a homogeneous space and its fundamental group is trivial. Any link in S^3 is contained inside a three-ball and any link in $\mathbb{R}^3$ is also contained inside a three-ball. The topological properties of links in S^3 and of links in $\mathbb{R}^3$ are equivalent.

Let us now consider the solution of the CS model in S^3. Clearly, the short-distance behaviour of the field theory in S^3 is equal to the behaviour in $\mathbb{R}^3$. Therefore, the theory in S^3 is renormalizable and finite. Since the model is defined by the same Lagrangian density (15.1) as in $\mathbb{R}^3$, the physical consequences of the renormalized Schwinger-Dyson equations in S^3 coincide with those in $\mathbb{R}^3$. To be more precise, the curl of the expression corresponding to the propagator of the vector fields must represent an intersection form of a one-dimensional manifold (for example, a given component of a link) and a two-dimensional manifold (the Seifert surface associated with a different link component). This structure of the propagator, which can easily be verified in $\mathbb{R}^3$ by direct inspection, is a consequence of the Schwinger-Dyson equations. Since S^3 is a homology sphere, the vector field propagator in S^3 must have precisely the same structure as the propagator in $\mathbb{R}^3$.

All the general properties of the expectation values of the observables in $\mathbb{R}^3$ are also valid in S^3. Consequently, the defining relations of the universal link polynomial determine the Wilson line expectation values in $\mathbb{R}^3$ and in S^3 as well.

There is a point, however, which we need to consider carefully. The numerical input of the defining relations **(1)**-**(5)** depends on the complex parameter q only. A change in the value of the coupling constant k does not modify rules **(1)**-**(5)**: it only modifies the numerical value of the deformation parameter q. Now, we have seen in Chapter 4 that, because of gauge invariance, the coupling constant k of the CS theory in S^3 must take integer values. Thus, the global structure of S^3 does not modify any of the results obtained in $\mathbb{R}^3$; it simply introduces the constraint that k must be an integer.

To sum up, the exact solution of the quantum CS theory in S^3 can be obtained from the solution in $\mathbb{R}^3$ by imposing the condition that k is an integer. Since a change in the sign of k is equivalent to a modification of the orientation of S^3, we need only to consider the case in which

$$k \geq 1 \qquad \text{and} \qquad k = \text{integer}. \tag{15.28}$$

Condition (15.28) implies that, in the expressions of the vacuum expectation values of Wilson line operators in S^3, the deformation parameter q is a root of unity. Constraint (15.28) does not introduce any singularities in the values of the universal link polynomial because $\{ E(L) \}$ are well defined for any non-trivial value of k. Yet, condition (15.28) drastically modifies the physics of the CS theory.

15.5 Reduced tensor algebra

We shall now describe the remarkable physical consequences of condition (15.28) in the case in which the gauge group is $G = SU(2)$.

Let us first assume that k is fixed and $k > 2$; the cases in which $k = 2$ and $k = 1$ will be analysed later. The values (15.19) of the unknots can be written as

$$E_0[J] = \frac{\sin\left[(2J+1)\pi/k\right]}{\sin(\pi/k)}. \tag{15.29}$$

This equation shows that $E_0[J]$ is a periodic function in J with period k. For increasing values of J, starting from $J = 0$, $E_0[J]$ vanishes for the first time,

$$E_0[J_0] = 0, \tag{15.30}$$

when

$$J_0 = \frac{k-1}{2}. \tag{15.31}$$

For the reasons we shall now explain, the corresponding element $\chi[J_0]$ in the tensor algebra $\mathcal{T}$ of $SU(2)$ is called the nontrivial ***null vector***.

Indeed, let us consider a generic link L in which one link component is associated with the element $\chi[J_0]$. According to property **G7**, we can write

$$E(L) = E_0[J_0]\ \mathcal{F}_{J_0}, \tag{15.32}$$

where $\mathcal{F}_{J_0} \in Z[q^{\pm 1/4}]$. Since $E_0[J_0] = 0$, we obtain $E(L) = 0$.

In conclusion, when one component (at least) of a generic link is labelled by $\chi[J_0]$, the associated observable $E(L)$ vanishes. Since the link invariants $\{E(L)\}$ represent the whole set of the gauge-invariant physical observables of the system, $\chi[J_0]$ is totally equivalent to the real null vector of the tensor algebra. In fact, as far as physics (which is defined by the set of observables) is concerned, there is no distinction at all between $\chi[J_0]$ and the real (trivial) null vector of $\mathcal{T}$:

$$\chi[J_0] \Longleftrightarrow 0. \tag{15.33}$$

Let us now prove the following general result [59]:

Property G8. *For fixed integer $k \geq 2$, and for any link L in which one component is labelled by the element $\chi = \chi[J_0]\chi[J]$, where J is arbitrary, the associated expectation value $E(L)$ vanishes.*

Proof. The element $\chi = \chi[J_0]\chi[J]$ of $\mathcal{T}$ corresponds to the tensor product of the irreducible representations (J_0) and (J) of $SU(2)$. Let the link component C of L be associated with the representation of $SU(2)$ which is equivalent to the tensor product $(J_0) \otimes (J)$. From the satellite relation (6.35), or (13.34), it follows that

$$E(C, \ldots; \chi = \chi[J_0]\chi[J], \ldots) = E(h^\circ(B), \ldots; \chi[J_0], \chi[J], \ldots), \tag{15.34}$$

where $h^\diamond(B)$ is the image of the pattern link B, shown in Fig. 6.6, under the homeomorphism $h^\diamond$ determined by the framing of C. In equation (15.34), the two components of $h^\diamond(B)$ are labelled by the representations (J_0) and (J) respectively. On the right-hand-side of equation (15.34), we have a satellite L' of L and one component of L' is associated with $\chi[J_0]$. According to equations (15.30) and (15.32), the associated value of the extended Jones polynomial vanishes. Therefore, we have shown that

$$E(C, \ldots; \chi = \chi[J_0]\,\chi[J], \ldots) = 0\,, \tag{15.35}$$

and this concludes the proof. □

In order to illustrate the consequences of property **G8**, let us consider a few examples. Suppose that the component C of a given link L is associated with

$$\chi[J_0]\,\chi[1/2] = \chi[J_0 + 1/2] + \chi[J_0 - 1/2]\,. \tag{15.36}$$

On the one hand, according to equation (15.16), we have

$$\begin{aligned} E(C, \ldots; \chi[J_0]\,\chi[1/2], \ldots) = {} & E(C, \ldots; \chi[J_0 + 1/2], \ldots) \\ & + E(C, \ldots; \chi[J_0 - 1/2], \ldots)\,. \end{aligned} \tag{15.37}$$

On the other hand, **G8** implies that $E(C, \ldots; \chi[J_0]\chi[1/2], \ldots) = 0$; therefore, we get

$$E(C, \ldots; \chi[J_0 + 1/2], \ldots) = E(C, \ldots; -\chi[J_0 - 1/2], \ldots)\,. \tag{15.38}$$

This equation, which holds for any link L, implies that the element $\chi[J_0 + 1/2]$ is physically equivalent to $-\chi[J_0 - 1/2]$:

$$\chi[J_0 + 1/2] \Longleftrightarrow -\chi[J_0 - 1/2]\,. \tag{15.39}$$

In fact, as far as the observables $\{E(L)\}$ are concerned, equation (15.38) shows that there is no difference between $\chi[J_0 + 1/2]$ and $-\chi[J_0 - 1/2]$.

Let us now consider the product

$$\chi[J_0]\,\chi[1/2]\,\chi[1/2] = \chi[J_0 + 1] + 2\,\chi[J_0] + \chi[J_0 - 1]\,. \tag{15.40}$$

One has

$$\begin{aligned} E(C, \ldots; \chi[J_0]\,\chi[1/2]\,\chi[1/2], \ldots) = {} & E(C, \ldots; \chi[J_0 + 1], \ldots) \\ & + 2\,E(C, \ldots; \chi[J_0], \ldots) + E(C, \ldots; \chi[J_0 - 1], \ldots)\,, \end{aligned} \tag{15.41}$$

and then, by using property **G8**, one obtains

$$E(C, \ldots; \chi[J_0 + 1], \ldots) = E(C, \ldots; -\chi[J_0 - 1], \ldots). \tag{15.42}$$

Therefore, as far as the gauge-invariant observables are concerned, the element $\chi[J_0 + 1]$ is equivalent to $-\chi[J_0 - 1]$:

$$\chi[J_0 + 1] \Longleftrightarrow -\chi[J_0 - 1]. \tag{15.43}$$

In conclusion, for fixed integer k with $k > 2$, not all the elements $\{\chi[J]\}$ of the standard basis of $\mathcal{T}$ correspond to physical inequivalent quantum numbers. Indeed, a recursive use of the argument presented above shows that only $(k-1)$ elements of $\mathcal{T}$ can be taken to represent the independent and physically inequivalent quantum numbers of the CS theory in S^3.

Let us now introduce the invariant subalgebra $\mathcal{S}$ of $\mathcal{T}$ associated with $\chi[J_0]$; all the elements which can be written in the form $\chi[J_0]\,\chi$, with $\chi \in \mathcal{T}$, belong to $\mathcal{S}$. Property **G8** implies that, at the level of gauge-invariant observables, any element of the subalgebra $\mathcal{S}$ is physically equivalent to the trivial null vector. Therefore, in order to determine the physically inequivalent quantum numbers of the CS theory, we need to factorize $\mathcal{S}$ out of $\mathcal{T}$.

In fact, the tensor algebra $\mathcal{T}$ can be decomposed, according to $\mathcal{S}$, into a set of equivalence classes $\{\Psi\}$; the elements χ and $\chi + \chi'$, with $\chi' \in \mathcal{S}$, belong to the same class. From **G8** it follows that, for fixed integer k, the observables $\{E(L)\}$ define a set of multilinear functions on such equivalence classes.

The addition and multiplication operations in $\mathcal{T}$ define the corresponding operations for the equivalence classes $\{\Psi\}$. Consequently, the set of the classes $\{\Psi\}$ has an algebra structure: this algebra is called [59] the ***reduced tensor algebra*** and is indicated by $\mathcal{T}_{(k)}$.

To sum up, we have shown that, for fixed integer k with $k > 2$, the set of the physical quantum numbers of the CS theory in S^3 is described by the k-reduced tensor algebra $\mathcal{T}_{(k)}$.

Let us now consider the main properties of $\mathcal{T}_{(k)}$. By using the structure constants (15.18) of $\mathcal{T}$ and property **G8**, it is easy to see that the elements $\chi[J]$ of $\mathcal{T}$, with $(J_0 - 1/2) \geq J \geq 0$, are not influenced by the existence of the nontrivial null vector $\chi[J_0]$. These elements can be chosen to represent the inequivalent nontrivial classes of $\mathcal{T}$ modulo $\mathcal{S}$. On the other hand, each element $\chi[J]$, with $J \geq J_0$, belongs to $\mathcal{S}$ or belongs to one of the nontrivial equivalence classes associated with $\chi[J]$ for $(J_0 - 1/2) \geq J \geq 0$.

Let us introduce a standard basis $\{\Psi[J]\}$ in $\mathcal{T}_{(k)}$. Each class $\Psi[J]$ is represented by

$$\Psi[J] \Longleftrightarrow \chi[J], \quad \text{with} \quad (k/2) - 1 \geq J \geq 0. \tag{15.44}$$

Thus, for $G = SU(2)$, the reduced tensor algebra $\mathcal{T}_{(k)}$ is of order $(k-1)$ and is generated by $\{\Psi[J]\}$ with $(k/2)-1 \geq J \geq 0$. Each element $\Psi[J]$ is represented by $\chi[J]$ which is associated with the irreducible representation (J) of $SU(2)$.

The structure constants of $\mathcal{T}_{(k)}$ can easily be obtained from those of $\mathcal{T}$. Clearly, one has to analyse equation (15.18) when both J_1 and J_2 belong to the set of values shown in equation (15.44); i.e.

$$\begin{aligned} (k/2)-1 &\geq J_1 \geq 0\,, \\ (k/2)-1 &\geq J_2 \geq 0\,, \end{aligned} \tag{15.45}$$

If $J_1 + J_2 \leq k/2 - 1$, the structure constants given in (15.18) coincide with those of $\mathcal{T}_{(k)}$. When $J_1 + J_2 > k/2 - 1$, one has to note that the contributions $\chi[J_0 + p]$ and $\chi[J_0 - p]$ cancel in the sum (15.18). Therefore, one finds

$$\Psi[J_1]\,\Psi[J_2] = \sum_{J=|J_1-J_2|}^{b(J_1,J_2)} \Psi[J]\,, \tag{15.46}$$

where

$$b(J_1, J_2) = \min\{J_1 + J_2, k-2-J_1-J_2\}\,. \tag{15.47}$$

The structure constants (15.46) of $\mathcal{T}_{(k)}$ coincide with the fusion rules [60] of the $SU(2)_\ell$ WZNW model with level ℓ given by

$$\ell = k - 2\,. \tag{15.48}$$

It is important to note that property **G8** can also be used to simplify the computation of the vacuum expectation values. Remember that each projector, introduced in the defining relations **(1)**-**(5)**, selects the corresponding irreducible representation of the gauge group running along the core of a given solid torus. Property **G8** implies that, for fixed integer k, two inequivalent representations of $SU(2)$ do not necessarily represent physically inequivalent quantum numbers. Since any element of the subalgebra $\mathcal{S}$ is equivalent to the null vector, there are consequently many cancellations between different representations of $SU(2)$ in the computation of $\{E(L)\}$.

Thus, rules **(1)**-**(5)** can be slightly modified in such a way that the consequences of property **G8** are taken into account from the very beginning. As we have seen, the physically inequivalent states, associated with a solid torus, are described by the elements $\{\Psi[J]\}$ of the standard basis of $\mathcal{T}_{(k)}$. Therefore, for fixed integer k, the different link components can be directly labelled by the elements of $\mathcal{T}_{(k)}$. This simply means that each link component can be labelled by an element of the tensor algebra which represents the corresponding class. Then, for each element $\Psi[J]$ one can introduce the corresponding projector and all the rules and the general properties that we have discussed in the previous chapters remain valid. The only modification is that, instead of the structure constants (15.18), we can use the structure constants (15.46). The satellite relations, in particular, can be constructed by using equation (15.46). Of course, for fixed integer k, this simplified method gives final results for the observables which are identical to the results obtained with the standard defining relations.

In this formalism, based on the physical quantum numbers of the reduced tensor algebra $\mathcal{T}_{(k)}$, the correspondence between $\{\Psi[J]\}$ and the conformal blocks [61] of the conformal two-dimensional theories is manifest. This is in agreement with the fact that some of the braid group representations identified by $E(L)$ appear also as monodromy representations [46] in the correlation functions (analytic building blocks) of several two-dimensional systems. In conformal theories, the braiding matrices act on the space of the conformal blocks [46,61]; whereas, in the CS theory, the braiding operations involve Wilson lines which are characterized by the irreducible representations of the gauge group. The tensor algebra cannot be identified with the fusion rules. However, for certain integer values of the coupling constant k, the observables $\{ E(L) \}$ define a set of multilinear functions on the elements of the k-reduced tensor algebra. As a result, the quantum numbers associated with the observables describe a finite number of physical states which are in correspondence with the conformal blocks. Therefore, for certain values of k, the reduced tensor algebra $\mathcal{T}_{(k)}$ can be interpreted as the fusion algebra.

When q is a root of unity, the effective reduction of the quantum $SU(2)_q$ modules to the set of the so-called integrable representations [60] has been analysed in [55,62,63]. In our approach, only the properties of the Wilson line expectation values and of the ordinary tensor algebra of $SU(2)$ have been used. The arguments presented above are closely related to those produced in Ref.[68]. Instead of the bracket polynomial, we have considered the extended Jones polynomial which is the invariant defined by the CS theory. Moreover, the role of the $SU(2)_q$ irreducible modules in Ref.[68] is played, in our case, by the irreducible representations of $SU(2)$.

15.6 Roots of unity

Let us now consider the relation between the tensor algebra and the reduced tensor algebra of $SU(2)$. As we have seen in the previous section, the presence of the nontrivial null vector $\chi[J_0]$ and the periodic structure of $\{ E(L) \}$ are encoded in the expression (15.29) of the values of the unknots. Equation (15.29) can also be written as

$$E_0[J] = \operatorname{Im} \exp\left[i\frac{\pi}{k}(2J+1) \right] \Big/ \operatorname{Im} \exp\left(i\frac{\pi}{k} \right) . \qquad (15.49)$$

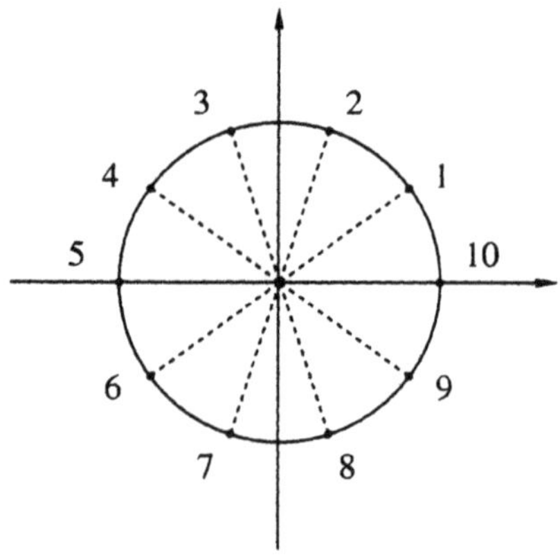

Figure 15.1. Root diagram for $k = 5$

This expression shows that $\eta(J)$, defined by

$$\eta(J) = \exp\left[i\frac{\pi}{k}(2J+1) \right], \tag{15.50}$$

is the relevant variable which must be considered. Let us represent the values of $\eta(J)$ in the complex plane. For fixed integer k, one has

$$[\eta(J)]^{2k} = 1, \tag{15.51}$$

and, by varying J, the values of $\eta(J)$ coincide with the different $(2k)$ roots of unity. When $k = 5$, the possible values of $\eta(J)$ are shown in Fig. 15.1, whereas the root diagram for $k = 6$ is shown in Fig. 15.2.

Clearly, $\eta(J)$ is a periodic function of J with period k. In Fig. 15.1 and Fig. 15.2, the values of J in the interval $(2J_0 + 1/2) \geq J \geq 0$ have been considered and the integer numbers, which label the different roots of unity, refer to the dimension $D(J) = 2J + 1$ of the irreducible representation (J) of $SU(2)$.

The values of the unknots are proportional to the imaginary part of the roots. The nontrivial null vector $\chi[J_0]$ corresponds to the negative root on the real axis. The $(k-1)$ roots on the upper half-plane are in one-to-one correspondence with the nontrivial classes $\{\Psi[J]\}$.

Let us consider a generic element $\chi[J]$. If the corresponding root $\eta(J)$ is on the real axis, then $\chi[J] \in \mathcal{S}$; if not, $\chi[J]$ belongs to the class determined by the real part of $\eta(J)$. On the one hand, when the value of $\eta(J)$ coincides with $\eta(J^*)$, with $(k/2) - 1 \geq J^* \geq 0$, then $\chi[J]$ is physically equivalent to $\chi[J^*]$. On the other hand, if $\eta(J)$ belongs to the lower half-plane and is related to $\eta(J^*)$, with

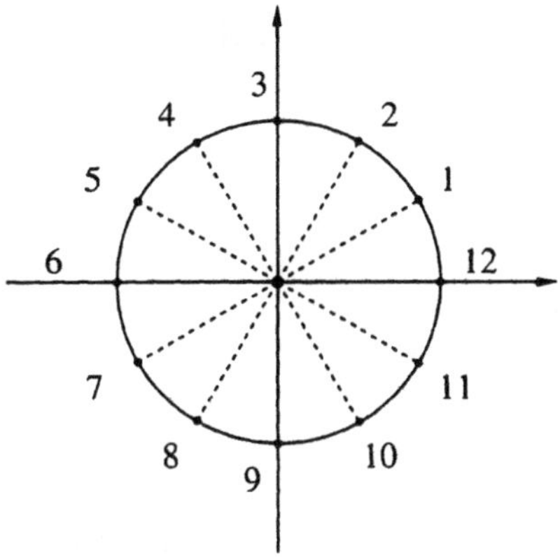

Figure 15.2. Root diagram for $k = 6$

$(k/2) - 1 \geq J^* \geq 0$, by a reflection on real axis, then $\chi[J]$ is physically equivalent to $-\chi[J^*]$.

The class $\Psi[(k-1)/2 - p]$, with $(k-1)/2 \geq p \geq 1$, can be represented by $\chi[(k-1)/2 - p]$ or by $-\chi[(k-1)/2 + p]$. In general, $\chi[J]$ is physically equivalent to $\chi[J + nk]$ for arbitrary (non-negative) integer n. It is easy to verify that

$$q^{Q((k-1)/2-p)} = q^{Q((k-1)/2+p)}, \tag{15.52}$$

and that

$$E_0[(k-1)/2 - p] = -E_0[(k-1)/2 + p]. \tag{15.53}$$

Let us consider one example of reduced tensor algebra. The four elements of the standard basis of $\mathcal{T}_{(5)}$ are $\Psi[0]$, $\Psi[1/2]$, $\Psi[1]$ and $\Psi[3/2]$. The element $\Psi[0]$ corresponds to the identity and the nontrivial structure constants of $\mathcal{T}_{(5)}$ are given by

$$\Psi[1/2]\,\Psi[1/2] = \Psi[0] + \Psi[1]; \quad \Psi[1/2]\,\Psi[3/2] = \Psi[1]; \tag{15.54}$$

$$\Psi[1/2]\,\Psi[1] = \Psi[1/2] + \Psi[3/2]; \quad \Psi[1]\,\Psi[3/2] = \Psi[1/2]; \tag{15.55}$$

$$\Psi[1]\,\Psi[1] = \Psi[0] + \Psi[1]; \quad \Psi[3/2]\,\Psi[3/2] = \Psi[0]. \tag{15.56}$$

In general, the element $\Psi[(k/2) - 1]$ generates the transformation which corresponds to a reflection on the vertical axis of the root diagram. This means that

$$\Psi[(k/2) - 1]\,\Psi[J] = \Psi[(k/2) - 1 - J], \tag{15.57}$$

where $(k/2) - 1 \geq J \geq 0$.

15.7 Special cases

When the CS coupling constant is equal to $k = 2$ or $k = 1$, one cannot use the known braiding properties of the WZNW model to conjecture what the solution of the quantum CS theory is. Nevertheless, when $k = 2$ or $k = 1$, the CS theory exists and is well defined. The point is that the three-dimensional topological field theory is not equivalent to (and should not be confused with) the two-dimensional conformal models. In this section, we shall discuss these two special cases in which $k = 2$ and $k = 1$.

The general results that we have obtained for integer k, with $k > 2$, are also valid for $k = 2$. When $k = 2$, the nontrivial null vector in $\mathcal{T}$ is $\chi[1/2]$. Consequently, the reduced tensor algebra $\mathcal{T}_{(2)}$ is of order equal to unity and its standard basis contains only one element: $\Psi[0]$. The class $\Psi[0]$ can be represented in $\mathcal{T}$ by

$$\chi[J = 2n] \iff \Psi[0] \iff -\chi[J = 2n+1], \qquad (15.58)$$

where n is an arbitrary natural number. According to **G8**, any element $\chi[J]$, with $J \neq$ integer, is equivalent to the trivial null vector. Therefore, the invariant defined by $E(L)$ is quite simple in this case. In fact, if one of the components of a generic link L is associated with a representation (J) with $J \neq$ integer, then the corresponding link invariant vanishes, $E(L) = 0$. On the other hand, when all the components $\{ C_i \}$ of L are labelled by $\{ \chi[J_i] \}$ with integers $\{ (J_i) \}$, one has simply

$$E(L) = (-1)^{\sum_i J_i}. \qquad (15.59)$$

Equation (15.59) can easily be understood. Since $\Psi[0]$ generates $\mathcal{T}_{(2)}$, it follows that $|E(L)| = 1$. The presence of the (-1) factors in (15.59) is due to the relation (15.58). Since all the elements of $\mathcal{T}_{(2)}$ are proportional to $\Psi[0]$, $E(L)$ is in this case invariant under a modification of the framing of any component of L. As we shall see, this implies that, for $k = 2$, the expression (15.59) represents the vacuum expectation values of the Wilson line operators in any (closed, connected and orientable) three-manifold $\mathcal{M}$.

The case in which $k = 1$ is more interesting. First of all, the values (5.19) of the unknots become

$$E_0[J] = (-1)^{2J}(2J+1). \qquad (15.60)$$

Consequently, among the elements of the standard basis of $\mathcal{T}$, there is no nontrivial null vector. Since $k = 1$, one has $q = q^{-1} = 1$ and $q^{1/2} = q^{-1/2} = -1$. For integer values of J, one gets

$$q^{Q(J)} = q^{Q(J)/2} = 1; \qquad (15.61)$$

whereas for $J = (2n+1)/2$ (with $n = 0, 1, 2, \ldots$) one finds

$$q^{Q(J)} = i, \quad q^{-Q(J)} = -i\,. \tag{15.62}$$

Equations (15.61) and (15.62), combined with the defining relations **(1)**-**(5)**, lead to the crossing relations shown in Fig. 15.3 and Fig. 15.4.

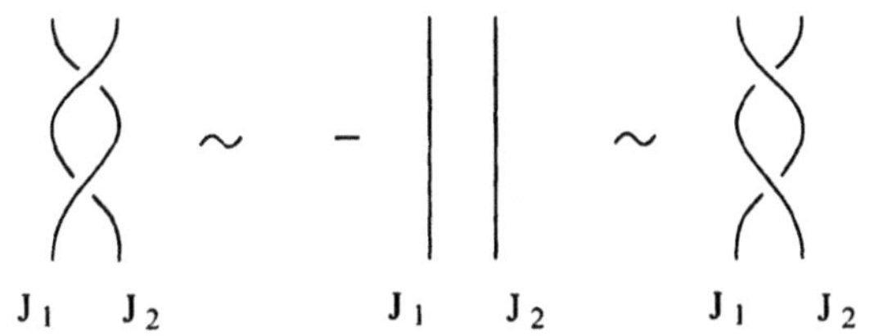

Figure 15.3. Crossing relations for $k = 1$ when both J_1 and J_2 are not integers

The relations shown in Fig. 15.3 are valid when both J_1 and J_2 are not integer numbers; in all the remaining cases, the relations shown in Fig. 15.4 hold.

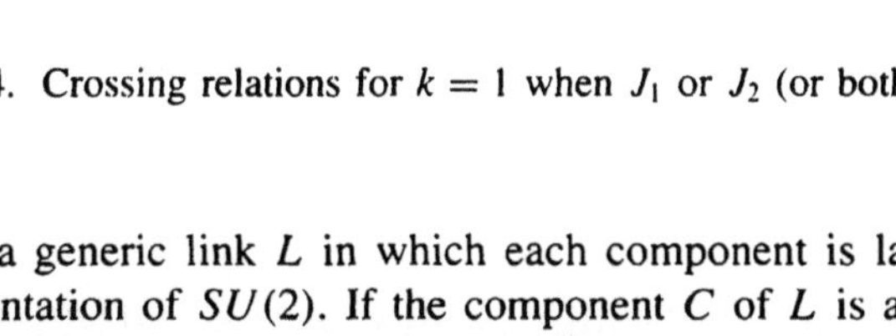

Figure 15.4. Crossing relations for $k = 1$ when J_1 or J_2 (or both) are integers

Consider now a generic link L in which each component is labelled by an irreducible representation of $SU(2)$. If the component C of L is associated with the (J)-representation with integer J, then, by using the relations shown in Fig. 15.4, this component can be unlinked and separated from the remaining components of L. At this point, the component C can be unknotted and reduced to a simple circle. So, when the component C is eliminated, it simply gives a factor $E_0[J]$. The same procedure can be repeated for all the components of L which are labelled by $\chi[J]$ with integer values of J.

The components of L which are labelled by non-integer values of J now remain to be considered. Let us denote these components by $\{C_1, \ldots, C_t\}$. By using the relations shown in Fig. 15.3, C_1 can be unlinked and separated from all the remaining link components. In doing this, we get a factor

$$(-1)^{\sum_{a=2}^{t} \chi(C_1, C_a)} = \exp\left[\left(\frac{i\pi}{2}\right) 2 \sum_{a=2}^{t} \chi(C_1, C_a)\right], \tag{15.63}$$

where $\chi(C_1, C_a)$ is the linking number of C_1 and C_a. Then, the component C_1 can be unknotted and transformed into a simple circle; according to equation (15.62) and the relations shown in Fig. 15.3, this gives a factor

$$(i)^{w(C_1)} = \exp\left[\left(\frac{i\pi}{2}\right) w(C_1)\right], \tag{15.64}$$

where $w(C_1)$ is the writhe of C_1. Finally, C_1 is ambient isotopic with the unknot (with zero writhe) labelled by $\chi[J_1]$. Therefore, it gives a contribution $E_0[J_1]$. Again, the same procedure can now be repeated for all the remaining components of L.

In conclusion, let us denote by $\{ C_1, \ldots, C_t, \ldots, C_m \}$ the components of L. The component C_i is labelled by $\chi[J_i]$ and J_i is an integer only for $t + 1 \leq i \leq m$. Then

$$E(L) = \left(\prod_{i=1}^{m} E_0[J_i]\right) \cdot \exp\left[\frac{i\pi}{2}\left(\sum_a w(C_a) + 2 \sum_{a<b} \chi(C_a, C_b)\right)\right], \tag{15.65}$$

where $1 \leq a, b \leq t$. Equation (15.65) gives the general expression of the expectation values of Wilson line operators in S^3 when $k = 1$.

The expression (15.65) shows that the number of independent and physically inequivalent states, associated with a solid torus in the $SU(2)$ CS theory, is drastically reduced also for $k = 1$. Indeed, when $k = 1$, the relevant reduced tensor algebra $\mathcal{T}_{(1)}$ is of order two. The construction of this algebra is slightly different from the case in which $k \geq 2$.

Let us introduce the two elements $\{ \Psi[0], \Psi[1/2] \}$ of the standard basis of $\mathcal{T}_{(1)}$ with structure constants determined by

$$\Psi[0]\,\Psi[0] = \Psi[0], \quad \Psi[0]\,\Psi[1/2] = \Psi[1/2], \tag{15.66}$$

and

$$\Psi[1/2]\,\Psi[1/2] = \Psi[0]. \tag{15.67}$$

Clearly, $\mathcal{T}_{(1)}$ coincides with the group algebra of the center Z_2 of $SU(2)$. When $k = 1$, each element of $\mathcal{T}$ corresponds to an element of $\mathcal{T}_{(1)}$ according to

$$\chi[J] \Longrightarrow (2J + 1)\,\Psi[0], \tag{15.68}$$

for integer values of J; whereas for values of J which are not integers

$$\chi[J] \implies (2J+1)\,\Psi[1/2]\,. \tag{15.69}$$

In agreement with equations (15.61) and (15.62), the values of the twist variable associated with $\Psi[0]$ and $\Psi[1/2]$ are given by

$$\alpha(0) = 1\,, \tag{15.70}$$

and

$$\alpha(1/2) = e^{i\pi/2}\,. \tag{15.71}$$

As in the case $k \geq 2$, rules **(1)**-**(5)** can now be simplified and written directly in terms of $\{\,\Psi[0],\,\Psi[1/2]\,\}$. The satellite relations are determined by the structure constants (15.66) and (15.67) and the values of the twist variable are shown in equation (15.70) and (15.71). It is easy to verify that the relations shown in Fig. 15.3 and Fig. 15.4 are indeed satisfied and, consequently, the expression (15.65) is recovered.

Chapter 16

Surgery on three-manifolds

A manifold $\mathcal{M}$ is closed if it is a compact manifold without boundary, i.e. $\partial\mathcal{M} = \emptyset$; $\mathcal{M}$ is called an open manifold if it is non-compact and $\partial\mathcal{M} = \emptyset$. In general, a manifold $\mathcal{M}$ can be understood as the union of several components glued together by some given identification of the points on their boundaries. If these components are glued in a different way, one may find a new manifold $\mathcal{M}'$. In this case, we say that $\mathcal{M}'$ can be obtained from $\mathcal{M}$ by means of surgery.

It is known [12,64] that any closed, orientable and connected three-manifold $\mathcal{M}$ can be obtained by surgery from the three-sphere S^3. In this chapter, we shall briefly review the description of surgery operations in S^3. We shall concentrate on Dehn surgery performed along knots or links in S^3.

16.1 Mapping class group of the torus

We shall begin our discussion by recalling the structure of the mapping class group of the two-dimensional torus $T^2 = S^1 \times S^1$. If we represent S^1 by the unit complex numbers, the points of T^2 have coordinates $(e^{i\theta_1}, e^{i\theta_2})$. The map

$$e^{i\phi} \;\rightarrow\; (e^{i\phi}, 1)\,, \tag{16.1}$$

from S^1 to T^2, defines a longitude and the map

$$e^{i\phi} \;\rightarrow\; (1, e^{i\phi}) \tag{16.2}$$

defines a meridian. With a given orientation for S^1, the longitude (16.1) and the meridian (16.2) are oriented and can be taken to represent the generators of the fundamental group of T^2. The homotopy class $[\,f\,]$ of a map $f\,:\,S^1 \rightarrow T^2$ is written in the longitude-meridian basis as $[\,f\,] = (a, b)$. The longitude class is $(1, 0)$ and the meridian class is $(0, 1)$. The class (a, b) of a knot in T^2 has either $a = 0 = b$ or a and b are relatively prime.

The longitudinal and meridinal twists h_L and h_M are the self-homeomorphisms of T^2 defined by

$$h_L\,(e^{i\theta_1}, e^{i\theta_2}) \;\to\; (e^{i(\theta_1+\theta_2)}, e^{i\theta_2}), \tag{16.3}$$

$$h_M\,(e^{i\theta_1}, e^{i\theta_2}) \;\to\; (e^{i\theta_1}, e^{i(\theta_1+\theta_2)}). \tag{16.4}$$

The action induced by h_L and h_M in $\pi_1(T^2)$ can be represented by the matrices

$$h_L \;\to\; \begin{pmatrix} 1 & 0 \\ 1 & 1 \end{pmatrix}, \tag{16.5}$$

and

$$h_M \;\to\; \begin{pmatrix} 1 & 1 \\ 0 & 1 \end{pmatrix}, \tag{16.6}$$

acting on $[\,f\,] = (a, b)$ as

$$h_L\,(a, b) \;=\; (a, b)\cdot\begin{pmatrix} 1 & 0 \\ 1 & 1 \end{pmatrix} \;=\; (a+b, b), \tag{16.7}$$

$$h_M\,(a, b) \;=\; (a, b)\cdot\begin{pmatrix} 1 & 1 \\ 0 & 1 \end{pmatrix} \;=\; (a, a+b). \tag{16.8}$$

Let us recall that a homeomorphism h of a space X is called an ambient isotopy if h is the end map h_1 of a homotopy $h_t\,:\,X \to X$ such that h_0 is the identity and h_t is a homeomorphism $\forall\, t \in [0, 1]$. The group of orientation-preserving self-homeomorphisms of T^2, modulo ambient isotopy, is generated precisely by the longitudinal and meridinal twists h_L and h_M and, in view of the representation (16.5) and (16.6), it is easy to see that this group is isomorphic with $SL(2, Z)$.

Two knots C_1 and C_2 in T^2 are ambient isotopic if and only if $[C_1] = \pm[C_2]$. On the other hand, for any knot C in T^2 with $[\,C\,] \neq (0, 0)$, a self-homeomorphism of T^2 exists such that $h(C)$ is a meridian.

16.2 Solid tori

In order to describe Dehn surgery operations, we need to consider solid tori. A solid torus V is a three-dimensional space homeomorphic with $S^1 \times D^2$, where D^2 is the two-dimensional disc (or the unit ball in $\mathbb{R}^2$ centered at the origin). A meridian of V is a simple closed curve in ∂V which is essential in ∂V (i.e., its class is nontrivial in ∂V) and which is homotopically trivial in V. A longitude of V is a closed curve on ∂V intersecting some meridian of V (transversally) in a single point. Any two meridians are ambient isotopic and, in this sense, the meridian is an intrinsic part of V. On the other hand, there is an infinite number of ambient isotopy classes of longitudes; in fact, the specification of a longitude, modulo ambient isotopy, involves a choice.

A given homeomorphism $h : S^1 \times D^2 \to V$ is called a framing of V. For some specified framing h of V, $h(1 \times \partial D^2)$ is a meridian and $h(S^1 \times 1)$ is a longitude.

A tubular neighbourhood N of a knot C in S^3 is an embedding $f : C \times D^2 \to S^3$ such that $f(x, 0) = x$ for $x \in C$. Therefore, a tubular neighbourhood of C is a solid torus in S^3 whose core is C. Given a tubular neighbourhood N of C and a framing h of N, the longitude $h(S^1 \times 1)$ of N defines a framing C_f of C.

On the other hand, a given framing C_f of a knot C defines a framing h of a tubular neighbourhood of C; with a given choice of the orientation of C, this framing h is unique up to ambient isotopy transformations. C_f is a preferred framing for C if the linking number of C and C_f is equal to zero. Similarly, we say that h is a preferred framing of N if the longitude $h(S^1 \times 1)$ of N has a vanishing linking number with the core of N (i.e. the knot C).

Let us now define a homotopy basis in ∂N. If the knot C is not oriented, we shall introduce an orientation for it. The longitude $\lambda = h(S^1 \times 1)$, defined by a preferred framing h of N, is oriented as C and the meridian $\mu = h(1 \times \partial D^2)$ is oriented in such a way that its linking number with C is equal to $+1$. We will say that the longitude λ and the meridian μ of N are the homotopy generators of a ***Rolfsen basis*** in ∂N and any class $[f] \in \pi_1(\partial N)$ is expressed in this basis as

$$[f] = a \cdot [\lambda] + b \cdot [\mu] = (a, b) . \tag{16.9}$$

A knot in ∂N is a meridian of N if and only if its class is $(0, \pm 1)$; this knot is a longitude of N if and only if its class is $(\pm 1, p)$ for some integer p.

A given homeomorphism $h : \partial N \to \partial N$ extends to a self-homeomorphism of the solid torus N if and only if h takes a meridian to a meridian. Any such homeomorphism h of the torus ∂N is characterized by an $SL(2, Z)$ matrix of the form

$$h \leftrightarrow \begin{pmatrix} 1 & n \\ 0 & 1 \end{pmatrix} , \tag{16.10}$$

where n is an integer and, therefore, h is ambient isotopic with $(h_M)^n$, h_M being a meridinal twist of ∂N.

In conclusion, the group of self-homeomorphisms of the solid torus N, modulo ambient isotopy, is Abelian and its generator is called a ***twist*** of N. A self-homeomorphism τ_+ of N, which is the extension in N of a meridinal twist $h_M : \partial N \to \partial N$ represented by the matrix (16.6) in a Rolfsen basis, is called a ***right-handed twist*** of N. The inverse homeomorphism τ_- is called a ***left-handed twist*** of N.

If we represent the disc D^2 in the complex plane, the points of $S^1 \times D^2$ have coordinates $(e^{i\theta_1}, r\, e^{i\theta_2})$ with $0 \leq r \leq 1$. The longitude $(e^{i\phi}, 1)$ and the meridian $(1, e^{i\phi})$ of $S^1 \times D^2$ are oriented in the usual anti-clockwise direction. The self-homeomorphisms $\widetilde{\tau}_{\pm}$ of $S^1 \times D^2$, defined by

$$\widetilde{\tau}_{\pm}\,(e^{i\theta_1}\,,\,r\,e^{i\theta_2}\,) \;=\; (e^{i\theta_1}\,,\,r\,e^{i(\theta_2 \pm \theta_1)}\,)\,, \tag{16.11}$$

are the basic twists of $S^1 \times D^2$. Consider then a tubular neighbourhood N of a knot C and let $h\,:\,S^1 \times D^2 \to N$ be a preferred framing of N such that $h(S^1 \times 1)$ and $h(1 \times \partial D^2)$ define a Rolfsen basis in ∂N. The right-handed τ_+ and the left-handed τ_- (meridinal) twists of N are given, up to ambient isotopy, by

$$\tau_{\pm} \;=\; h \cdot \widetilde{\tau}_{\pm} \cdot h^{-1}\,. \tag{16.12}$$

It is important to consider the effects produced by the twists $\tau_{\pm}$ on the links defined in the solid torus; we shall now elaborate on this subject. Our description is expressed in a form which is convenient for our subsequent discussion. A tubular neighbourhood N of the unknot U (simple circle) in S^3 is shown in Fig. 16.1.

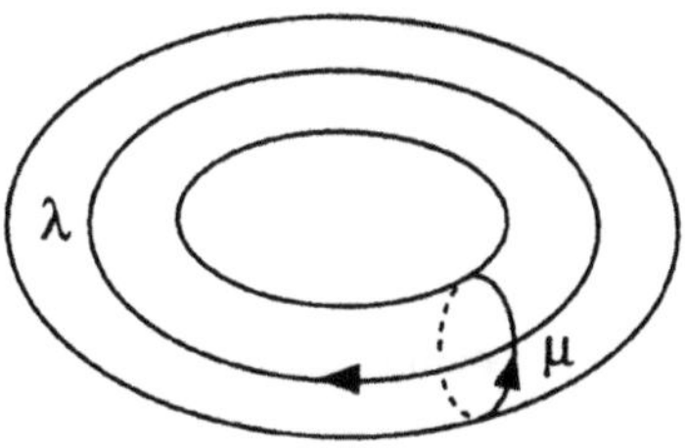

Figure 16.1.

The longitude λ and the meridian μ of N, shown in Fig. 16.1, are the generators of a Rolfsen basis in ∂N. Up to ambient isotopy, there are only two possible Rolfsen bases: one of them is shown in Fig. 16.1 and the other one is obtained by reversing both orientations of λ and μ.

The space $S^3 - \mathring{N}$, obtained by removing the interior $\mathring{N}$ of N from S^3, is a solid torus called the complement solid torus of U in S^3. The oriented curves λ and μ define a Rolfsen basis also in $\partial(S^3 - \mathring{N})$ but their roles are exchanged; indeed, λ is a meridian of $S^3 - \mathring{N}$ and μ is a longitude of $S^3 - \mathring{N}$. We are now interested in twist homeomorphisms $\tau_{\pm}$ of the solid torus $S^3 - \mathring{N}$.

A two-component link (which is not framed) in the complement solid torus of U in S^3 is shown in Fig. 16.2. Under a right-handed twist τ_+ of $S^3 - \mathring{N}$, Fig. 16.2 is transformed into Fig. 16.3, whereas the effects produced by τ_- are shown in Fig. 16.4.

Roughly speaking, $\tau_{\pm}$ generate a pure-braid group (over/under)-crossing between the two components running along the core of the solid torus. This effect does

U

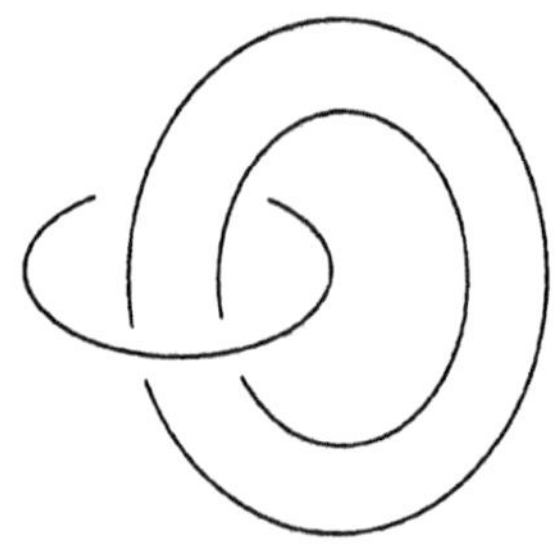

Figure 16.2.

U

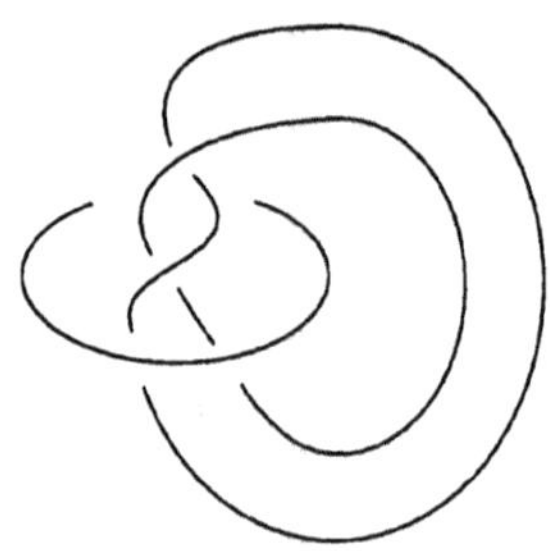

Figure 16.3.

U

Figure 16.4.

not depend on the particular choice of the Rolfsen basis. One can also imagine that, instead of a two-component and not framed link, Fig. 16.2 represents a single framed knot. In this case, Fig. 16.3 and Fig. 16.4 describe the effects of τ_+ and τ_- on the framing.

On the other hand, if Fig. 16.2 represents a framed two component link in $S^3 - \mathring{N}$ (with the vertical framing convention), then the effects of a right-handed

Figure 16.5.

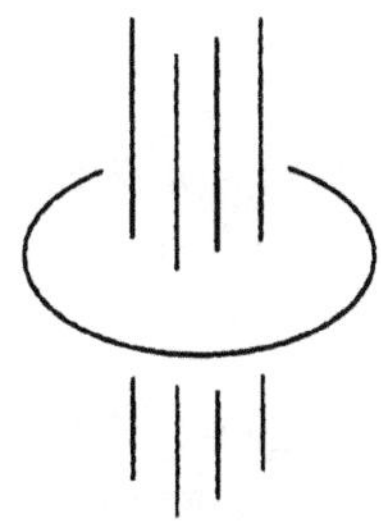

Figure 16.6.

twist τ_+ are shown in Fig. 16.5. The link shown in Fig. 16.5 has been considered in Chapters 13 and 14.
There is a simple method which can be used to describe the effects produced by $\tau_\pm$ on a generic link L in $S^3 - \mathring{N}$. Consider a disc D bounded by U, as shown in Fig. 16.6; some of the components of L will intersect the disc D. Each component may intersect the disc D several times.

One can always imagine that these components intersect D perpendicularly. In a neighbourhood of the disc D, they are represented by parallel segments which lie on a piece of band; $\tau_\pm$ simply twist the band by a $\pm 2\pi$ rotation.

For example, the action of τ_+ on the picture of Fig. 16.6 is shown in Fig. 16.7. In general, if there are n segments intersecting D, the picture representing the action of τ_+ is described by the element $(g_{n-1} \cdots g_2 g_1)^n$ of the braid group B_n, $\{ g_i \}$ being the generators.
It is clear that $\tau_\pm$ do not modify the number of components of a link L in the solid torus, but the linking numbers between the different components can be modified. Let us consider, in particular, the behaviour of the framings. Let C be a framed

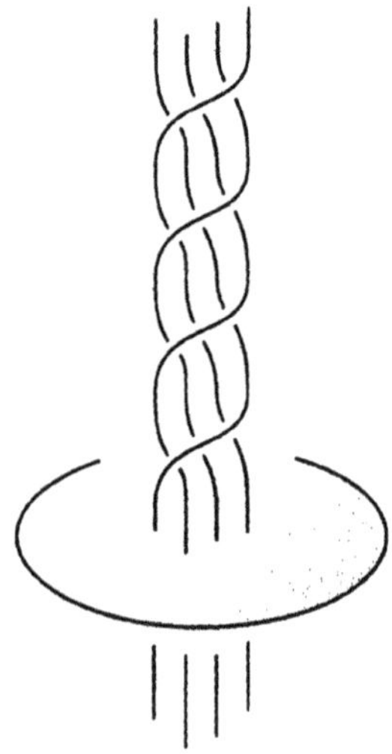

Figure 16.7.

component of a link in the complement solid torus $S^3 - \mathring{N}$ of the unknot U in S^3. $\chi(C, C_f)$ is the linking number of C and of its framing C_f, whereas $\chi(C, U)$ is the linking number of C and U. Under a twist homeomorphism $\tau_\pm$ of $S^3 - \mathring{N}$, C is transformed into C' and its framing is modified according to the relation [12]

$$\chi(C', C'_f) = \chi(C, C_f) \pm [\chi(C, U)]^2 . \tag{16.13}$$

16.3 Dehn surgery

Let us now consider surgery operations in S^3. A Dehn surgery performed along a knot $\mathcal{Z}$ in S^3 consists of

(1) removing the interior $\mathring{N}$, of a tubular neighbourhood N of the knot $\mathcal{Z}$, from S^3;
(2) considering $S^3 - \mathring{N}$ and N as distinct spaces whose (distinct) boundaries $\partial(S^3 - \mathring{N})$ and ∂N are tori;
(3) gluing back N and $S^3 - \mathring{N}$ by identifying the points on their boundaries ∂N and $\partial(S^3 - \mathring{N})$ according to a given homeomorphism $h : \partial N \to \partial(S^3 - \mathring{N})$.

The knot $\mathcal{Z}$ and the "gluing" homeomorphism h completely specify the surgery operation and the resulting manifold is indicated by

$$\mathcal{M} = (S^3 - \overset{\circ}{N}) \bigcup_h N . \tag{16.14}$$

Actually, the manifold (16.14) depends [12], up to homeomorphism, only on the homotopy class of $h(\mu)$ in $\partial(S^3 - \overset{\circ}{N})$, where μ is a meridian of N. The surgery is characterized then by the knot $\mathcal{Z}$ and by a closed curve $Y \in \partial N$ representing $h(\mu)$. The convention introduced by Rolfsen [12], which is used to codify the surgery instruction, is the following: the class $[Y] \in \pi_1(\partial N)$ is written as

$$[Y] = a \cdot [\lambda] + b \cdot [\mu] , \tag{16.15}$$

where the generators λ and μ are the longitude and the meridian of a Rolfsen basis in ∂N. Since Y is a knot in ∂N, the integer coefficients a and b appearing in equation (16.15) are relatively prime. The ratio

$$r = \frac{b}{a} \tag{16.16}$$

is called the ***surgery coefficient***. In conclusion, the surgery instruction is specified simply by the knot $\mathcal{Z}$ in S^3 and by the rational surgery coefficient r.

The knot $\mathcal{Z}$, along which surgery is performed, is not oriented and Y is not oriented either. Therefore, in a fixed Rolfsen basis, the coefficients a and b appearing in equation (16.15) possess an overall ambiguity in their signs, depending on the choice of the orientation of Y. Similarly, for fixed orientation of Y, a different choice of the Rolfsen basis modifies both signs of the coefficients a and b. This ambiguity does not affect the resulting manifold obtained by surgery and, in fact, this ambiguity disappears in the surgery coefficient r.

When $a = 0$, b must be ± 1 and the surgery coefficient is indicated by $r = \infty$. In this case, Y is a meridian of N. This means that the image, under the gluing homeomorphism h, of the meridian μ of N is ambient isotopic with μ itself. Therefore, the resulting manifold is simply S^3. In conclusion, the surgery instruction specified by an arbitrary knot $\mathcal{Z}$ with surgery coefficient $r = \infty$ corresponds to the identity.

Clearly, the surgery operation of removing and sewing a solid torus can be repeated several times. Therefore, a general ***surgery instruction*** consists of an unoriented link $\mathcal{L}$ in S^3, called the ***surgery link***, with given surgery coefficients $\{r_i\}$ assigned to its components $\{\mathcal{L}_i\}$.

For example, when $\mathcal{L}$ is the unknot with surgery coefficient $r = b/a$, the resulting space is homeomorphic with the lens space L(b,a). In particular, the unknot with surgery coefficient $r = 0$ corresponds to $S^2 \times S^1$. If $\mathcal{L}$ coincides with the Borromean rings with all the surgery coefficients equal to $+1$, the resulting manifold is the icosahedral space or Poincaré manifold $\mathcal{P}$. The Borromean Rings with surgery coefficients $r_i = 0$ (for $i = 1, 2, 3$) represent $S^1 \times S^1 \times S^1$.

Different surgery instructions do not necessarily correspond to different manifolds. To be more precise, two manifolds associated with different surgery instructions are homeomorphic if and only if the two surgery instructions are related [12,65] by a finite sequence of ***Rolfsen moves***.

A Rolfsen move of the first type is quite obvious: it states that one can add or eliminate a component of the surgery link $\mathcal{L}$ with surgery coefficient $r = \infty$. A Rolfsen move of the second type describes the effects of an appropriate twist homeomorphism $\tau_\pm$ acting on a solid torus which contains part of the surgery link.

Let $\mathcal{L}$ be a surgery link in which one of its components, say $\mathcal{L}_1$, is the unknot with surgery coefficient r_1. This means that all the remaining components $\{\mathcal{L}_j\}$ (with $j \neq 1$) of $\mathcal{L}$ belong to the complement solid torus $\mathcal{V}_1$ of $\mathcal{L}_1$ in S^3. Under a twist homeomorphism $\tau_\pm$ of $\mathcal{V}_1$, the component $\mathcal{L}_1$ is not modified, i.e. $\mathcal{L}'_1 = \mathcal{L}_1$. On the other hand, the remaining components $\{\mathcal{L}_j\}$ are transformed, $\tau_\pm : \mathcal{L}_j \to \mathcal{L}'_j$, according to the rules illustrated in the previous section.

Furthermore, the surgery coefficients also are modified. One can show [12] that the new surgery coefficients are

$$r'_1 = \frac{1}{(1/r_1) \pm 1}, \tag{16.17}$$

and

$$r'_j = r_j \pm \left[\chi(\mathcal{L}_j, \mathcal{L}_1)\right]^2 \quad \text{for} \quad j \neq 1, \tag{16.18}$$

where $\chi(\mathcal{L}_j, \mathcal{L}_1)$ is the linking number of $\mathcal{L}_j$ and $\mathcal{L}_1$. Thus, we have two surgery instructions which are described, respectively, by

(i) the surgery link $\mathcal{L}$ with surgery coefficients $\{r_i\}$;
(ii) the surgery link $\mathcal{L}'$ with surgery coefficients $\{r'_i\}$.

The surgery instructions (i) and (ii) are said to be related by a Rolfsen move of the second kind and describe homeomorphic manifolds. The connecting homeomorphism is just the image under surgery of $\tau_\pm$, of course.

The three surgery instructions shown in Fig. 16.8 describe homeomorphic manifolds: $S^2 \times S^1$. The first and the second are related by a Rolfsen move of the second kind, whereas the second and the third are related by a Rolfsen move of the first kind.

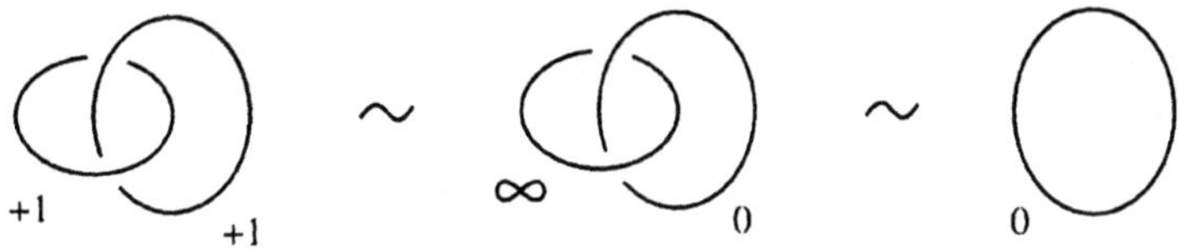

Figure 16.8.

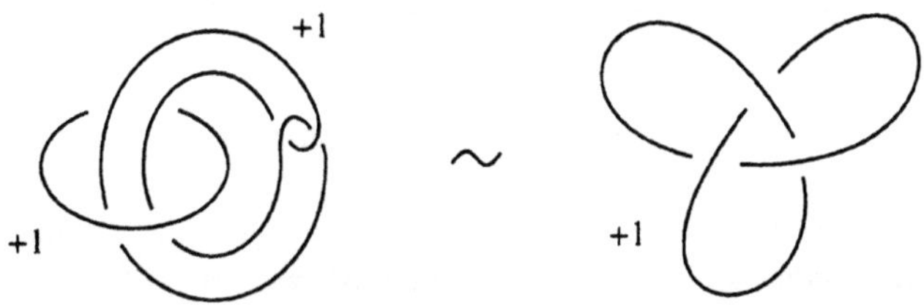

Figure 16.9.

The surgery instructions, shown in Fig. 16.9, are related by Rolfsen moves and correspond to the Poincaré manifold $\mathcal{P}$.

Two different surgery instructions, which are related by a finite sequence of Rolfsen moves, are called equivalent. The set of all possible surgery instructions can be decomposed into classes of equivalent instructions and a function defined on these equivalence classes is called a ***three-manifold invariant***.

It should be noted that the finite sequences of Rolfsen moves actually correspond to the set of orientation-preserving self-homeomorphisms of a given three-manifold. If we wish to include self-homeomorphisms which are not orientation-preserving, we simply need to add the inversion of the orientation of S^3 (or mirror-reflection) as an admissible move.

16.4 Links in three-manifolds

Any orientable, closed and connected three-manifold $\mathcal{M}$ has a presentation in terms of Dehn surgery in S^3. In the CS theory, one is interested in framed links defined in $\mathcal{M}$. We need to find a way of representing them. We already know that links defined in S^3 can easily be described by link diagrams; each string component of the picture simply represents the projection on a plane of the corresponding component of the link $L \subset S^3$.

Link diagrams can also be used to represent links in a closed orientable connected three-manifold $\mathcal{M}$. In this case, not all the components of the picture are associated with the components of the links in the three-manifold $\mathcal{M}$.

In order to represent a given link L in $\mathcal{M}$, we shall use a link diagram corresponding to a link in S^3 in which

(1) some of the components of this link coincide with the components of the surgery link $\mathcal{L}$: $\mathcal{L}$ describes the surgery instructions corresponding to a presentation of the manifold $\mathcal{M}$ in terms of Dehn surgery in S^3;
(2) the remaining components of this link describe how the real link L is placed in $\mathcal{M}$.

Let us now consider the ambient isotopy classes of links $\{ L \}$ in a generic manifold $\mathcal{M}$.

In the previous chapters we have seen that the observables of the CS theory in S^3 are not invariant under a generic homeomorphism (acting on the links) which is not smoothly connected with the identity. In general, the expectation values of the Wilson line operators in the CS theory are invariant under isotopy transformations acting on the links in a given ambient space. Since any surgery instruction defines a three-manifold up to homeomorphism, there does not seem to be sufficient information in the surgery link $\mathcal{L}$ to characterize the ambient isotopy classes of links in the corresponding three-manifold $\mathcal{M}$.

The method that we have introduced above to represent links in $\mathcal{M}$ is based on the surgery link $\mathcal{L}$. Consequently, we need to discuss the consistency of this method. We must specify, in particular, the correspondence between the link diagrams, considered in points (1) and (2), and the ambient isotopy classes of links in $\mathcal{M}$.

Let us consider a surgery instruction described by the knot $\mathcal{Z}$ with surgery coefficient r. As explained in Section 16.3, the resulting manifold $\mathcal{M}$ is the union of the two spaces $S^3 - \mathring{N}$ and N, where N is a tubular neighbourhood of $\mathcal{Z}$. Without loss of generality, the ambient isotopy class of any link L in $\mathcal{M}$ can be represented by a link in $S^3 - \mathring{N}$. In fact, suppose that at the beginning L is not in $S^3 - \mathring{N}$. In this case, by means of an isotopy we can smoothly deform L so that its parts contained in N reach the boundary ∂N and then, through the gluing homeomorphism h, these parts are transported in $S^3 - \mathring{N}$.

Thus, we can always represent a generic link L in $\mathcal{M}$ by means of a link diagram in which:

(i) one of the components describes the knot $\mathcal{Z}$ in S^3 with a given surgery coefficient r attached to it;
(ii) the remaining components simply describe the link L which belongs to the complement of $\mathcal{Z}$ in S^3.

For example, the diagram shown in Fig. 16.10 describes a two component link, with components C_1 and C_2, in $S^2 \times S^1$.

It should be noted that the orientation of S^3 is kept fixed; moreover, we will use the standard right-handed convention to compute linking numbers. We shall now show that the prescription illustrated above is consistent and can be used to describe arbitrary (oriented and framed) links $\{L\}$ in $\mathcal{M}$.

Remember that the knot $\mathcal{Z}$ and the gluing homeomorphism $h : \partial N \to \partial(S^3 - \mathring{N})$ completely specify the surgery operation. The surgery coefficient r, however, does not identify h uniquely (up to ambient isotopy). The freedom which is left is precisely the origin of the statement that "the manifold $\mathcal{M}$ depends, up to homeomorphism, only on the homotopy class of $h(\mu)$ in $\partial(S^3 - \mathring{N})$".

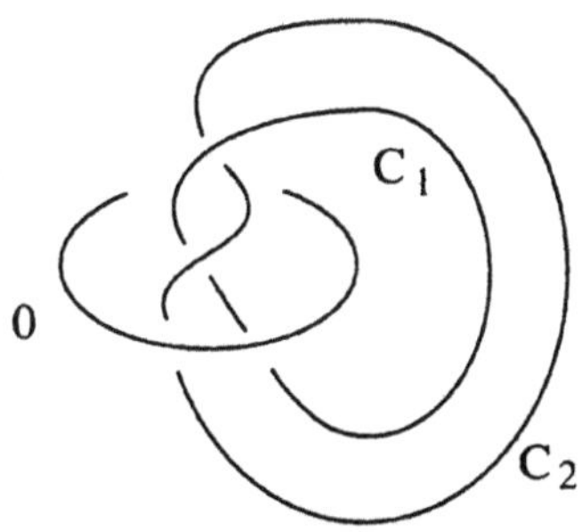

Figure 16.10.

Let us now consider, for fixed $r = b/a$, all the ambiguity which remains in the definition of h. Let h and h' be two homeomorphisms such that $[h(\mu)] = [h'(\mu)] = (a, b)$. Then, the self-homeomorphism $\tau = h^{-1} \cdot h'$ of ∂N takes a meridian to a meridian. Therefore, as we have seen in Section 16.3, τ is ambient isotopic with a certain power of a meridinal twist h_M of ∂N which, in turn, extends to a twist homeomorphism of N. In conclusion, all the ambiguity that we have in the description of $\mathcal{M}$, and which is left by specifying the surgery coefficient r, consists of self-homeomorphisms of $\mathcal{M}$ which, up to ambient isotopy, act trivially in $S^3 - \mathring{N}$ and coincide with twist homeomorphisms in the solid torus N. Since we represent the ambient isotopy classes of links in $\mathcal{M}$ by links in $S^3 - \mathring{N}$, this ambiguity is completely harmless and the prescription based on the points (i) and (ii) is consistent.

Clearly, the same argument can also be used in the case in which the surgery link $\mathcal{L}$ has several components. Therefore, the description of links $\{L\}$ in $\mathcal{M}$, specified in (1) and (2), is consistent. Needless to say, this also shows that any surgery instruction itself, as described in Section 16.3, is in fact well defined.

To sum up, the surgery instruction corresponding to a three-manifold $\mathcal{M}$ is described by a surgery link $\mathcal{L}$ with given surgery coefficients $\{r_i\}$ assigned to its components $\{\mathcal{L}_i\}$. A generic link L in $\mathcal{M}$ is represented by a diagram which shows the surgery link $\mathcal{L}$ in S^3 and which shows, at the same time, how L is embedded in the complement of $\mathcal{L}$ in S^3.

16.5 Elementary surgeries

Dehn surgery provides a simple and direct description of three-manifolds. A basic surgery operation consists of removing and sewing a solid torus. Depending on the embedding of this solid torus in S^3 (or in the complement of a link in S^3) and on the surgery coefficient r, an infinite number of different basic surgery operations can be constructed. As we have already mentioned in Section 16.3, however, different surgery instructions may describe homeomorphic manifolds. Thus, it is natural to ask whether, by means of Rolfsen moves, any surgery instruction can be expressed in some kind of standard form. The answer can be formulated [12,64,65] according to the following

Lickorish (or Fundamental) Theorem. *Every closed, orientable and connected three-manifold can be obtained by surgery in S^3. Moreover, one may always find such a surgery presentation in which the surgery coefficients are all ± 1 and the individual components of the surgery link are unknotted.*

This theorem represents one of the most important results in the theory of three-manifolds and provides a connection between Dehn surgery and Heegaard decomposition of three-manifolds. For our purposes, Lickorish Theorem is crucial because it permits us to solve the quantum CS field theory.

Two different surgery instructions, expressed in the form specified by the Fundamental Theorem, may still correspond to homeomorphic manifolds. The consequences of the Fundamental Theorem, which are particularly important in the context of the topological field theory, can be summarized as follows:

(1) any Dehn surgery can be understood as a suitable combination of "elementary" surgery operations;
(2) only two kinds of such elementary surgeries exist: their corresponding surgery instructions coincide with the unknot in S^3 with surgery coefficient ± 1.

Let us denote by S_+ and S_- the elementary surgery operations in S^3 corresponding to the instructions described by the unknot U in S^3 with surgery coefficient $r = +1$ and $r = -1$ respectively.

The three-manifold obtained according to a single surgery instruction S_+ or S_- is homeomorphic with S^3. By combining several surgeries $S_\pm$, any three-manifold $\mathcal{M}$ can be obtained.

We would now like to describe the effects produced by the single elementary surgery S_+ or S_-. Let us denote by $\mathcal{V}$ the complement solid torus of U in S^3. Under a twist homeomorphism $\tau_\mp$ of the solid torus $\mathcal{V}$, the surgery coefficient $r = \pm 1$ of U is transformed, according to equation (16.17), into $r \to r' = \infty$. But any surgery instruction with surgery coefficient equal to ∞ corresponds to the identity; therefore,

(3) The action of the elementary surgery S_+ (S_-) is simply equivalent to a left-handed (right-handed) twist on the complement solid torus $\mathcal{V}$ of U in S^3.

Thus, a remarkable consequence of the Fundamental Theorem is that the theory of closed, orientable and connected three-manifolds can be reduced to the study of the twist homeomorphisms of a solid torus standardly embedded in S^3.

In order to clarify the meaning of $S_\pm$, it is useful to present a few examples. First of all it should be clear that, in any given situation, the general strategy is always the same: we shall twist the complement solid torus $\mathcal{V}$ of U in S^3 in such a way that the original value (± 1) of the surgery coefficient of U is transformed into ∞. In other words, we shall now illustrate the meaning of the Rolfsen moves of the second kind in the particular case in which, with the notations of Section 16.3, $r_1 = \pm 1$ and in which $\mathcal{V}_1$ may also contain "true" links, in addition to links corresponding to surgery instructions.

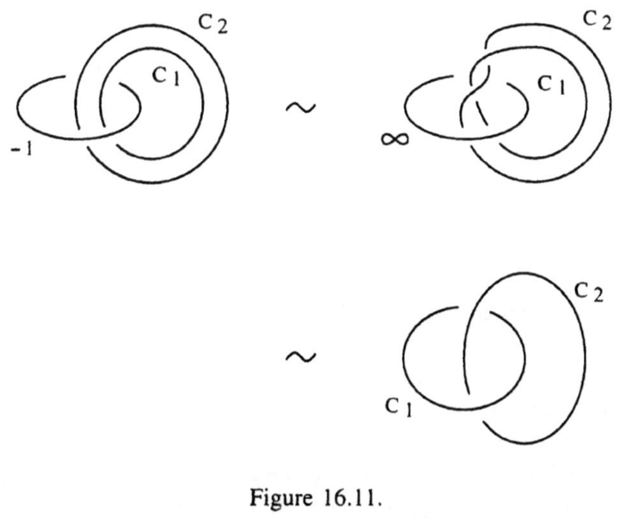

Figure 16.11.

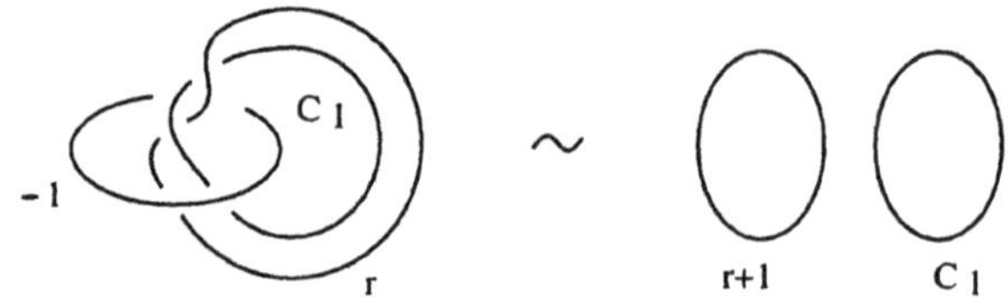

Figure 16.12.

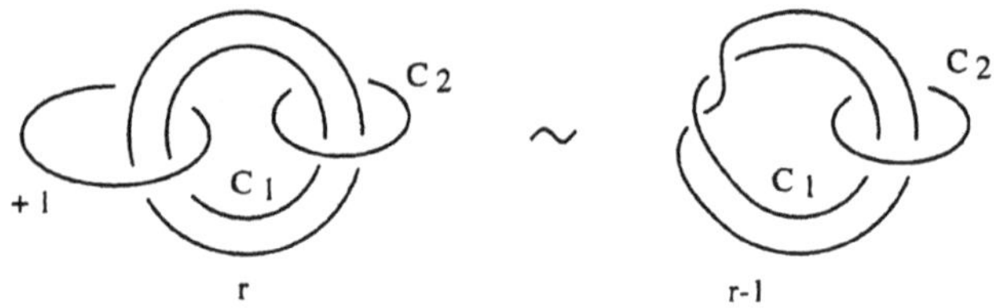

Figure 16.13.

In the first example, the effects of S_- on a two-component link, contained in $\mathcal{V}$, are shown in Fig. 16.11.
In the example of Fig. 16.12, one of the components of the link in $\mathcal{V}$ represents a surgery instruction with surgery coefficient r. The remaining component C_1 of the link represents a real knot. In addition to the right-handed (2π-rotation) twist introduced by S_- on these two components, the surgery coefficient also is modified according to equation (16.18).

Finally, in the example of Fig. 16.13, the effects of S_+ on a three-component link are shown. Needless to say, all these simple examples just illustrate the general rules, introduced in sects.16.2 and 16.3, which describe twist homeomorphisms of solid tori.

16.6 Physical interpretation

Let us summarize the main aspects of surgery on three-manifolds which are relevant in the topological field theory context.

We have seen that a surgery presentation of a generic three-manifold (which is orientable, connected and closed) can be produced by removing and sewing solid tori. Therefore, by means of Dehn surgery, the description and the study of three-manifolds is reduced to knot theory in S^3. This is a good starting point because, in S^3, we know the exact solution of the quantum CS theory.

At this stage, our problem consists of finding the field theory rules which correspond to the abstract surgery instructions. In order to solve this problem, we shall give a physical interpretation of surgery in terms of symmetries. Such an interpretation is the natural consequence of two basic results in the theory of three-manifolds:

(i) the Fundamental Theorem;
(ii) the homeomorphism equivalence of surgeries under Rolfsen moves.

Lickorish Theorem states that any three-manifold has a surgery presentation which is obtained by combining two elementary surgery operations $S_{\pm}$. The surgery instruction corresponding to S_+ (S_-) is described by unknot U in S^3 with surgery coefficient $r = +1$ ($r = -1$). By definition, each single surgery operation $S_{\pm}$ consists of removing from S^3 a tubular neighbourhood of U and sewing it back in such a way that its meridian is mapped into a longitude which has a linking number with U equal to ± 1. The whole effect of the surgery operations $S_{\pm}$ has been analysed by using the homeomorphism equivalence of surgeries under Rolfsen moves. The result is summarized in point (3) of the previous section; in other words,

(3′) the elementary surgeries $S_{\pm}$ can be understood as the generators of symmetry transformations; these transformations are twist homeomorphisms $\tau_{\mp}$ of the complement solid torus of a tubular neighbourhood of the unknot U in S^3.

At this point, the problem of finding the field theory rules which correspond to the surgery instructions can be formulated in terms of symmetries in quantum field theory.

Since $S_{\pm}$ connect S^3 with S^3 and since the CS theory in S^3 has been solved exactly, we shall study how the symmetry transformations $\tau_{\mp}$ act on the observables of the theory. Then, we shall construct the field theory operators $\widetilde{S}_{\pm}$ which represent $S_{\pm}$. By means of these operators $\widetilde{S}_{\pm}$, we shall provide an explicit operator realization of any surgery which is expressed in the form specified by the Fundamental Theorem. Therefore, by means of the operator surgery method, the solution of the quantum CS field theory in any three-manifold $\mathcal{M}$ will be derived from the solution in S^3.

We shall conclude this section by introducing the so-called ***"honest" surgeries*** [12,64,65]. It is not difficult to prove that one can always find an "honest" surgery presentation of a given three-manifold: this simply means that all the surgery coefficients $\{r_i\}$ are integers.

When the ratio r appearing in equation (16.16) is an integer, one can take $a = 1$ and $b = r$ in equation (16.15). In this case, the curve Y is a longitude of N and can be interpreted as a framing of the surgery knot $\mathcal{Z}$. The linking number $\chi(\mathcal{Z}, Y)$ of $\mathcal{Z}$ and Y is given precisely by the coefficient b, see equation (16.15), and then

$$\chi(\mathcal{Z}, Y) = r . \tag{16.19}$$

Therefore, a surgery link $\mathcal{L}$ with integer surgery coefficients $\{r_i\}$ can be represented by a framed link $\mathcal{L}$ in which the linking number of the component $\mathcal{L}_i$ and its framing $\mathcal{L}_{if}$ is equal to r_i.

Let us now consider the set of all possible instructions corresponding to "honest" surgeries. Two such different instructions describe homeomorphic manifolds if and only if they are related [65] by a finite sequence of ***Kirby moves***. A Kirby move is the analogue of a Rolfsen move and can be defined as follows.

Suppose that one component of $\mathcal{L}$ is the unknot. Let $\mathcal{L}_1$ be this component with framing $\mathcal{L}_{1f}$ such that $\chi(\mathcal{L}_1, \mathcal{L}_{1f}) = \pm 1$. Then, this component can be eliminated provided that we perform a twist $\tau_{\mp}$ on the complement solid torus $\mathcal{V}_1$ of $\mathcal{L}_1$ in S^3. Of course, this means that all the remaining components of $\mathcal{L}$, which belong to $\mathcal{V}_1$, must be transformed according to the rules explained in Section 16.2 and Section 16.3. In general, let us consider a Kirby move when a real link L is also present in the manifold $\mathcal{M}$. In this case, the twist $\tau_{\mp}$ of solid torus $\mathcal{V}_1$ clearly acts on all the remaining components of $\mathcal{L}$ and, simultaneously, on the link L.

By using equations (16.17) and (16.18) and the transformation property (16.13) of framings under twist homeomorphisms, it is easy to prove that the invariance under Kirby moves is in fact equivalent to the invariance under Rolfsen moves. Actually, we have already considered Kirby moves, as particular cases of Rolfsen moves, in the previous section.

16.7 The fundamental group

Let C be a knot in S^3; the fundamental group of $S^3 - C$ is simply called the group of the knot C. Clearly, $\pi_1(S^3 - C)$ is also equal to $\pi_1(S^3 - N)$, where N is a tubular neighbourhood of C. Given a diagram description of C, one can easily derive a presentation of $\pi_1(S^3 - C)$ called the Wirtinger presentation. We shall now briefly describe the procedure [12] which can be used to determine $\pi_1(S^3 - C)$.

If the knot C is not oriented, we shall introduce an orientation. The associated diagram consists of an ordered set of arcs $\{\alpha_i\}$, with $i = 1, 2, \ldots, n$. Each α_i is connected to α_{i-1} and α_{i+1} by arcs which are placed below the plane of the knot diagram. Let us introduce, for each α_i, an arrow x_i as shown in Fig. 16.14.
Each arrow x_i represents an oriented closed path in $S^3 - C$. Consider a fixed base point X^* in $S^3 - C$; for example, X^* is placed somewhere over the plane of the knot diagram. The loop associated with x_i consists of three parts: a path connecting

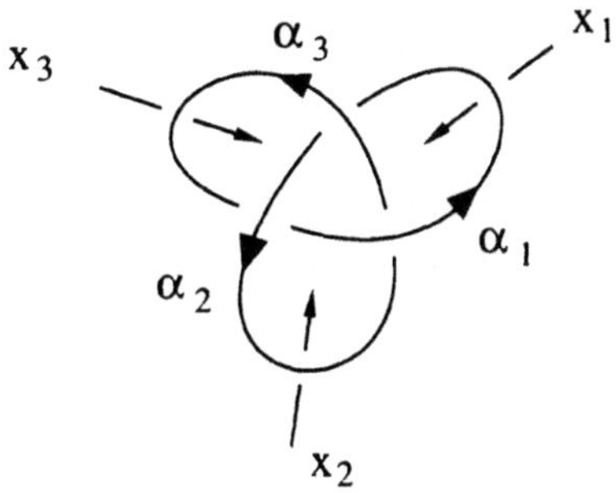

Figure 16.14.

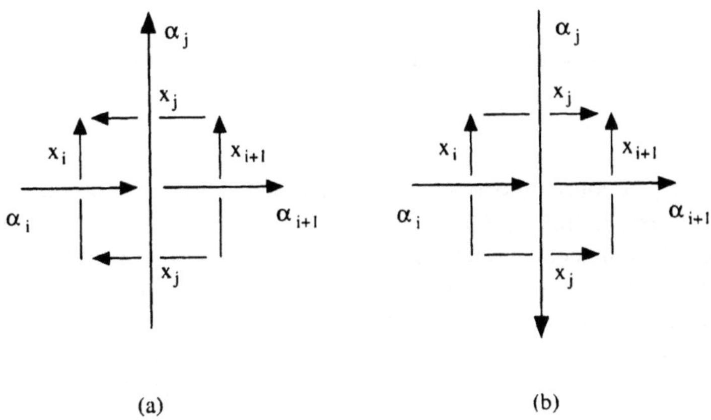

Figure 16.15.

X^* with the tail of x_i, the arrow x_i itself and, finally, a path connecting the head of x_i with X^*. The orientation of this loop is given by the orientation of x_i.

Thus, $\{x_i\}$ can be taken to represent a set of generators for the group $\pi_1(S^3 - C)$. In general, $\pi_1(S^3 - C)$ is not freely generated by $\{x_i\}$; when $n > 1$, at each crossing of the knot diagram one has a relation. At the crossing point in which α_i is connected with α_{i+1}, the two possible dispositions of the arrows are shown in Fig. 16.15.

In the case of the configuration (a), the associated relation $\mathcal{R}_i$ reads [12]

$$\mathcal{R}_i \quad : \quad x_j x_i = x_{i+1} x_j ; \tag{16.20}$$

whereas, in the case (b), the relation $\mathcal{R}_i$ is given by

$$\mathcal{R}_i \quad : \quad x_i \, x_j \;=\; x_j \, x_{i+1} \, . \tag{16.21}$$

Clearly, with $n > 1$ arcs $\{\alpha_i\}$, there are n relations $\{\mathcal{R}_i\}$. The group $\pi_1(S^3 - C)$ is generated by x_i and has the presentation

$$\pi_1(S^3 - C) \;=\; (\, x_1, \ldots, x_n \; ; \; \mathcal{R}_1, \ldots, \mathcal{R}_n \,) \, . \tag{16.22}$$

One of the relations $\{\mathcal{R}_i\}$ can always be eliminated because it is a consequence of the remaining ones. For the trefoil knot shown in Fig. 16.14, the group is generated by $\{x_1, x_2, x_3\}$ with relations

$$x_1 \, x_3 \;=\; x_3 \, x_2 \, , \tag{16.23}$$

$$x_2 \, x_1 \;=\; x_1 \, x_3 \, , \tag{16.24}$$

$$x_3 \, x_2 \;=\; x_2 \, x_1 \, . \tag{16.25}$$

The Wirtinger procedure, illustrated above, can also be used to compute $\pi_1(S^3 - L)$ where L is a generic link in S^3.

Each closed, connected and orientable three-manifold $\mathcal{M}$ is characterized by its ***fundamental group*** $\pi_1(\mathcal{M})$. Let the surgery link $\mathcal{L}$ in S^3 be the surgery instruction corresponding to a Dehn surgery description of $\mathcal{M}$. A presentation of $\pi_1(\mathcal{M})$ can easily be derived from $\mathcal{L}$. We shall now describe how to compute $\pi_1(\mathcal{M})$.

Let $\{\mathcal{L}_i\}$, with $i = 1, 2, \ldots, m$, be the components of the surgery link $\mathcal{L}$; we shall denote by N_i a tubular neighbourhood of $\mathcal{L}_i$. Each component $\mathcal{L}_i$ is characterized by the surgery coefficient $r_i = b_i / a_i$. Let the coprime integers a_i and b_i specify the class $[Y_i] = (a_i, b_i)$ of a simple curve $Y_i \subset \partial N_i$ with respect to the homotopy generators of a Rolfsen basis in ∂N_i.

As we have seen in Section 16.3, $\mathcal{M}$ admits the following surgery presentation. The manifolds $S^3 - \{\mathring{N}_1 \cup \cdots \cup \mathring{N}_m\}$ and $\{N_1 \cup \cdots \cup N_m\}$, considered as distinct spaces, have homeomorphic boundaries. The three-manifold $\mathcal{M}$ is the union

$$\mathcal{M} \;=\; \left(S^3 - \{\mathring{N}_1 \cup \cdots \cup \mathring{N}_m\} \right) \bigcup_h \, (\, N_1 \cup \cdots \cup N_m \,) \, , \tag{16.26}$$

in which the points on the two boundaries are identified through the "gluing" homeomorphism h

$$h \; : \; \partial \, (\, N_1 \cup \cdots \cup N_m \,) \;\Longrightarrow\; \partial \left(S^3 - \{\mathring{N}_1 \cup \cdots \cup \mathring{N}_m\} \right) . \tag{16.27}$$

The homeomorphism h must be consistent with the given values $\{ r_i \}$ of the surgery coefficients. This means that, for each component $\mathcal{L}_i$, the restriction of h ,

$$h \Big|_i = h_i \; : \; \partial (N_i) \Longrightarrow \partial \left(S^3 - \mathring{N}_i \right) , \tag{16.28}$$

must be a homeomorphism which sends a meridian μ_i of N_i into the curve $h_i(\mu_i)$ which corresponds to Y_i.

In order to determine the fundamental group of the manifold $\mathcal{M}$, shown in equation (16.26), let us first consider the space $S^3 - \{ \mathring{N}_1 \cup \cdots \cup \mathring{N}_m \}$. Its fundamental group is isomorphic with the group $\pi_1(S^3 - \mathcal{L})$ of the surgery link $\mathcal{L}$ in S^3. Then, let us take into account the sewing of the solid tori $(N_1 \cup \cdots \cup N_m)$. As a consequence of this operation, any homotopy can now be extended from $S^3 - \{ \mathring{N}_1 \cup \cdots \cup \mathring{N}_m \}$ into a certain solid torus. Any simple closed curve, which is essential on the boundary of a given solid torus, is homotopically trivial in the solid torus if and only if this curve is a meridian. Therefore, the fundamental group $\pi_1(\mathcal{M})$ can be obtained from the Wirtinger presentation of $\pi_1(S^3 - \mathcal{L})$ simply by imposing the new relations

$$\widetilde{\mathcal{R}}_i \quad : \quad \text{class of } \; Y_i \; = \; 1 \quad , \quad \text{for } \; i = 1, 2, \ldots , m \, , \tag{16.29}$$

where the class of Y_i has to be written with respect to the generators of $\pi_1(S^3 - \mathcal{L})$.

Chapter 17

Surgery and field theory

In this chapter, we shall derive the surgery rules in the topological quantum CS field theory. We shall concentrate on the case in which the gauge group is $G = SU(2)$.

We shall first construct the surgery operators which represent the elementary surgeries $S_\pm$ in S^3. Then, we shall give the expression of the surgery operator which corresponds to a generic " honest" surgery instruction and we shall define a field theory procedure which represents Dehn surgery in S^3. Finally, we will show that the results obtained according to this procedure are invariant under Kirby moves.

17.1 Basic pairing

Let us analyse, in the field theory context, the elementary surgery operations $S_\pm$ described by the unknot U in S^3 with surgery coefficient $r = \pm 1$. Surgery in field theory can be easily understood in terms of path-integral.

Consider a quantum field theory in S^3 defined by means of a set of fields, that we denote simply by $\phi(x)$, and a given action $S = S[\phi]$. In the CS theory, the components of ϕ represent the vector gauge fields and all the other fields (auxiliary and ghost fields) which are required to fix the gauge. In principle, the computation of the whole path-integral in S^3 can be divided into three parts. First, one can fix the values of the fields on the boundary ∂N of a tubular neighbourhood N of the unknot U in S^3. Then, one integrates over the fields with fixed boundary conditions in ∂N. Finally, one integrates over all possible values of the fields in ∂N.

The path-integral over the fields defined in N, with fixed values ϕ_0 on ∂N, is a functional $\Psi_{in}[\phi_0]$ of ϕ_0. Similarly, the path-integral over the fields defined in $S^3 - \mathring{N}$, with the same fixed boundary conditions ϕ_0 on $\partial(S^3 - \mathring{N})$, is denoted by $\Psi_{out}[\phi_0]$. The functionals $\Psi_{in}[\phi_0]$ and $\Psi_{out}[\phi_0]$ of course can be considered to be the wave functions of two vectors belonging to the state spaces H_{in} and H_{out} and the last path-integration over ϕ_0 simply defines a pairing between them. In conclusion, the whole path-integral can be understood as a pairing between vectors belonging to two state spaces.

As we have already mentioned in Section 16.4, we are also interested in the case in which Wilson line operators are associated with links defined in $S^3 - \mathring{N}$. So, depending on the particular link which is present in $S^3 - \mathring{N}$, several different functionals $\{ \Psi_{out}[\phi_0] \}$ can be obtained.

Clearly, in the above procedure one identifies the points of the boundaries ∂N and $\partial(S^3 - \mathring{N})$ according to the canonical decomposition of S^3 described by the embedding N in S^3. Instead of doing this, suppose that we now identify the points of ∂N and of $\partial(S^3 - \mathring{N})$ (considered as distinct spaces) according to a given homeomorphism $h : \partial N \to \partial(S^3 - \mathring{N})$. The resulting new pairing will describe precisely the effects of surgery in the field theory context. Our problem consists of determining this new pairing in the quantum CS field theory. In order to solve this problem, we need to introduce the Hopf matrix and study its properties.

In this section, we shall give the expression of the standard pairing between the states of H_{in} and H_{out} in the quantum CS theory in S^3. Standard pairing means the pairing between the states when the points of the boundaries ∂N and $\partial(S^3 - \mathring{N})$ are identified according to the canonical decomposition of S^3 (i.e. no surgery operation has been introduced).

The unknot U is taken to be a circle in S^3; U is the core of its tubular neighbourhood N. By definition, N is a solid torus standardly embedded in S^3; its complement $S^3 - \mathring{N}$ is also a solid torus standardly embedded in S^3. Therefore, the two-dimensional torus ∂N separates S^3 into two solid tori which are completely equivalent. Thus, the path-integral in N, with given boundary conditions in ∂N, gives the same result as the path-integral in $S^3 - \mathring{N}$ with the same boundary conditions in $\partial(S^3 - \mathring{N})$. The pairing between H_{in} and H_{out} must be symmetric under exchange of H_{in} and H_{out} because there is no distinction at all between N and $S^3 - \mathring{N}$.

Let us consider the state space of the quantum CS model. With the introduction of the appropriate gauge-fixing terms in the action, the states of the field theory in $\mathbb{R}^3$ are associated with the values of the fields (which play the role of "coordinates") on a fixed-time two-dimensional plane. As in any field theory, the associated linear space is infinite-dimensional and, as in any gauge theory, this linear space is not a Hilbert space. Since the states corresponding to pure-gauge degrees of freedom are not physical, the linear space must contain "negative-norm" states which cancel out the contributions coming from the longitudinal part of the vector fields in the computation of the observables.

The linear state space of any gauge field theory does not coincide with the space of the physical states of the system. In order to describe the physics of a gauge model and produce meaningful results, we must consider gauge-invariant quantities.

Let us now reconsider the CS model and the states of the field theory associated with the boundary ∂N of a solid torus N standardly embedded in $\mathbb{R}^3$. We are really interested in the ***physical*** subspace, i.e. in the space of states modulo the action of gauge (or BRS) transformations. Because of general covariance, the thickness of the solid torus N is completely irrelevant. Therefore, the physical (gauge-invariant) states in which we are interested can be understood as the gauge-invariant states associated with the core U of N.

As we have seen in Chapter 7, a complete set of gauge-invariant observables associated with links is given by the Wilson line operators. Therefore, the gauge-invariant states associated with ∂N correspond to the possible labels of the Wilson line operator $W(U)$. As we have seen in Chapter 15, these labels are the elements of the tensor algebra $\mathcal{T}$ associated with the gauge group.

As a linear space, $\mathcal{T}$ is infinite-dimensional. Therefore, in the quantum CS theory in $\mathbb{R}^3$, with non-rational values of k, the space of the physical gauge-invariant states associated with a two-dimensional torus is infinite-dimensional. This is in agreement with the results of harmonic analysis for compact simple Lie groups.

When the CS theory is defined in S^3, the coupling constant k takes integer values and, as we have seen in Chapter 15, the space of gauge-invariant states associated with the boundary of a solid torus is described by the reduced tensor algebra $\mathcal{T}_{(k)}$.

As a linear space, $\mathcal{T}_{(k)}$ is finite-dimensional. For the quantum CS theory in S^3, the space of the physical (gauge-invariant) states associated with a two-dimensional torus has dimension $(k-1)$ when $k \geq 2$ and has dimension 2 when $k = 1$. Thus, the space of all possible vectors in H_{in} (or in H_{out}) modulo the action of gauge transformations is equivalent to $\mathcal{T}_{(k)}$.

Any physical state $\Psi[J_1]$ associated with ∂N, and which is the result of the path-integral in N, can be described by the Wilson line operator $W(U;\, \Psi[J_1])$ where U is the core of N.

Similarly, any physical state $\Psi[J_2]$ associated with $\partial(S^3 - \mathring{N})$, and which is the result of the path-integral in $S^3 - \mathring{N}$, is described by the Wilson operator $W(C;\, \Psi[J_2])$ where C is the core of $S^3 - \mathring{N}$.

The ***pairing*** $(\, \Psi[J_1]\, ,\ \Psi[J_2]\,)$ between $\Psi[J_1]$ and $\Psi[J_2]$ is given by the whole path-integral of the quantum CS theory in S^3; this path-integral is precisely the vacuum expectation value $E(U, C;\, \Psi[J_1],\, \Psi[J_2])$ of the Hopf link with components U and C, shown in Fig. 17.1.

When $k \geq 2$, each element $\Psi[J]$ can be represented in $\mathcal{T}$ according to equation (15.44). Therefore, from equation (12.23) we obtain

$$(\, \Psi[J_1]\, ,\ \Psi[J_2]\,) \;=\; \frac{q^{(2J_1+1)(2J_2+1)/2} - q^{-(2J_1+1)(2J_2+1)/2}}{q^{1/2} - q^{-1/2}} \, . \tag{17.1a}$$

When $k = 1$, the elements $\Psi[0]$ and $\Psi[1/2]$ of $\mathcal{T}_{(1)}$ can be represented in $\mathcal{T}$ according to equations (15.68) and (15.69). Therefore, from equation (15.65) we get

$$(\, \Psi[0]\, ,\ \Psi[0]\,) \;=\; 1\, ; \tag{17.1b}$$

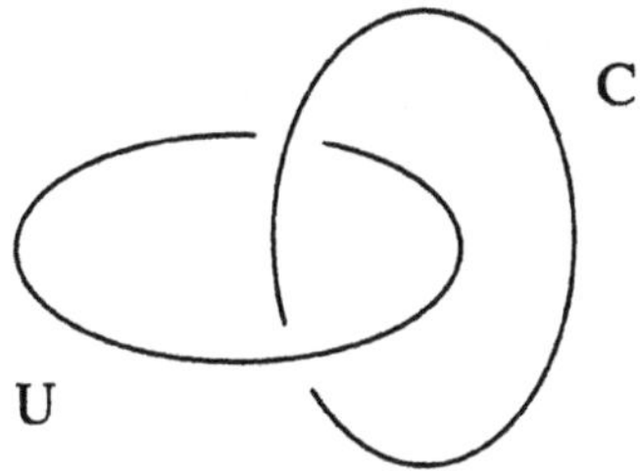

Figure 17.1.

$$(\Psi[0]\,,\ \Psi[1/2]) \;=\; -1 \;=\; (\Psi[1/2]\,,\ \Psi[0])\,; \tag{17.1c}$$

$$(\Psi[1/2]\,,\ \Psi[1/2]) \;=\; -1\,. \tag{17.1d}$$

The pairing, given by equations (17.1a), (17.1b), (17.1c) and (17.1d), is symmetric under an exchange of $\Psi[J_1]$ and $\Psi[J_2]$, as it should be. The construction of the surgery operators will be based exclusively on the properties of the pairing (17.1).

17.2 Properties of the Hopf matrix

When $k = 1$, the properties of the pairing (17.1) can easily be derived because we only need to analyse the (two by two) matrix whose matrix elements are shown in equations (17.1b), (17.1c) and (17.1d). When $k \geq 2$, the properties of the resulting pairing (17.1a) are not completely obvious and, in this section, we will concentrate on them.

Let us denote by $H_{J_1J_2}$ the value $E(U, C;\, \Psi[J_1],\, \Psi\,[J_2])$ of the Hopf link in S^3 with components U and C shown in Fig. 17.1,

$$H_{J_1J_2} \;=\; \frac{q^{(2J_1+1)(2J_2+1)/2} - q^{-(2J_1+1)(2J_2+1)/2}}{q^{1/2} - q^{-1/2}}\,. \tag{17.2}$$

As we have seen in the previous section, $H_{J_1J_2}$ represents the value of the pairing $(\Psi[J_1]\,,\ \Psi[J_2])$ between the physical states of the CS theory in S^3 when $k \geq 2$.

For fixed integer k (with $k \geq 2$), the reduced tensor algebra $\mathcal{T}_{(k)}$ is of order $(k-1)$. As we have seen in Chapter 15, the elements $\{\Psi[J]\}$ of the standard basis of $\mathcal{T}_{(k)}$ are labelled by J which takes values in the interval

$$\frac{k}{2} - 1 \;\geq\; J \;\geq\; 0\,. \tag{17.3}$$

For the interesting values of J_1 and J_2 (satisfying equation (17.3)), the entries shown in equation (17.2) represent the matrix elements of a $(k-1) \times (k-1)$ matrix H called the ***Hopf matrix***. Clearly, H is real and symmetric. From now on, it is understood that the indices (for example, J and I) take the values shown in equation (17.3) with the natural order from 0 to $(k/2)-1$.

As we have explained in Chapter 15, for fixed integer k, the defining relations **(1)-(5)** can be simplified so that, even in the intermediate steps of the computations, the structure of the reduced tensor algebra $\mathcal{T}_{(k)}$ is manifestly shown. In our case, the matrix elements of H can also be written as

$$H_{J_1 J_2} = q^{-Q(J_1)} q^{-Q(J_2)} \sum_{J=|J_1-J_2|}^{b(J_1,J_2)} q^{Q(J)} E_0[J] , \tag{17.4}$$

where

$$b(J_1, J_2) = \min \{ J_1 + J_2 , \; k-2-J_1-J_2 \} . \tag{17.5}$$

In the expression (17.4) one can easily recognize the structure constants of the reduced tensor algebra.

Property G9. *The Hopf matrix satisfies*

$$\omega_J = \sum_{I=0}^{k/2-1} E_0[I] \, H_{IJ} = \binom{k}{2} \frac{1}{\sin^2(\pi/k)} \, \delta_{J0} . \tag{17.6}$$

Proof. By using the explicit expressions (15.19) and (17.2) of the unknots and of the Hopf matrix, one finds

$$\omega_J = \left(q^{1/2} - q^{-1/2}\right)^{-2} \sum_{I=0}^{k/2-1} \Big\{ q^{(2I+1)(J+1)} - q^{-(2I+1)J} - q^{(2I+1)J} + q^{-(2I+1)(J+1)} \Big\} ,$$

and, putting $x = 2I+1$, one can write

$$\omega_J = \left(q^{1/2} - q^{-1/2}\right)^{-2} \sum_{x=0}^{k-1} \Big\{ q^{(J+1)x} - q^{-Jx} - q^{Jx} + q^{-(J+1)x} \Big\} . \tag{17.7}$$

The explicit computation of the geometric series, contained in the expression (17.7), immediately gives the result (17.6). □

Let us now consider the link $\mathcal{G}$ in S^3, shown in Fig. 17.2, in which the three components $\{C_1, C_2, C_3\}$ have preferred framings. The components C_1 and C_2 are associated with the elements $\Psi[J_1]$ and $\Psi[J_2]$ respectively. The component C_3 is associated with the element

$$\widetilde{\Psi}_0 = \sum_{I=0}^{k/2-1} E_0[I]\,\Psi[I] \tag{17.8}$$

of $\mathcal{T}_{(k)}$. The expectation value $E(\mathcal{G})$ of the corresponding Wilson line operator is

$$E(\mathcal{G}) = E(C_1, C_2, C_3; \Psi[J_1], \Psi[J_2], \widetilde{\Psi}_0) = \left(\frac{k}{2}\right) \frac{1}{\sin^2(\pi/k)}\,\delta_{J_1 J_2}. \tag{17.9}$$

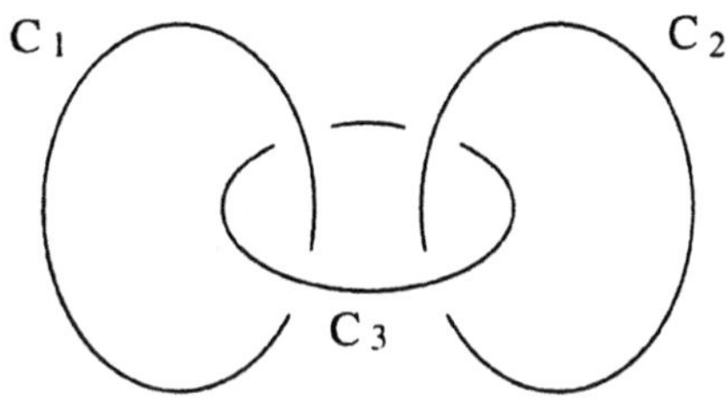

Figure 17.2. Link $\mathcal{G}$ in S^3

In fact, the link $\mathcal{G}$ is a satellite of the Hopf link constructed with the pattern link B shown in Fig. (14.2). The components C_1 and C_2 represent the two components of $h^\diamond(B)$. Therefore, by using the satellite relations and the structure constants of the reduced tensor algebra, one gets

$$E(\mathcal{G}) = \sum_{J=|J_1-J_2|}^{b(J_1,J_2)} \omega_J\,, \tag{17.10}$$

where $b(J_1, J_2)$ is given in equation (17.5) and ω_J is shown in equation (17.6). Since the decomposition of the product $\Psi[J_1]\,\Psi[J_2]$ contains $\Psi[0]$ if and only if $J_1 = J_2$, equation (17.9) follows from equation (17.6).

Property G10. *The Hopf matrix satisfies*

$$(H^2)_{IJ} = \left(\frac{k}{2}\right) \frac{1}{\sin^2(\pi/k)}\,\delta_{IJ}\,. \tag{17.11}$$

Proof. The link $\mathcal{G}$ shown in Fig. 17.2 is the connected sum of the two Hopf links shown in Fig. 17.3; the connected sum is performed by acting on the components

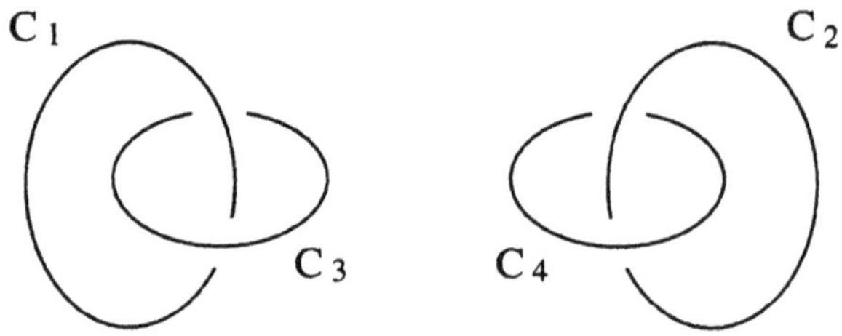

Figure 17.3.

C_3 and C_4. Let both components C_3 and C_4 be labelled by the element $\Psi[I\,]$. The components C_1 and C_2 are associated with $\Psi[J_1]$ and $\Psi[J_2]$ respectively. According to the general formula (13.49) of the connected sum of links, one has

$$\begin{aligned} E_0[I\,]\, E((C_1,C_3)\#(C_2,C_4);\, \Psi[J_1], \Psi[J_2], \Psi[I\,]) \;&= \\ = \; E(C_1,C_3;\, \Psi[J_1], \Psi[I\,])\; E(C_2,C_4;\, \Psi[J_2], \Psi[I\,])\,. \end{aligned} \tag{17.12}$$

By taking the sum over I of both sides of equation (17.12), one obtains

$$E(\mathcal{G}) \;=\; \sum_{I=0}^{k/2-1} H_{J_1 I}\, H_{I J_2}\,. \tag{17.13}$$

Equation (17.13) combined with equation (17.9) show that equation (17.11) holds. Therefore, the statement of property **G10** is proved. By using the explicit form (17.2) of the Hopf matrix, or the expression (17.4), equation (17.11) can also be obtained by means of a direct computation. □

Let $E = \{\, E_{IJ} \,\}$ be the (invertible) matrix

$$E_{IJ} \;=\; E_0[J]\, \delta_{IJ}\,, \tag{17.14}$$

and $\Lambda = \{\, \Lambda_{JI} \,\}$ the real matrix

$$\Lambda_{JI} \;=\; \{\, H \cdot E^{-1} \,\}_{JI} \;=\; H_{JI}\, E_0^{-1}[I\,]\,. \tag{17.15}$$

Equation (17.11) implies that the Hopf matrix H is invertible; therefore, the matrix Λ also is invertible. Equation (17.15) can be rewritten as

$$H_{JI} \;=\; \Lambda_{JI}\, E_0[I\,] \qquad , \tag{17.16}$$

and the matrix elements $\{ \Lambda_{JI} \}$ are given by

$$\Lambda_{JI} = \frac{q^{(2J+1)(2I+1)/2} - q^{-(2J+1)(2I+1)/2}}{q^{(2I+1)/2} - q^{-(2I+1)/2}} . \tag{17.17}$$

Property G11. *The matrix* Λ *satisfies*

$$\sum_{J=0}^{k/2-1} q^{Q(J)} E_0[J] \Lambda_{JI} = Z_{(1)} q^{-Q(I)} , \tag{17.18}$$

where

$$Z_{(1)} = \sqrt{\frac{k}{2}} \frac{1}{\sin(\pi/k)} \exp\left(-i\frac{\pi}{4} \frac{3(k-2)}{k}\right) . \tag{17.19}$$

Proof. Let us denote by $f(I)$ the sum

$$f(I) = \sum_{J=0}^{k/2-1} q^{Q(J)} E_0[J] \Lambda_{JI} . \tag{17.20}$$

From equation (17.15), it follows that

$$f(I) = \sum_{J=0}^{k/2-1} q^{Q(J)} E_0[J] H_{JI} E_0^{-1}[I] , \tag{17.21}$$

and, by using the expression (17.4), we have

$$f(I) = \sum_{J=0}^{k/2-1} E_0[J] \sum_{M=|J-I|}^{b(J,I)} q^{Q(M)} E_0[M] q^{-Q(I)} E_0^{-1}[I] . \tag{17.22}$$

At this stage, it is convenient to use the following identity [68]

$$\sum_{J=0}^{k/2-1} \sum_{M=|J-I|}^{b(J,I)} = \sum_{M=0}^{k/2-1} \sum_{J=|M-I|}^{b(M,I)} . \tag{17.23}$$

Since the satellite relations imply

$$\sum_{J=|M-I|}^{b(M,I)} E_0[J] = E_0[M] E_0[I] , \tag{17.24}$$

we obtain

$$f(I) = \sum_{M=0}^{k/2-1} q^{Q(M)} E_0^2[M] \, q^{-Q(I)} . \tag{17.25}$$

This expression has the correct dependence on $q^{-Q(I)}$. It remains to be shown that the value of $Z_{(1)}$, given by

$$Z_{(1)} = \sum_{J=0}^{k/2-1} q^{Q(J)} E_0^2[J] , \tag{17.26}$$

coincides with the expression (17.19).

From equations (15.52) and (15.53), it follows that we can write

$$Z_{(1)} = (1/4) \sum_{J=0}^{2k-1} q^{Q(J)} E_0^2[J] . \tag{17.27}$$

By inserting in this expression the values of the unknots and putting $x = 2J + 1$, we get

$$Z_{(1)} = \frac{q^{-1/4}}{4(q^{1/2} - q^{-1/2})^2} \sum_{x=1}^{4k-1} \Big\{ q^{-1} q^{(x+2)^2/4} + q^{-1} q^{(x-2)^2/4} - 2 q^{x^2/4} \Big\} . \tag{17.28}$$

With a suitable redefinition of the summation indices, equation (17.28) takes the form

$$Z_{(1)} = - \frac{q^{-3/4}}{2(q^{1/2} - q^{-1/2})} \sum_{x=0}^{4k-1} q^{x^2/4} . \tag{17.29}$$

The series

$$\sum_{x=0}^{4k-1} q^{x^2/4} = \sum_{x=0}^{4k-1} \exp \left(-i2\pi \cdot \frac{x^2}{4k} \right) \tag{17.30}$$

is known in mathematics as the complex conjugate of the Gauss sum $S(1, 4k)$. Let us recall that, in general, the Gauss sum $S(n, m)$ is defined by

$$S(n, m) = \sum_{x=0}^{m-1} \exp \left(2\pi i \frac{n}{m} x^2 \right) , \tag{17.31}$$

and it is known [66] that

$$S(1,m) = \begin{cases} (1+i)\sqrt{m} & \text{if } m \equiv 0 \ (\bmod 4); \\ \sqrt{m} & \text{if } m \equiv 1 \ (\bmod 4); \\ 0 & \text{if } m \equiv 2 \ (\bmod 4); \\ i\sqrt{m} & \text{if } m \equiv 3 \ (\bmod 4). \end{cases} \tag{17.32}$$

Therefore, from equations (17.29) and (17.32) it follows that

$$Z_{(1)} = \sqrt{\frac{k}{2}} \frac{1}{\sin(\pi/k)} \exp\left(-i\frac{\pi}{4}\frac{3(k-2)}{k}\right). \tag{17.33}$$

This expression coincides with the value of $Z_{(1)}$ given in equation (17.19); this concludes the proof of property **G11**. □

Finally, by taking the complex conjugates of equations (17.18) and (17.19), one finds

$$\sum_{J=0}^{k/2-1} q^{-Q(J)} E_0[J] \Lambda_{JI} = Z_{(-1)} q^{Q(I)}, \tag{17.34}$$

where

$$Z_{(-1)} = \sqrt{\frac{k}{2}} \frac{1}{\sin(\pi/k)} \exp\left(i\frac{\pi}{4}\frac{3(k-2)}{k}\right). \tag{17.35}$$

17.3 Elementary surgery operators

The surgery instruction corresponding to S_+ (S_-) is the unknot U in S^3 with surgery coefficient $r=+1$ $(r=-1)$. As we have seen in Section 16.6, $S_\pm$ connect S^3 with S^3 and can be understood as the generators of twist homeomorphisms $\tau_\mp$ of the complement solid torus of U in S^3. Therefore, for any given framed link L in the complement of U in S^3, the action of $S_\pm$ can be represented by

$$S_\pm : L \rightarrow L^{(\mp)}, \tag{17.36}$$

where $L^{(\mp)}$ is the image of L under a twist homeomorphism $\tau_\mp$ of the complement solid torus of U in S^3. So, under surgery $S_\pm$, the observables of the theory transform as

$$S_\pm : \langle W(L) \rangle \Big|_{S^3} \rightarrow \langle W(L^{(\mp)}) \rangle \Big|_{S^3}. \tag{17.37}$$

Let us recall that $\langle\, W(L)\,\rangle\big|_{S^3}$ denotes the normalized field theory expectation value

$$\begin{aligned}\langle\, W(L)\,\rangle\Big|_{S^3} &\equiv \frac{\langle 0|\; W(L)\; |0\rangle\Big|_{S^3}}{\langle 0\,|\,0\rangle\Big|_{S^3}} \\ &= \frac{\langle 0|\; W_{\rho_1}(C_1)\cdots W_{\rho_m}(C_m)\; |0\rangle\Big|_{S^3}}{\langle 0\,|\,0\rangle\Big|_{S^3}}\,.\end{aligned} \tag{17.38}$$

Under surgery operations $S_\pm$, both path-integrals $\langle 0|\; W(L)\; |0\rangle\big|_{S^3}$ and $\langle 0\,|\,0\rangle\big|_{S^3}$ are modified and, as a result of this modification, we know that equation (17.37) must be satisfied. Let us denote by $\widetilde{S}_\pm$ the operators of the field theory representing the surgery operations $S_\pm$; this means that equation (17.37) can be written as

$$\langle\, W(L^{(\mp)})\,\rangle\Big|_{S^3} = \frac{\langle 0\,|\; W(L)\; \widetilde{S}_\pm\; |\,0\rangle\Big|_{S^3}}{\langle 0\,|\; \widetilde{S}_\pm\; |\,0\rangle\Big|_{S^3}}\,. \tag{17.39}$$

With respect to the standard vacuum normalization of the theory in S^3, equation (17.39) takes the form

$$\begin{aligned}\langle\, W(L^{(\mp)})\,\rangle\Big|_{S^3} &= \frac{\langle 0\,|\; W(L)\; \widetilde{S}_\pm\; |\,0\rangle\Big|_{S^3}}{\langle 0\,|\,0\rangle\Big|_{S^3}} \Bigg/ \frac{\langle 0\,|\; \widetilde{S}_\pm\; |\,0\rangle\Big|_{S^3}}{\langle 0\,|\,0\rangle\Big|_{S^3}} \\ &= \langle\; W(L)\; \widetilde{S}_\pm\;\rangle\Big|_{S^3} \Bigg/ \langle\; \widetilde{S}_\pm\;\rangle\Big|_{S^3}\,.\end{aligned} \tag{17.40}$$

The operators $\widetilde{S}_\pm$ must be associated with the surgery link which, in our case, is the unknot U in S^3. Furthermore, since surgery operations connect gauge-invariant observables with themselves, $\widetilde{S}_\pm$ must be gauge invariant. The Wilson line operators, computed for the different inequivalent irreducible representations of the gauge group, represent a complete set of gauge-invariant operators associated with a given knot C. Therefore, the operators $\widetilde{S}_\pm$ necessarily have the form of appropriate linear combinations $\widetilde{W}(U;\pm 1)$ of Wilson line operators associated with the unknot U in different representations:

$$\widetilde{S}_\pm = \widetilde{W}(U;\pm 1)\,. \tag{17.41}$$

Since Wilson line operators are defined on framed links, we must introduce a framing for U and, for any given choice of the framing, the numerical coefficients appearing in the linear combination $\widetilde{W}(U;\pm 1)$ are fixed. The quantum holonomy is defined for oriented knots and, therefore, an orientation of U will be introduced. The particular choice of this orientation is irrelevant. Indeed, a change in the orientation of a knot is equivalent to substituting the associated representation for its complex conjugate and each representation of $SU(2)$ is equivalent to its complex conjugate.

The elementary surgery operators $\widetilde{W}(U;\pm 1)$ will be determined by imposing that equation (17.40) holds for any framed link L in the complement of U in S^3 (with arbitrary representations of $SU(2)$ associated with its components). We shall first consider the case in which $k \geq 2$ and then we shall analyse the case $k = 1$.

For fixed integer k, with $k \geq 2$, a complete basis of knot observables is characterized by the $(k-1)$ elements $\{\Psi[J]\}$ of the reduced tensor algebra. Each class $\Psi[J]$ can be represented by $\chi[J]$. In practice, this means that it is sufficient to consider the values of J in the interval $0 \leq J \leq (k/2) - 1$. Therefore, one has

$$\widetilde{W}(U;\pm 1) = \sum_{J=0}^{k/2-1} \phi_{\pm}(J)\, W(U;\Psi[J])\,, \tag{17.42}$$

where $\phi_{\pm}(J)$ are complex numbers and $W(U;\Psi[J])$ is the Wilson line operator associated with the unknot U which is labelled by $\Psi[J]$ and which has some specified framing. In order to simplify the exposition, we shall choose a ***preferred framing*** for U. (In the next section, we shall see that there is a different and more natural choice for the framing of the knot U.)

Let us now consider a generic framed L which does not intersect U. This means that L belongs to the complement solid torus $\mathcal{V}$ of a tubular neighbourhood of U in S^3. According to equation (13.18), $W(L)$ can always be written as

$$W(L) = \sum_{I=0}^{(k/2)-1} \xi(I)\, W(C;\Psi[I])\,, \tag{17.43}$$

where $W(C;\Psi[I])$ is the Wilson line operator associated with the framed core C of $\mathcal{V}$ with preferred framing. The knots U and C are shown in Fig. 17.4.

The complex coefficients $\{\xi(I)\}$ depend on the structure of L and the sum over I corresponds to the sum over all the elements of the standard basis of the reduced tensor algebra $\mathcal{T}_{(k)}$ determined by the fixed integer value of k.
Let $L^{(\pm)}$ $(C^{(\pm)})$ be the image of L (C) under a twist homeomorphism $\tau_{\pm}$ of $\mathcal{V}$. From equation (17.43) it follows that

$$W(L^{(\pm)}) = \sum_{I=0}^{(k/2)-1} \xi(I)\, W(C^{(\pm)};\Psi[I])\,. \tag{17.44}$$

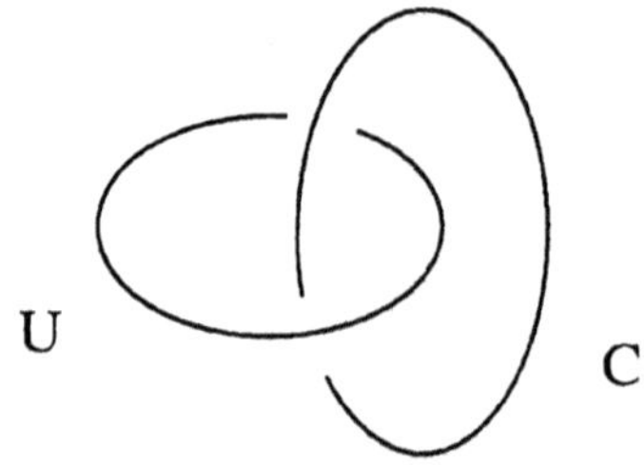

Figure 17.4.

For the CS theory in S^3, the vacuum expectation value of both sides of equation (17.44) gives

$$E(L^{(\pm)}) = \sum_{I=0}^{(k/2)-1} \xi(I)\, q^{\pm Q(I)}\, E_0[I\,]. \tag{17.45}$$

This expression has been obtained by taking into account the transformation properties of the extended Jones polynomial under a modification of the framing of the link components. By considering equations (17.41)-(17.45), equation (17.40) takes the form

$$\begin{aligned} \lambda_\pm \sum_{I=0}^{(k/2)-1} \xi(I)\, q^{\mp Q(I)}\, E_0[I\,] &= \\ = \sum_{I=0}^{(k/2)-1} \sum_{J=0}^{(k/2)-1} \xi(I)\, \phi_\pm(J)\, \langle\, W(C;\, \Psi[I\,])\, W(U;\, \Psi[J])\,\rangle \Big|_{S^3}, \end{aligned} \tag{17.46}$$

where $\lambda_\pm$ denote the expectation values

$$\lambda_\pm = \langle\, \widetilde{W}(U;\, \pm 1)\,\rangle \Big|_{S^3}. \tag{17.47}$$

Here we have assumed that $\lambda_\pm \neq 0$; this assumption is justified *a posteriori*. Indeed, we shall find that $\lambda_\pm \neq 0$. The coefficients $\{\, \phi_\pm(J)\,\}$ must be chosen in such a way that equation (17.46) holds for any link L. This means that equation (17.46) must hold for arbitrary coefficients $\{\, \xi(I)\,\}$. Therefore, for any $0 \leq I \leq (k/2) - 1$, $\{\, \phi_\pm(J)\,\}$ must satisfy

$$\sum_{J=0}^{(k/2)-1} \phi_\pm(J)\, E(U, C;\, J, I\,) = \lambda_\pm\, q^{\mp Q(I)}\, E_0[I\,], \tag{17.48}$$

where $E(U, C; J, I)$ is the value of the Hopf link in S^3 with components U and C,

$$E(U, C; J, I) = \langle W(C; \Psi[I]) \, W(U; \Psi[J]) \rangle \Big|_{S^3} . \tag{17.49}$$

Both components U and C of the Hopf link, shown in Fig. 17.4, have preferred framings and are labelled by $\Psi[J]$ and $\Psi[I]$ respectively. Let us recall that the values $E(U, C; J, I)$ coincide with the matrix elements of the Hopf matrix H_{JI} and, by taking into account the definition (17.16), equation (17.48) takes the form

$$\sum_{J=0}^{(k/2)-1} \phi_{\pm}(J) \, \Lambda_{JI} = \lambda_{\pm} \, q^{\mp Q(I)} . \tag{17.50}$$

Consider now the coefficients $\phi_{\pm}(J)$ given by [10,11]

$$\phi_{\pm}(J) = a \, q^{\pm Q(J)} \, E_0[J] , \tag{17.51}$$

where

$$a = \sqrt{\frac{2}{k}} \, \sin(\pi/k) , \tag{17.52}$$

From property **G11** and equations (17.34) and (17.45), it follows that equation (17.50) is satisfied for any $0 \leq I \leq (k/2) - 1$ and the value of $\lambda_{\pm}$ turns out to be

$$\lambda_{\pm} = e^{\mp i\varphi} , \tag{17.53}$$

where

$$e^{i\varphi} = \exp\left[\left(\frac{i\pi}{4} \right) \frac{3(k-2)}{k} \right] . \tag{17.54}$$

Therefore, with the coefficients $\{ \phi_{\pm}(J) \}$ given in equation (17.51), for any link L one has

$$\langle W(L) \, \widetilde{W}(U; \pm 1) \rangle \Big|_{S^3} = e^{\mp i\varphi} \langle W(L^{(\mp)}) \rangle \Big|_{S^3} , \tag{17.55}$$

where $L^{(\mp)}$ is the image of the framed link $L \subset \mathcal{V}$ under a left-handed (right-handed) twist of $\mathcal{V}$. Equation (17.55) implies, in particular, that

$$\langle \widetilde{W}(U; \pm 1) \rangle \Big|_{S^3} = e^{\mp i\varphi} . \tag{17.56}$$

Finally, by combining equations (17.55) and (17.56), one gets

$$\langle W(L^{(\mp)}) \rangle \Big|_{S^3} = \frac{\langle W(L) \, \widetilde{W}(U; \pm 1) \rangle \Big|_{S^3}}{\langle \widetilde{W}(U; \pm 1) \rangle \Big|_{S^3}} . \tag{17.57}$$

Consequently, we have proved the following

Theorem R6. *When* $k \geq 2$, *the field theory operators* $\widetilde{W}(U; \pm 1)$, *defined by* equations (17.42) *and* (17.51), *represent the elementary surgery operations* $S_\pm$ *in the sense specified by* equation (17.57).

Clearly, the operators $\widetilde{W}'(U; \pm 1) = \eta_\pm \widetilde{W}(U; \pm 1)$, with $\eta_\pm \neq 0$, can also be used to represent $S_\pm$ because the constants $\eta_\pm$ cancel out in equation (17.57). Our choice of the normalization of the coefficients $\phi_\pm(J)$, shown in equation (17.51), just simplifies our subsequent discussion on the Kirby moves.

Note that the coefficients $\phi_\pm(J)$ are fixed by equation (17.50). Since the matrix Λ is invertible, this means that, apart from an overall non-zero multiplicative constant, the elementary surgery operators $\widetilde{W}(U; \pm 1)$ are uniquely determined.

When $k = 2$, the reduced tensor algebra is of order equal to unity and the elementary surgery operators $\widetilde{W}(U; \pm 1)$ are equal to the identity operator. This means that, for $k = 2$, the expectation values of Wilson line operators are given by the expression (15.59) in any three-manifold $\mathcal{M}$.

Let us now consider the case in which $k = 1$. Equations (15.61) and (15.62) show that, when $k = 1$, the twist variable can take only two inequivalent values: 1 or i. This means that, under an elementary modification of the framing of a link component, there are only two possible inequivalent behaviours of $E(L)$. This is in agreement with the fact that $\mathcal{T}_{(1)}$ is of order two.

Consequently, the operators $\widetilde{W}(U; \pm 1)$ can be constructed by using only two elements of the standard basis of $\mathcal{T}$. One of these elements must have integer J and the second must have non-integer J. Equivalently, one could use the two elements of the standard basis of $\mathcal{T}_{(1)}$. In order to simplify the exposition, we shall use the elements of $\mathcal{T}$.

Theorem R7. *When* $k = 1$, *the elementary surgery operators can be taken to be*

$$\widetilde{W}(U; \pm 1) = \frac{1}{\sqrt{2}} W(U; \chi[0]) \mp \frac{i}{2\sqrt{2}} W(U; \chi[1/2]) \, . \tag{17.58}$$

Proof. Let the framed knot C of Fig. 17.4 be labelled by $\chi[I]$. Then

$$\begin{aligned} \langle\, W(C; \chi[I])\, \widetilde{W}(U; \pm 1) \,\rangle\Big|_{S^3} &= \frac{1}{\sqrt{2}} \langle\, W(C; \chi[I]) \,\rangle\Big|_{S^3} \\ &\mp \frac{i}{2\sqrt{2}} \langle\, W(C; \chi[I])\, W(U; \chi[1/2]) \,\rangle\Big|_{S^3} \, . \end{aligned} \tag{17.59}$$

If I is an integer, from equations (15.65) and (15.61) it follows that

$$\begin{aligned} \langle\, W(C;\chi[I\,])\;\widetilde{W}(U;\pm 1)\,\rangle\Big|_{S^3} &= \frac{1\pm i}{\sqrt{2}}\,\langle\, W(C;\chi[I\,])\,\rangle\Big|_{S^3} \\ &= e^{\pm i\pi/4}\,\langle\, W(C^{(\mp)};\chi[I\,])\,\rangle\Big|_{S^3}\,. \end{aligned} \tag{17.60}$$

When I is not an integer, by using equations (15.65) and (15.62) we find

$$\begin{aligned} \langle\, W(C;\chi[I\,])\;\widetilde{W}(U;\pm 1)\,\rangle\Big|_{S^3} &= \frac{1\mp i}{\sqrt{2}}\,\langle\, W(C;\chi[I\,])\,\rangle\Big|_{S^3} \\ &= e^{\pm i\pi/4}\,\langle\, W(C^{(\mp)};\chi[I\,])\,\rangle\Big|_{S^3}\,. \end{aligned} \tag{17.61}$$

Equations (17.60) and (17.61) imply that, for any arbitrary link L, one has

$$\langle\, W(L)\;\widetilde{W}(U;\pm 1)\,\rangle\Big|_{S^3} = e^{\pm i\pi/4}\,\langle\, W(L^{(\mp)})\,\rangle\Big|_{S^3}\,; \tag{17.62}$$

therefore,

$$\langle\, W(L^{(\mp)})\,\rangle\Big|_{S^3} = \frac{\langle\, W(L)\;\widetilde{W}(U;\pm 1)\,\rangle\Big|_{S^3}}{\langle\, \widetilde{W}(U;\pm 1)\,\rangle\Big|_{S^3}}\,.$$

This equation shows that $\widetilde{W}(U;\pm 1)$, given in equation (17.58), are the elementary surgery operators for $k=1$. □

17.4 Surgery operator

Let us summarize the results of Section 17.3. For $k \geq 2$, the elementary surgery operators $\widetilde{W}(U;\pm 1)$ are given by

$$\widetilde{W}(U;\pm 1) = W(U;\Psi_{\pm})\,, \tag{17.63}$$

where the unknot U has ***preferred framing*** and

$$\Psi_{\pm} = \sqrt{\frac{2}{k}}\,\sin(\pi/k)\sum_{J=0}^{(k/2)-1} q^{\pm Q(J)}\,E_0[J]\,\Psi[J]\,. \tag{17.64}$$

The insertion of the operator $\widetilde{W}(U;\pm 1)$ in the vacuum expectation value $\langle W(L)\rangle\big|_{S^3}$ of a given observable,

$$\widetilde{W}(U;\pm 1)\ :\ \langle\, W(L)\,\rangle\Big|_{S^3} \longrightarrow \langle\, W(L)\ \widetilde{W}(U;\pm 1)\,\rangle\Big|_{S^3}, \tag{17.65}$$

is called the action of $\widetilde{W}(U;\pm 1)$. The result (17.55) can be restated by saying that the action of $\widetilde{W}(U;\pm 1)$ on the element $\Psi[I]$ of the reduced tensor algebra $\mathcal{T}_{(k)}$, associated with the core of the complement solid torus of U in S^3, is given by

$$\widetilde{W}(U;\pm 1)\ :\ \Psi[I] \to e^{\mp i\varphi}\, q^{\mp Q(I)}\, \Psi[I], \tag{17.66}$$

where the value of $e^{i\varphi}$ is shown in equation (17.54). When $k=1$, one has

$$\widetilde{W}(U;\pm 1) = W(U;\chi_{\pm}), \tag{17.67}$$

where the unknot U has ***preferred framing*** and

$$\chi_{\pm} = \frac{1}{\sqrt{2}}\left(\chi[0] - \frac{1}{2}\, q^{\pm Q(1/2)}\, \chi[1/2]\right). \tag{17.68}$$

The action of $\widetilde{W}(U;\pm 1)$ on the element $\chi[I]$ of the tensor algebra $\mathcal{T}$, associated with the core of the complement solid torus of U in S^3, is given by

$$\widetilde{W}(U;\pm 1)\ :\ \chi[I] \to e^{\pm i\pi/4}\, q^{\mp Q(I)}\, \chi[I]. \tag{17.69}$$

Equations (17.66) and (17.69) show that, apart from the overall phase factors $e^{\pm i\varphi}$ or $e^{\pm i\pi/4}$, an insertion of $\widetilde{W}(U;\pm 1)$ in the expectation values generates a twist homeomorphism $\tau_{\mp}$ of the complement solid torus of U in S^3. Since these phase factors are constants (i.e., they do not depend on the representation under consideration), their presence in equations (17.66) and (17.69) is completely harmless. In fact, by introducing the correct vacuum normalization for the expectation values, as shown in equation (17.40), these phase factors cancel out. Actually, we could eliminate them simply by redefining the elements $\Psi_{\pm}$ and $\chi_{\pm}$. The normalization of the elements $\Psi_{\pm}$ and $\chi_{\pm}$, shown in equations (17.64) and (17.68), is convenient for our discussion on the Kirby moves contained in the next section.

Since the elementary surgery operators $\widetilde{W}(U;\pm 1)$ are known, we can now compute the expectation values of Wilson line operators for the CS theory defined in any orientable, closed connected three-manifold $\mathcal{M}$. In this section, we shall consider the surgery rules in the case in which the surgery instructions are of the form specified by the Fundamental Theorem.

According to Lickorish Theorem, $\mathcal{M}$ has a presentation in terms of Dehn surgery in S^3. Moreover, it is always possible to find a surgery description of $\mathcal{M}$ corresponding to a surgery link $\mathcal{L}$ in which all its components $\{\mathcal{L}_i\}$ (with $1 \leq i \leq p$) are simple circles and have surgery coefficients $\{r_i\}$ equal to $+1$ or -1. Since we know how to represent in the field theory the elementary surgery $S_\pm$, associated with each single component of $\mathcal{L}$, we can construct the operator $\widetilde{W}(\mathcal{L})$ representing the whole surgery.

Indeed, for each component $\mathcal{L}_i$, we shall introduce a ***preferred framing*** $\mathcal{L}_{if}$, and we shall consider the operator $\widetilde{W}(\mathcal{L}_i\,;r_i)$, as defined in equations (17.63) and (17.67). The field theory operator $\widetilde{W}(\mathcal{L})$, associated to the whole surgery described by $\mathcal{L}$, is

$$\widetilde{W}(\mathcal{L}) = \prod_{i=1}^{p} \widetilde{W}(\mathcal{L}_i\,;r_i)\,. \tag{17.70}$$

Let L be a given framed link in $\mathcal{M}$. As we have seen in Chapter 16, the isotopy class of $L \subset \mathcal{M}$ can be described by a link (that we indicate by the same symbol L) in the complement of $\mathcal{L}$ in S^3. The expectation value $\langle\, W(L)\,\rangle\big|_{\mathcal{M}}$ is simply

$$\langle\, W(L)\,\rangle\Big|_{\mathcal{M}} = \frac{\langle\, W(L)\,\widetilde{W}(\mathcal{L})\,\rangle\Big|_{S^3}}{\langle\, \widetilde{W}(\mathcal{L})\,\rangle\Big|_{S^3}}\,. \tag{17.71}$$

The physical interpretation of equation (17.71) is clear. Surgery operations on the three-manifold can be understood as transformations acting on the system described by the topological CS model. The field theory operator $\widetilde{W}(\mathcal{L})$ of equation (17.70) represents the surgery on S^3 corresponding to the instruction given by $\mathcal{L}$. Under surgery operations, the propagators and the interaction vertices of the theory are modified. All the effects, produced by this modification of the propagators and of the vertices in the Feynman diagrams, are described by an insertion of $\widetilde{W}(\mathcal{L})$. Since the contribution of the vacuum-to-vacuum diagrams also is modified, $\widetilde{W}(\mathcal{L})$ must appear in the numerator and in the denominator of the ratio shown in equation (17.71).

By means of the surgery rules (17.70) and (17.71), one can compute the expectation values $\{\,\langle\, W(L)\,\rangle\big|_{\mathcal{M}}\,\}$ in any orientable, closed connected three-manifold $\mathcal{M}$.

Quite often, the surgery instructions are not given in the form specified by the Fundamental Theorem. This is not a problem because, by means of an appropriate sequence of Rolfsen moves, we can always find an equivalent surgery instruction of the desired form. In order to simplify the whole procedure, however, in the next section we shall give the surgery rules which directly correspond to "honest" surgeries in general.

17.5 Surgery rules and Kirby moves

Any "honest" surgery is described by a surgery link $\mathcal{L}$ whose components $\{\mathcal{L}_i\}$ have integer surgery coefficients $\{r_i\}$. Each single component $\mathcal{L}_i$ is not necessarily ambient isotopic with a simple circle, of course.

For each component $\mathcal{L}_i$, we shall introduce a ***framing*** $\mathcal{L}_{if}$ such that the linking number of $\mathcal{L}_i$ and $\mathcal{L}_{if}$ is given by

$$\chi(\mathcal{L}_i, \mathcal{L}_{if}) = r_i . \tag{17.72}$$

It should be noted that the framing $\mathcal{L}_{if}$ that we have introduced here is different from the preferred framing used in Section 17.4.

Then, when $k \geq 2$, for each component $\mathcal{L}_i$ we shall consider the Wilson line operator

$$W(\mathcal{L}_i \, ; \Psi_0) , \tag{17.73}$$

which is associated to the framed component $\mathcal{L}_i$ with framing $\mathcal{L}_{if}$ satisfying equation (17.72). The element Ψ_0 is given by

$$\Psi_0 = \sqrt{\frac{2}{k}} \sin(\pi/k) \sum_{J=0}^{k/2-1} E_0[J] \, \Psi[J] . \tag{17.74}$$

When $k = 1$, we shall consider the Wilson operator

$$W(\mathcal{L}_i \, ; \chi_0) , \tag{17.75}$$

with

$$\chi_0 = \frac{1}{\sqrt{2}} \left(\chi[0] - \frac{1}{2} \chi[1/2] \right) . \tag{17.76}$$

The field-theory ***surgery rules*** corresponding to the "honest" surgery described by the surgery link $\mathcal{L}$ in the three-sphere S^3 are given by the following

Theorem R8. *The surgery operator, corresponding to the "honest" surgery described by* $\mathcal{L} = \{\mathcal{L}_i\}$ *and by the integer surgery coefficients* $\{r_i\}$ *(with* $1 \leq i \leq p$*), is given by*

$$\widetilde{W}(\mathcal{L}) = \prod_{i=1}^{p} W(\mathcal{L}_i \, ; \Psi_0), \quad \text{for} \quad k \geq 2 , \tag{17.77a}$$

and

$$\widetilde{W}(\mathcal{L}) = \prod_{i=1}^{p} W(\mathcal{L}_i \, ; \chi_0), \quad \text{for} \quad k = 1 . \tag{17.77b}$$

The elements Ψ_0 *and* χ_0 *are shown in* equations (17.74) *and* (17.76). *Consequently, the expectation value* $\langle W(L) \rangle \big|_{\mathcal{M}}$ *can be written as*

$$\langle W(L) \rangle \Big|_{\mathcal{M}} = \frac{\langle W(L) \, \widetilde{W}(\mathcal{L}) \rangle \Big|_{S^3}}{\langle \widetilde{W}(\mathcal{L}) \rangle \Big|_{S^3}} . \tag{17.78}$$

Proof. We need to prove that the surgery rules (17.77) and (17.78) are consistent. The proof contains two parts. First, we will see that

(i) when all the components $\{\mathcal{L}_i\}$ are simple circles and the surgery coefficients $\{r_i\}$ are equal to $+1$ or -1, equation (17.78) coincides with equation (17.71);

secondly, we will show that

(ii) the expectation values $\{ \langle W(L) \rangle \big|_{\mathcal{M}} \}$, which are obtained according to equations (17.77) and (17.78), are invariant under Kirby moves.

Let us consider point (i). Each operator $W(\mathcal{L}_i \; ; \Psi_0)$ is defined for the framed $\mathcal{L}_i$ with framing specified in equation (17.72). Suppose that $r_i = \pm 1$ and $\mathcal{L}_i$ is the unknot. Remember that the operators $\widetilde{W}(\mathcal{L}_i \; ; \pm 1)$ are defined for $\mathcal{L}_i$ with preferred framing. By taking into account the effects produced by a change of framing, one finds, for $k \geq 2$,

$$W(\mathcal{L}_i \; ; \Psi_0) = \widetilde{W}(\mathcal{L}_i \; ; \pm 1) , \tag{17.79}$$

and for $k = 1$

$$W(\mathcal{L}_i \; ; \chi_0) = \widetilde{W}(\mathcal{L}_i \; ; \pm 1) . \tag{17.80}$$

Therefore, when all the components $\{\mathcal{L}_i\}$ are simple circles and the surgery coefficients $\{r_i\}$ are equal to $+1$ or -1, equation (17.78) coincides with equation (17.71).

Let us now consider point (ii). Let $\mathcal{L}$ be the instruction corresponding to a given "honest" surgery. Suppose that one component of $\mathcal{L}$, say $\mathcal{L}_1$, is a simple circle with surgery coefficient $r_1 = \pm 1$. All the remaining components of $\mathcal{L}$ and the given link L belong to the complement solid torus $\mathcal{V}_1$ of $\mathcal{L}_1$ in S^3.

As we have seen above, $W(\mathcal{L}_1; \Psi_0)$ (or $W(\mathcal{L}_1; \chi_0)$) is equivalent to $\widetilde{W}(\mathcal{L}_1; \pm 1)$.

Consider now the numerator and, separately, the denominator of the expression (17.78). From equation (17.66) (or equation (17.69)), it follows that $\mathcal{L}_1$ can be eliminated provided that we perform a twist homeomorphism $\tau_\mp$ of $\mathcal{V}_1$ and, simultaneously, that we multiply by the phase factor $e^{\mp i\varphi}$ (or $e^{\pm i\pi/4}$). This phase factor cancels out in the ratio (17.78); therefore, only the effects of the twist homeomorphism $\tau_\mp$ are relevant.

Under this twist, L is accordingly modified into its homeomorphic image, as it should be. The same happens with the remaining components of $\mathcal{L}$. It remains to be verified that the new surgery coefficients $\{r'_j\}$ (with $j \neq 1$) have the correct values shown in equation (16.18). In fact, as a consequence of the transformation properties (16.13) of framings under twist homeomorphisms, this is indeed the case. In conclusion, the results obtained by means of the surgery rules (17.77) and (17.78) are invariant under Kirby moves. On the other hand, with an appropriate sequence of Kirby moves, any "honest" surgery instruction can be transformed into an equivalent surgery instruction of the type specified by the Fundamental Theorem. In this case, we know that the expressions (17.78) and (17.71) coincide and this concludes the proof of the validity of equations (17.77) and (17.78). □

With our definition of the elements $\Psi_\pm$ and $\chi_\pm$, shown in equations (17.64) and (17.68), Ψ_+ can be obtained from Ψ_- by means of two elementary changes of framing (the same happens with $\chi_\pm$). This is why we have chosen their normalization as shown in equations (17.64) and (17.68).

From the previous discussion on the Kirby moves, it is clear that the presence of the phase factors $e^{\pm i\varphi}$ or $e^{\pm i\pi/4}$ cannot be avoided. To be more precise, whatever the normalization of $\Psi_\pm$ or $\chi_\pm$ is, the numerator (and, separately, the denominator) of the expression (17.78) is ***not*** invariant under Kirby moves. (The ratio (17.78) is invariant, of course.) This means that a projective representation of the homeomorphism group is realized on the state space of the CS theory. Some consequences of this fact, connected with the construction of three-manifold invariants, will be discussed in Chapter 19.

When $\langle \widetilde{W}(\mathcal{L}) \rangle \big|_{S^3} \neq 0$ the expression (17.78) is well defined. In this case, if L and the surgery link $\mathcal{L}$ are disjoint then, according to equation (11.4),

$$\langle W(L)\, \widetilde{W}(\mathcal{L}) \rangle \Big|_{S^3} = \langle W(L) \rangle \Big|_{S^3} \langle \widetilde{W}(\mathcal{L}) \rangle \Big|_{S^3} , \tag{17.81}$$

and, therefore,

$$\langle W(L) \rangle \Big|_{\mathcal{M}} = \langle W(L) \rangle \Big|_{S^3} . \tag{17.82}$$

This conclusion is in perfect agreement with the predictions of the topological quantum field theory. Indeed, if L and $\mathcal{L}$ are disjoint, then L belongs to the interior of a three-ball in $\mathcal{M}$ and, because of general covariance, we know that equation (17.82) must hold.

The expectation value $\langle \widetilde{W}(\mathcal{L}) \rangle \big|_{S^3}$ gives information on the three-manifold $\mathcal{M}$, associated with the surgery link $\mathcal{L}$ in S^3, and depends on the value of the coupling constant k. For fixed $\mathcal{M}$, $\langle \widetilde{W}(\mathcal{L}) \rangle \big|_{S^3}$ may vanish when k takes values on a certain set $\mathcal{F}$ of integers; in this case, equation (17.78) is not well defined, of course. Let us recall that, in general, the internal consistency of the quantum CS theory defined in $\mathcal{M}$ puts some restrictions [4] on the possible values of k. It is natural to expect that $\langle \widetilde{W}(\mathcal{L}) \rangle \big|_{S^3}$ vanishes for precisely those values of k for which the quantum CS theory is not well defined in $\mathcal{M}$.

Chapter 18

Observables in three-manifolds

In this chapter, we shall consider several examples of closed, connected and orientable three-manifolds. The expectation values of the Wilson line operators associated with framed links in a generic three-manifold $\mathcal{M}$ will be determined by using the surgery rules discussed in Chapter 17 and by computing the values of the extended Jones polynomial for the relevant links.

18.1 The manifold $S^2 \times S^1$

The manifold $S^2 \times S^1$ has a surgery presentation described by the unknot U in S^3 with surgery coefficient $r = 0$.

The fundamental group of $S^2 \times S^1$ can easily be derived from the surgery link by means of the Wirtinger construction illustrated in Chapter 16. The unknot U and the curve Y, which is the image $h(\mu)$ of the meridian of a tubular neighbourhood of U, are represented in Fig. 18.1.

Figure 18.1.

The generator g of the fundamental group of the complement of U in S^3 is also shown; X is a base point. Since the class of Y is already trivial, there is no new constraint to be introduced. Therefore, $\pi_1(S^2 \times S^1)$ is the free group of rank 1 , $\pi_1(S^2 \times S^1) \cong Z$.

Framed links in $S^2 \times S^1$ are described by framed links in the complement of the surgery link (which in this case is the unknot U with surgery coefficient $r = 0$) in S^3. An example of knot in $S^2 \times S^1$ is shown in Fig. 18.2.

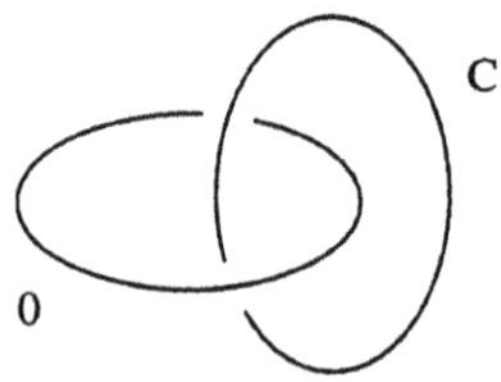

Figure 18.2.

With the introduction of a suitable orientation, the knot C shown in Fig. 18.2 is homotopic with the generator g of $\pi_1(S^2 \times S^1)$. An alternative description of the knot C can be obtained in the following way. As shown in Fig. 18.3, the torus T^2 can be represented by an annulus with the identification of the points (of the internal and external circles) which have the same angular coordinates.

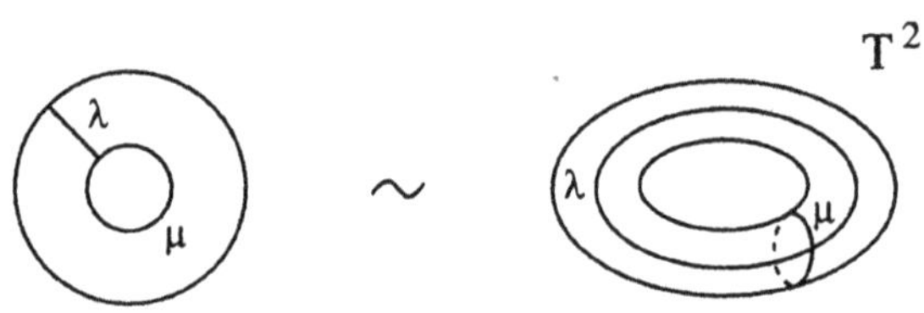

Figure 18.3.

A solid torus V can be represented then by the region of $\mathbb{R}^3$ delimited by an annulus and by two half-spheres, with different radii, as shown in Fig. 18.4. The points which belong to the two half-sphere surfaces and which have the same angular coordinates are identified.

According to the surgery instruction described by the unknot U with surgery coefficient $r = 0$, $S^2 \times S^1$ is equivalent to the union of two solid tori with their boundaries identified in such a way that the two meridians coincide. This means that $S^2 \times S^1$ is homeomorphic with the region of $\mathbb{R}^3$ delimited by two spherical surfaces Σ_1 and Σ_2. The points on Σ_1 and Σ_2 are identified when they have

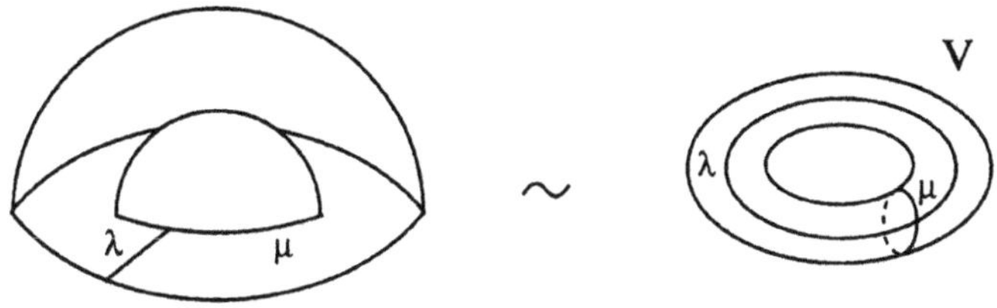

Figure 18.4.

the same angular coordinates. Therefore, an equivalent description of the knot C, shown in Fig. 18.2, is given in Fig. 18.5.

For $k \geq 2$, the surgery operator is

$$W(U;\Psi_0) = \sqrt{\frac{2}{k}}\,\sin(\pi/k)\sum_{J=0}^{k/2-1} E_0[J]\,W(U;\Psi[J])\,, \tag{18.1}$$

where the Wilson line operators are defined for the unknot U with preferred framing. The sum of the expectation values

$$\sum_{I=0}^{k/2-1} E_0[I]\,\langle\, W(U;\Psi[I])\,\rangle\Big|_{S^3} \tag{18.2}$$

is given by the expression (17.6) when $J=0$. Therefore, from equation (18.1) it follows that

$$\langle\, W(U;\Psi_0)\,\rangle\Big|_{S^3} = \sqrt{\frac{k}{2}}\,\frac{1}{\sin(\pi/k)}\,. \tag{18.3}$$

Let us consider the knot C in $S^2 \times S^1$, shown in Fig. 18.2 or, equivalently, in Fig. 18.5. Let C be labelled by $\Psi[J]$ and have preferred framing. From equation (17.6), it follows that

$$\langle\, W(C;\Psi[J])W(U;\Psi_0)\,\rangle\Big|_{S^3} = \sqrt{\frac{k}{2}}\,\frac{1}{\sin(\pi/k)}\,\delta_{J0}\,. \tag{18.4}$$

Therefore

$$\langle\, W(C;\Psi[J])\,\rangle\Big|_{S^2\times S^1} = \delta_{J0}\,. \tag{18.5}$$

The result (18.5) has a simple physical interpretation. In a canonical approach to the CS theory in $S^2 \times S^1$, S^1 can be interpreted as a (compactified) time interval. At any fixed time, C intersects S^2 in a single point, see Fig. 18.5. The "electric" flux, originated by this puncture on S^2, cannot be conserved unless the puncture

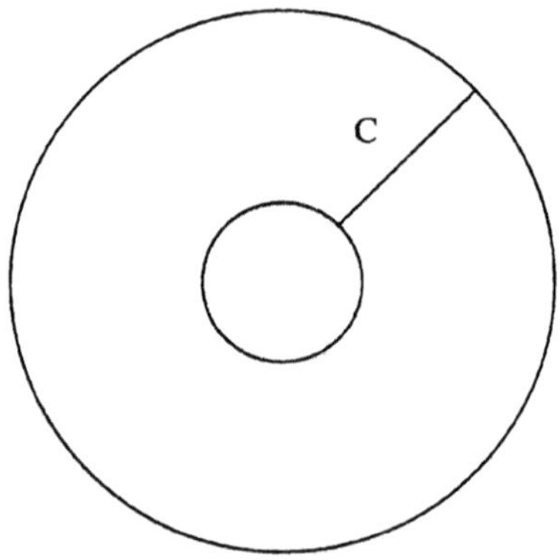

Figure 18.5.

has vanishing non-abelian charge (i.e., the element $\Psi[0]$ in $\mathcal{T}_{(k)}$). But in the case in which $\Psi[J] = \Psi[0]$, the presence of the puncture on S^2 is completely illusory (in other words, the puncture can be removed) and the associated state vector coincides with the vacuum; so, equation (18.5) is obtained. Clearly, the result (18.5) is invariant under an arbitrary modification of the framing of C.

The two components C_1 and C_2 of the link shown in Fig. 18.6 have preferred framings.

Figure 18.6.

By using the satellite formulae and equation (18.5), one obtains

$$\langle\, W(C_1; \Psi[I\,])\, W(C_2; \Psi[J])\,\rangle\Big|_{S^2\times S^1} = \delta_{IJ}\,. \tag{18.6}$$

Similarly, for the three-component link of Fig. 18.7 one gets

$$\langle\, W(C_1; \Psi[I\,])\, W(C_2; \Psi[J])\, W(C_3; \Psi[M])\,\rangle\Big|_{S^2\times S^1} = N_{IJM}\,, \tag{18.7}$$

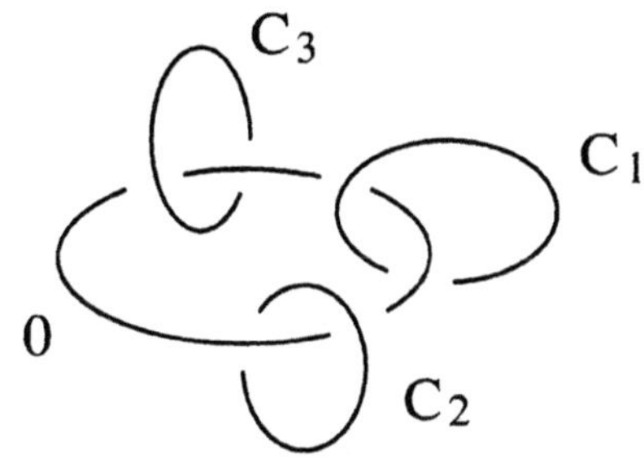

Figure 18.7.

Figure 18.8.

where

$$N_{IJM} = \sum_{N=|I-J|}^{b(I,J)} \delta_{NM} \tag{18.8}$$

vanishes if $\Psi[M]$ is not contained in the decomposition of $\Psi[I]\Psi[J]$; otherwise, it is equal to unity.

With the vertical framing convention, framed links in $S^2 \times S^1$ are represented by the regular isotopy classes of link diagrams in the complement of the surgery link in S^3. Consider the framed knot C shown in Fig. 18.8; the associated expectation value is

$$\langle W(C; \Psi[J]) \rangle \Big|_{S^2 \times S^1} = q^{-Q(J)} (-1)^{2J} . \tag{18.9}$$

For the framed link shown in Fig. 18.9, one obtains

$$\langle W(C_1; \Psi[I]) \, W(C_2; \Psi[J]) \rangle \Big|_{S^2 \times S^1} = E_0[J] \, \delta_{I0} . \tag{18.10}$$

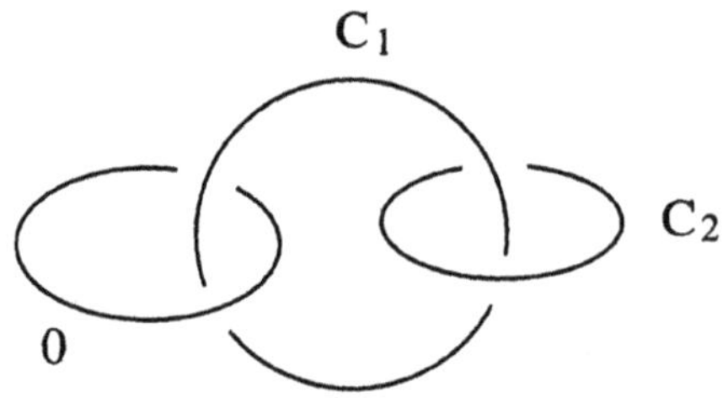

Figure 18.9.

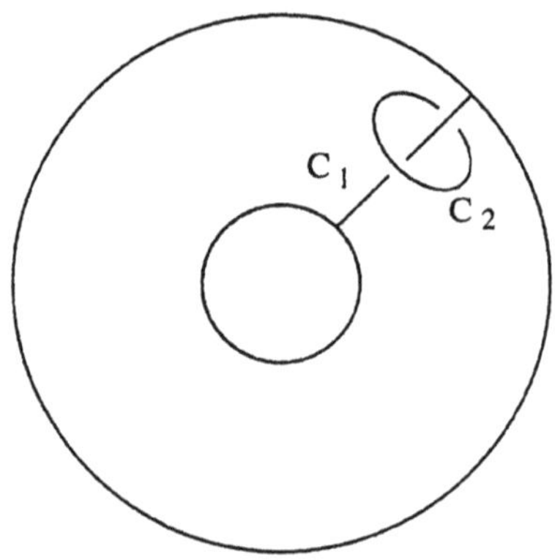

Figure 18.10.

Equation (18.10) is in perfect agreement with the ambient isotopy invariance in $S^2 \times S^1$. Indeed, the link shown in Fig. 18.9 can equivalently be described by the link shown in Fig. 18.10 which, in turn, is ambient isotopic with the link shown in Fig. 18.11. Since the component C_2 belongs to the interior of a three-ball in $S^2 \times S^1$, the result (18.10) is consistent with equation (18.5).

Let us now consider the framed link shown in Fig. 18.12. This link can also be represented in $S^2 \times S^1$ as shown in Fig. 18.13.

Let the components C_1 and C_2 of the link of Fig. 18.12 be labelled by $\Psi[I]$ and $\Psi[J]$ respectively. The surgery rules give

$$\langle W(C_1; \Psi[I])\, W(C_2; \Psi[J]) \rangle \Big|_{S^2 \times S^1} = q^{-2Q(J)}\, \delta_{IJ} \,. \tag{18.11}$$

Equation (18.11) confirms the validity of the surgery rules derived in Chapter 17. Indeed, as shown in Fig. 18.14, the framed link of Fig. 18.13 is ambient isotopic with the framed link shown in Fig. 18.14(c). It should be noted that, instead of drawing a band for each component, in Fig. 18.14 we have used the regular isotopy

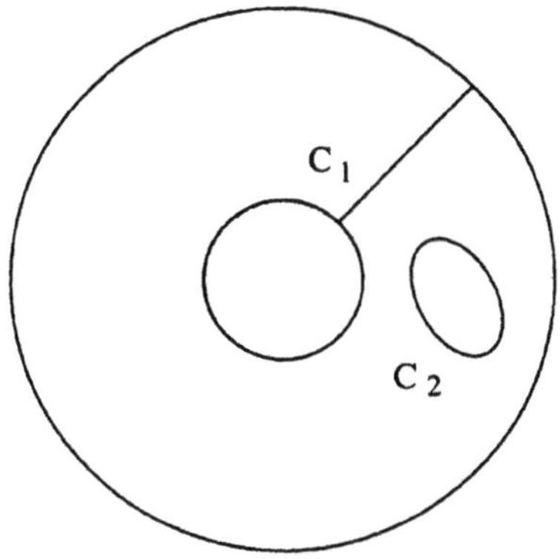

Figure 18.11.

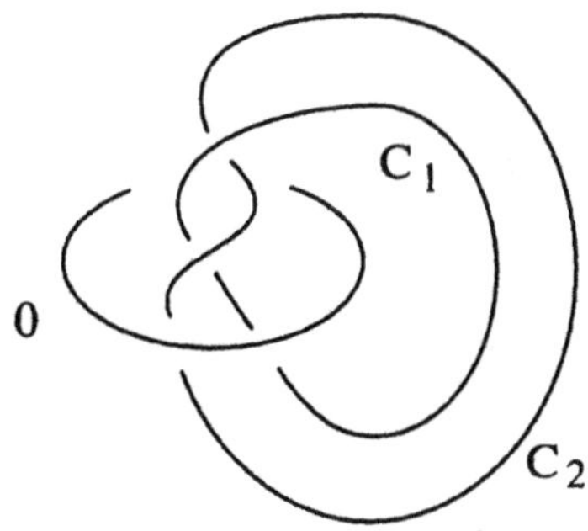

Figure 18.12.

convention for link diagrams (vertical framing). Thus, by taking into account the effects of a modification of the framing, equation (11.3), one finds that the result (18.11) is perfectly consistent with equation (18.6).

Finally, for the link in $S^2 \times S^1$ shown in Fig. 18.15, one obtains

$$\langle\, W(C_1;\, \Psi[J_1])\, W(C_2;\, \Psi[J_2])\, W(C_3;\, \Psi[J_3])\, W(C_4;\, \Psi[I\,])\,\rangle\Big|_{S^2\times S^1} =$$

$$= N_{J_2 J_3 I}\, q^{Q(I)-Q(J_2)-Q(J_3)}\, \frac{q^{(2J_1+1)(2I+1)/2} - q^{-(2J_1+1)(2I+1)/2}}{q^{(2I+1)/2} - q^{-(2I+1)/2}}\,. \qquad (18.12)$$

When $k = 1$, the surgery operator is

$$W(U;\chi_0) = \frac{1}{\sqrt{2}}\, W(U;\chi[0]) - \frac{1}{2\sqrt{2}}\, W(U;\chi[1/2])\,, \qquad (18.13)$$

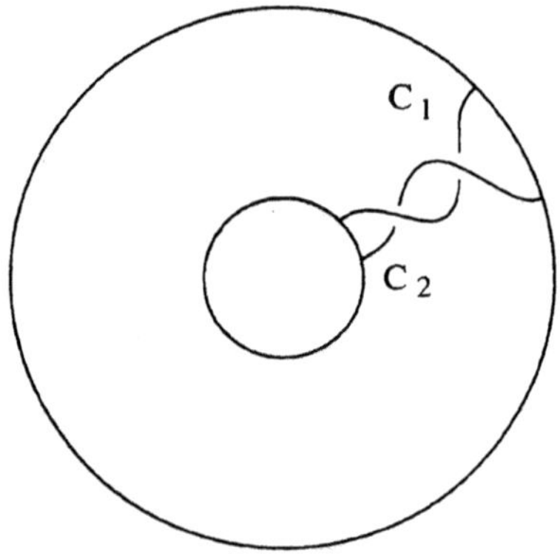

Figure 18.13.

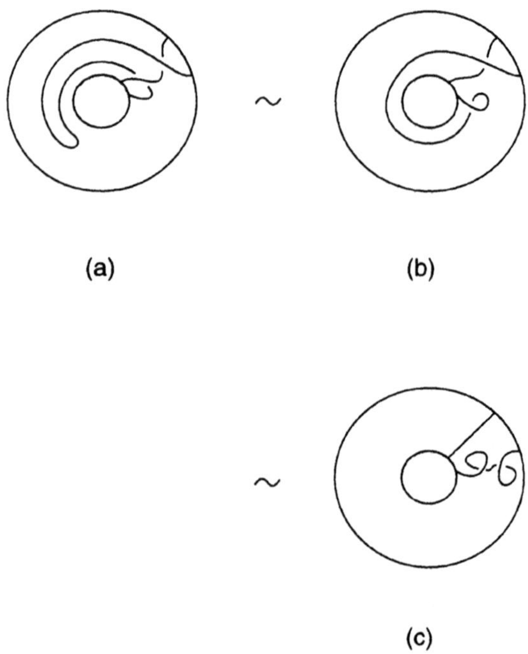

Figure 18.14.

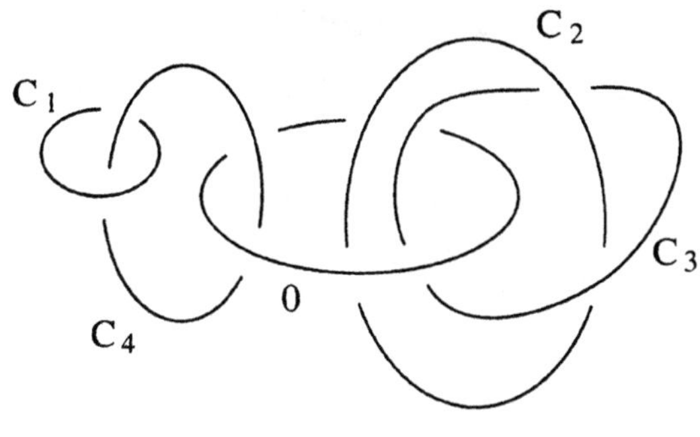

Figure 18.15.

where the Wilson line operators are defined for U with preferred framing. From the general expression (15.65) of the expectation values for $k = 1$, it follows that

$$\langle W(U; \chi_0) \rangle \Big|_{S^3} = \sqrt{2}. \tag{18.14}$$

By means of equation (15.65), the computation of the expectation values in $S^2 \times S^1$ is straightforward. For example, for the knot C shown in Fig. 18.2, one finds

$$\langle W(C; \chi[J]) \rangle \Big|_{S^2 \times S^1} = \frac{(2J+1)}{2} \left[1 + (-1)^{2J} \right]. \tag{18.15}$$

For the two-component link shown in Fig. 18.6, one has

$$\begin{aligned} \langle W(C_1; \chi[I])\, W(C_2; \chi[J]) \rangle \Big|_{S^2 \times S^1} &= \\ &= \frac{(2I+1)(2J+1)}{2} \left[1 + (-1)^{2(I+J)} \right]. \end{aligned} \tag{18.16}$$

Let us now consider the framed knot C, shown in Fig. 18.8, which has writhe equal to $+1$; we obtain

$$\langle W(C; \chi[J]) \rangle \Big|_{S^2 \times S^1} = \begin{cases} (2J+1) & \text{if } \; 2J = \text{ even;} \\ -i\,(2J+1) & \text{if } \; 2J = \text{ odd.} \end{cases} \tag{18.17}$$

18.2 The manifold RP^3

The lens space $L(2,1) \cong RP^3$ can be obtained from the sphere S^3 by identifying its antipodal points. RP^3 admits a surgery presentation [12] whose instruction is given by the unknot U in S^3 with surgery coefficient $r = +2$.

As shown in Fig. 18.16, the fundamental group of RP^3 can be described in terms of a generator g with the condition

$$[Y] = g^2 = 1 . \tag{18.18}$$

Therefore, $\pi_1(RP^3) \cong Z_2$.

Figure 18.16.

For $k \geq 2$, the surgery operator is

$$\widetilde{W}(\mathcal{L}) = W(U; \Psi_0) , \tag{18.19}$$

where the Wilson line operator $W(U; \Psi_0)$ is defined for U with framing U_f such that

$$\chi(U, U_f) = +2 . \tag{18.20}$$

Let us consider the denominator appearing in equation (17.78). We have

$$\langle \widetilde{W}(\mathcal{L}) \rangle \Big|_{S^3} = \sqrt{\frac{2}{k}} \sin(\pi/k)\, Z_{(2)} , \tag{18.21}$$

where $Z_{(2)}$ denotes the sum

$$Z_{(2)} = \sum_{J=0}^{k/2-1} q^{2Q(J)} E_0^2[J] . \tag{18.22}$$

By using equations (15.52) and (15.53), one can write

$$Z_{(2)} = (1/2) \sum_{J=0}^{k-1} q^{2Q(J)} E_0^2[J] . \tag{18.23}$$

By inserting the values of the unknots in this expression, one finds

$$\begin{aligned} Z_{(2)} = (1/2) \frac{q^{-1/2}}{(q^{1/2}-q^{-1/2})^2} \sum_{x=1}^{2k-1} \Big\{ & q^{-1/2} q^{(x+1)^2/2} + \\ & + q^{-1/2} q^{(x-1)^2/2} - 2 q^{x^2/2} \Big\} , \end{aligned} \tag{18.24}$$

where $x = 2J + 1$. Equation (18.24) can be written as

$$Z_{(2)} = - \frac{q^{-3/4}(q^{1/4} - q^{-1/4})}{(q^{1/2}-q^{-1/2})^2} \sum_{x=0}^{2k-1} q^{x^2/2} . \tag{18.25}$$

From equation (17.32) it follows that

$$\sum_{x=0}^{2k-1} q^{x^2/2} = \begin{cases} (1-i)\sqrt{2k} & \text{if } k = \text{even}; \\ 0 & \text{if } k = \text{odd}. \end{cases} \tag{18.26}$$

Therefore,

$$Z_{(2)} = \begin{cases} \sqrt{k} \, \frac{\sin(\pi/2k)}{\sin^2(\pi/k)} \exp\left(-i \frac{\pi}{4} \frac{3(k-2)}{k}\right) & \text{if } k = \text{even}; \\ 0 & \text{if } k = \text{odd}. \end{cases} \tag{18.27}$$

From equation (18.21) it follows that

$$\langle \widetilde{W}(\mathcal{L}) \rangle \Big|_{S^3} = \begin{cases} \left[\sqrt{2} \cos(\pi/2k)\right]^{-1} e^{-i(\pi/4)\, 3(k-2)/k} & \text{if } k = \text{even}; \\ 0 & \text{if } k = \text{odd}. \end{cases} \tag{18.28}$$

When $k = 1$, the surgery operator is

$$\widetilde{W}(\mathcal{L}) = W(U; \chi_0) , \tag{18.29}$$

where the Wilson line operators are defined for U with framing U_f satisfying equation (18.20). Equation (15.65) implies that, for $k = 1$,

$$\langle \widetilde{W}(\mathcal{L}) \rangle \Big|_{S^3} = 0 . \tag{18.30}$$

Equations (18.28) and (18.30) show that $\langle \widetilde{W}(\mathcal{L}) \rangle \Big|_{S^3}$ is different from zero if and only if k is even. Consequently, for the manifold RP^3, only for even values of k is the surgery rule (17.78) meaningful. This seems to suggest that k must be even for the CS field theory in RP^3 to be consistent. In order to clarify this point, we need to analyse the so-called large gauge transformations, i.e. gauge transformations which are not connected with the identity.

Let us reconsider the CS theory in S^3. As we have already mentioned in Chapter 4, under a gauge transformation

$$A_\mu(x) \;\rightarrow\; A^\Omega_\mu(x) \;=\; \Omega^{-1}(x)\, A_\mu(x)\, \Omega(x) \;-\; i\, \Omega^{-1}(x)\, \partial_\mu \Omega(x)\,, \tag{18.31}$$

the CS action transforms as

$$S_{CS}[A^\Omega] \;=\; S_{CS}[A] \;+\; 2\pi\, k\, \Gamma_t[\Omega]\,, \tag{18.32}$$

where the functional Γ_t is shown in equation (4.7). The integer $\Gamma_t[\Omega]$ coincides with the index of the class of $\Omega(x)$. Let us represent S^3 by $\mathbb{R}^3 + \{\infty\}$. For $G = SU(2)$, the elementary nontrivial map $\Omega^{(1)} : S^3 \rightarrow SU(2)$, given by

$$\Omega^{(1)}(\vec{x}\,) \;=\; \exp\left(i\, \pi\, \hat{x} \cdot \vec{\sigma}\, \frac{|\vec{x}\,|}{|\vec{x}\,| + r} \right), \tag{18.33}$$

satisfies $\Gamma_t[\Omega^{(1)}] = 1$. As we have already said, S^3 is the universal covering space of RP^3. Let the points of S^3 have coordinates $(z) = (z_1, z_2, z_3, z_4)$ with $\sum_i z_i^2 = 1$. The homeomorphism $\tau : S^3 \rightarrow S^3$, given by $\tau(z) = (-z)$, is the generator of the cyclic group of covering translations. The associated orbit space is homeomorphic with RP^3. Consider now a generic map $\omega : RP^3 \rightarrow SU(2)$ corresponding to a gauge transformation; we are interested in computing $\Gamma_t[\omega]$. Since the group of the homotopy classes of the maps $\{\omega\}$ is isomorphic with Z, it is sufficient to consider a generator $\omega^{(1)}$.

When S^3 is represented by $\mathbb{R}^3 + \{\infty\}$, the action of τ can be taken to be $\tau(\vec{x}\,) = -\vec{x} \cdot r^2/|\vec{x}\,|^2$ and, under this transformation, the matrix $\Omega^{(1)}(\vec{x}\,)$ goes into $-\Omega^{(1)}(\vec{x}\,)$. Since the integrand of the expression (4.7) is invariant under $\Omega \rightarrow -\Omega$ and since the gauge transformations (18.31) also have this property, we have

$$\Gamma_t[\omega^{(1)}] \;=\; \tfrac{1}{2}\, \Gamma_t[\Omega^{(1)}]\,. \tag{18.34}$$

By combining equation (18.34) with equation (18.32), we can conclude that gauge invariance of the CS theory in RP^3 implies that k must be an even integer. This is precisely the same conclusion obtained from our surgery construction.

The expectation values of the Wilson line operators in RP^3 can be computed by means of the surgery rules of Section 17.5. Let us consider, for example, the knot C in RP^3 shown in Fig. 18.17.

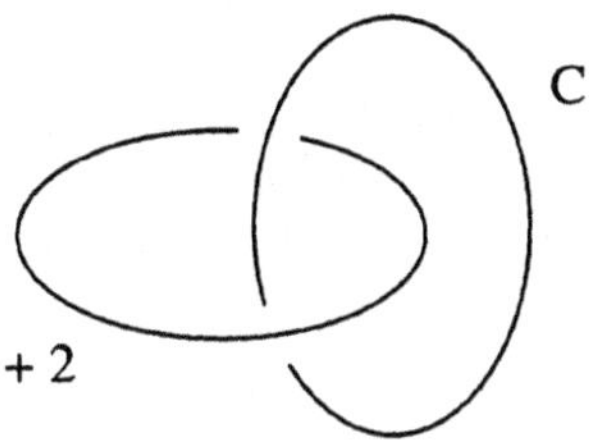

Figure 18.17.

When the knot C has preferred framing and is labelled by $\Psi[J]$, one has

$$\langle W(C; \Psi[J])\, \widetilde{W}(\mathcal{L}) \rangle\Big|_{S^3} = \sqrt{\frac{2}{k}}\, \sin(\pi/k) \sum_{I=0}^{k/2-1} q^{2Q(I)}\, E_0[I]\, H_{IJ}\,, \qquad (18.35)$$

where H_{IJ} is the Hopf matrix (17.2). Note that the general dependence of the vacuum expectation values on the variable k is not always easy to produce. Therefore, we shall now give the numerical values of $\langle W(C; \Psi[J]) \rangle\big|_{RP^3}$ when the integer k has been fixed. Let us consider, for instance, the values $k = 2, 4, 6$.

When $k = 2$, the surgery operator is the identity and, in any three-manifold $\mathcal{M}$, the expectation values are given in equation (15.59).

For $k = 4$, we obtain

$$\langle W(C; \Psi[1/2]) \rangle\Big|_{RP^3} = 0\,, \qquad (18.36)$$

$$\langle W(C; \Psi[1]) \rangle\Big|_{RP^3} = i\, \text{tang}\,(3\pi/8)\,. \qquad (18.37)$$

Finally, for $k = 6$ we find

$$\langle W(C; \Psi[1/2]) \rangle\Big|_{RP^3} = 0 = \langle W(C; \Psi[3/2]) \rangle\Big|_{RP^3}\,, \qquad (18.38)$$

$$\langle\, W(C;\, \Psi[1])\,\rangle\Big|_{RP^3} \;=\; \frac{1+i\sqrt{3}}{\sqrt{3}-1}\,, \tag{18.39}$$

$$\langle\, W(C;\, \Psi[2])\,\rangle\Big|_{RP^3} \;=\; -\,(2+\sqrt{3})\,. \tag{18.40}$$

18.3 Lens spaces

The lens space $L(p,1)$ (with integer p) corresponds to the surgery described [12] by the unknot U with surgery coefficient $r = p$. The fundamental group of $L(p,1)$ is $\pi_1(L(p,1)) = Z_p$. Any link in $L(p,1)$ can be described (in the surgery presentation) by a link which belongs to the complement solid torus of U in S^3. This being the case, the computation of any expectation value can be simplified by writing $W(L)$ in the form shown in equation (13.18), where the circle C is shown in Fig. 18.18. At this point, it is sufficient to compute $\langle\, W(C;\, \chi[I\,])\,\rangle\Big|_{L(p,1)}$.

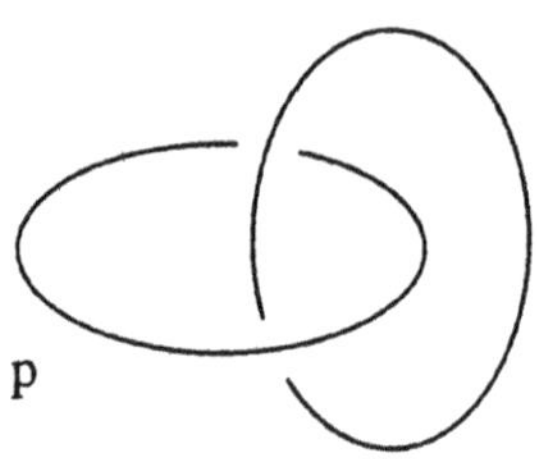

Figure 18.18.

Let us compute the expectation values of the observables when the integer k is fixed. We shall consider the values $1 \le k \le 6$. We have seen in Section 17.3 that, when $k = 2$, the surgery operator is the identity; therefore, this trivial case will not be considered.

When $k = 1$, the surgery operator is

$$\widetilde{W}(\mathcal{L}) \;=\; \frac{1}{\sqrt{2}}\, W(U;\, \chi[0]) - \frac{1}{2\sqrt{2}}\, W(U;\, \chi[1/2])\,, \tag{18.41}$$

and is defined for U with framing U_f such that $\chi(U, U_f) = p$. One has

$$\langle\, \widetilde{W}(\mathcal{L})\,\rangle\Big|_{S^3} \;=\; \frac{1}{\sqrt{2}}\,\big(1 + (i)^p\big)\;; \tag{18.42}$$

therefore, the value $k = 1$ is excluded for $p \equiv 2 \pmod 4$. In all the remaining cases, the expectation values are

$$\langle W(C; \chi[I]) \rangle \Big|_{L(p,1)} = 2I + 1 \qquad \text{if } I = \text{integer}, \tag{18.43}$$

and, for non-integer I,

$$\langle W(C; \chi[I]) \rangle \Big|_{L(p,1)} = \begin{cases} 0 & \text{if } p \equiv 0 \pmod 4; \\ i\,(2I+1) & \text{if } p \equiv 1 \pmod 4; \\ -i\,(2I+1) & \text{if } p \equiv 3 \pmod 4. \end{cases} \tag{18.44}$$

For $k \geq 2$, the surgery operator is

$$\widetilde{W}(\mathcal{L}) = \sqrt{\frac{2}{k}} \sin(\pi/k) \sum_{J=0}^{k/2-1} E_0[J]\, W(U; \Psi[J]), \tag{18.45}$$

and is defined for U with framing U_f such that $\chi(U, U_f) = p$. Clearly,

$$\langle \widetilde{W}(\mathcal{L}) \rangle \Big|_{S^3} = \sqrt{\frac{2}{k}} \sin(\pi/k) \sum_{J=0}^{k/2-1} q^{\,pQ(J)}\, E_0^2[J]. \tag{18.46}$$

In general, this expression is periodic in p with period $p \to p + 4k$. When $k = 3$, one finds

$$\langle \widetilde{W}(\mathcal{L}) \rangle \Big|_{S^3} = \frac{1}{\sqrt{2}} \left(1 + (-i)^p\right); \tag{18.47}$$

therefore, $k = 3$ must be excluded for $p \equiv 2 \pmod 4$. In the remaining cases,

$$\langle W(C; \Psi[1/2]) \rangle \Big|_{L(p,1)} = \frac{1 - (-i)^p}{1 + (-i)^p}. \tag{18.48}$$

When $k = 4$, the expression (18.46) takes the value

$$\langle \widetilde{W}(\mathcal{L}) \rangle \Big|_{S^3} = \xi(4) = \frac{1 + (-i)^p}{2} + e^{-i3\pi p/8}; \tag{18.49}$$

therefore, $k = 4$ is excluded for $p \equiv 8 \pmod{16}$. In the remaining cases,

$$\langle W(C; \Psi[1/2]) \rangle \Big|_{L(p,1)} = \xi^{-1}(4)\, \frac{1 - (-1)^p}{\sqrt{2}}, \tag{18.50}$$

$$\langle W(C; \Psi[1]) \rangle \Big|_{L(p,1)} = \xi^{-1}(4)\, \tfrac{1}{2} \left(1 + (-1)^p - 2e^{-i3\pi p/8}\right). \tag{18.51}$$

When $k=5$, one obtains

$$\langle \widetilde{W}(\mathcal{L}) \rangle\Big|_{S^3} = \xi(5) = \sqrt{\frac{2}{5}}\,\sin(\pi/5) \times \\ \left[1+(i)^p + 4e^{-i3\pi p/10}\,(1+(-i)^p)\,\cos^2(\pi/5)\right]. \tag{18.52}$$

For those values of p such that $\xi(5) \neq 0$, one has

$$\langle W(C;\Psi[1/2]) \rangle\Big|_{L(p,1)} = \xi^{-1}(5)\sqrt{\frac{2}{5}}\,\sin(2\pi/5) \times \\ \left[(1-(i)^p) + e^{-i3\pi p/10}\,(1-(-i)^p)\right], \tag{18.53}$$

$$\langle W(C;\Psi[1]) \rangle\Big|_{L(p,1)} = \xi^{-1}(5)\sqrt{\frac{2}{5}}\,\sin(2\pi/5) \times \\ \left[(1+(i)^p) - e^{-i3\pi p/10}\,(1+(-i)^p)\right], \tag{18.54}$$

$$\langle W(C;\Psi[3/2]) \rangle\Big|_{L(p,1)} = \xi^{-1}(5)\sqrt{\frac{2}{5}}\,\sin(\pi/5) \times \\ \left[1-(i)^p - 4e^{-i3\pi p/10}\,(1-(-i)^p)\,\cos^2(\pi/5)\right]. \tag{18.55}$$

Finally, let us consider $k=6$. In this case,

$$\langle \widetilde{W}(\mathcal{L}) \rangle\Big|_{S^3} = \xi(6) = \frac{1}{2\sqrt{3}}\Big[2 + 4e^{-i2\pi p/3} + \\ + 3(e^{-i\pi p/4} + e^{i3\pi p/4})\Big]; \tag{18.56}$$

and, for those values of p such that $\xi(6) \neq 0$,

$$\langle W(C;\Psi[1/2]) \rangle\Big|_{L(p,1)} = \xi^{-1}(6)\,\frac{\sqrt{3}}{2}\left[e^{-i\pi p/4} - e^{i3\pi p/4}\right], \tag{18.57}$$

$$\langle W(C;\Psi[1]) \rangle\Big|_{L(p,1)} = \xi^{-1}(6)\,\frac{2}{\sqrt{3}}\left[1 - e^{-i2\pi p/3}\right], \tag{18.58}$$

$$\langle W(C;\Psi[3/2]) \rangle\Big|_{L(p,1)} = \xi^{-1}(6)\,\frac{\sqrt{3}}{2}\left[e^{i3\pi p/4} - e^{-i\pi p/4}\right], \tag{18.59}$$

$$\langle W(C;\Psi[2]) \rangle\Big|_{L(p,1)} = \xi^{-1}(6)\,\frac{1}{2\sqrt{3}}\Big[2 + 4e^{-i2\pi p/3} - \\ - 3(e^{-i\pi p/4} + e^{i3\pi p/4})\Big]. \tag{18.60}$$

A generic lens space of type (p, t), where p and t are coprime integers, is denoted by $L(p, t)$. Provided an "honest" surgery presentation of $L(p, t)$ is given, the computation of the Wilson line operators in $L(p, t)$ is straightforward.

A surgery instruction corresponding to $L(p, t)$ is given [12] by the unknot with surgery coefficient $r = p/t$. An equivalent "honest" surgery can be obtained by using the following method [12]. Consider the link of Fig. 18.19(a). Two components of this link represent surgery instructions with integer surgery coefficients a_1 and a_2, whereas the remaining component C is a framed knot. By means of a finite sequence of Rolfsen moves, Fig. 18.19(a) can be transformed into Fig. 18.19(b) in which the surgery coefficient r is given by

$$r = a_2 - \frac{1}{a_1}. \tag{18.61}$$

A recursive use of this procedure provides a continued fraction decomposition of the rational p/t. For example, the manifold $L(18, 5)$ admits the "honest" surgery presentation shown in Fig. 18.20.

Figure 18.19.

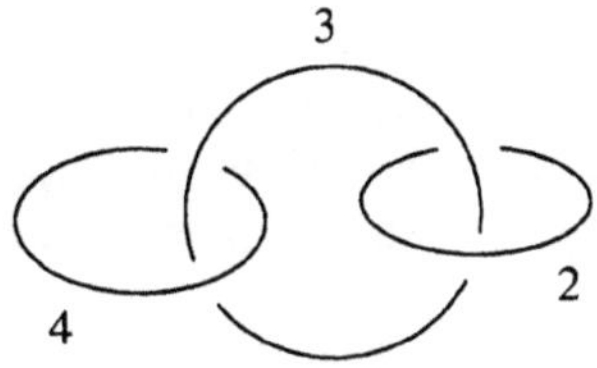

Figure 18.20.

18.4 The Poincaré manifold

The Poincaré manifold $\mathcal{P}$ corresponds [12] to the surgery described by the right-handed trefoil T with surgery coefficient $r = +1$. The fundamental group of $\mathcal{P}$ is isomorphic with the binary icosahedral group and has a presentation [12]

$$\pi_1(\mathcal{P}) \;=\; \{\, g_1, g_2\, ;\; (g_1 g_2)^2 = g_1^3 = g_2^5 \,\}, \tag{18.62}$$

where the generators g_1 and g_2 are shown in Fig. 18.21.

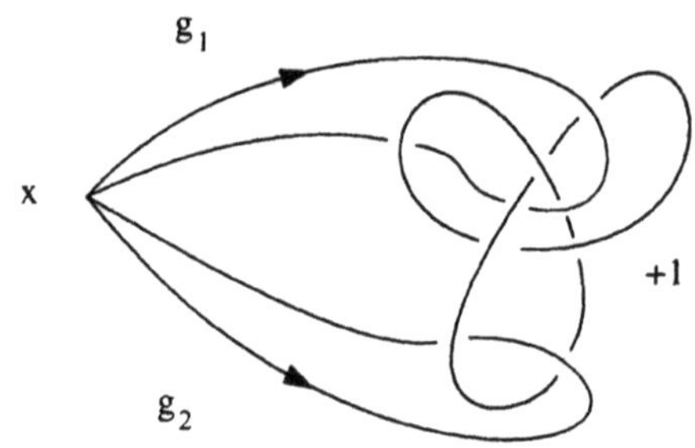

Figure 18.21.

For $k = 1$, the surgery operator is

$$\widetilde{W}(\mathcal{L}) \;=\; \frac{1}{\sqrt{2}}\, W(T;\chi[0]) - \frac{1}{2\sqrt{2}}\, W(T;\chi[1/2])\,, \tag{18.63}$$

and is defined for T with framing T_f such that $\chi(T, T_f) = +1$. One finds

$$\langle\, \widetilde{W}(\mathcal{L})\,\rangle\Big|_{S^3} \;=\; e^{i\pi/4}\,. \tag{18.64}$$

Let us consider the knots C_1 and C_2 in $\mathcal{P}$ which are related to the generators of $\pi_1(\mathcal{P})$ as shown in Fig. 18.22 and Fig. 18.23. Both C_1 and C_2 are homotopically nontrivial in $\mathcal{P}$; moreover, C_1 is not homotopically equivalent to C_2.
The expectation values associated with the knots C_1 and C_2 (both with preferred framings), are

$$\langle\, W(C_1;\chi[I\,])\,\rangle\Big|_{\mathcal{P}} \;=\; (-1)^{2I}\,(2I+1)\,, \tag{18.65}$$

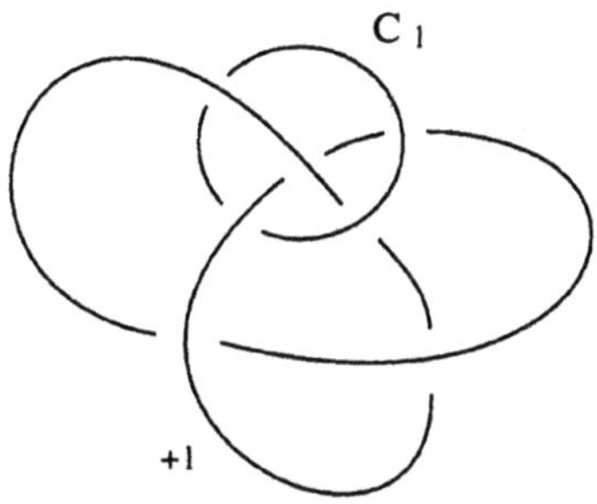

Figure 18.22.

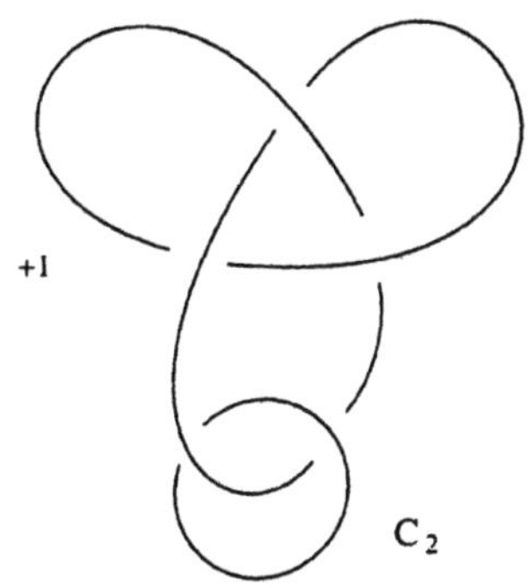

Figure 18.23.

and

$$\langle W(C_2; \chi[I])\rangle\Big|_{\mathcal{P}} = (-1)^{2I}\,(2I+1)\,\exp\left[(i\pi/4)\left((-1)^{2I}-1\right)\right]. \qquad (18.66)$$

For $k \geq 2$, the surgery operator is

$$\widetilde{W}(\mathcal{L}) = \sqrt{\frac{2}{k}}\,\sin(\pi/k) \sum_{J=0}^{k/2-1} E_0[J]\, W(T;\, \Psi[J])\,, \tag{18.67}$$

with framing T_f such that $\chi(T, T_f) = +1$.

When $k = 3$, one has

$$\langle\, \widetilde{W}(\mathcal{L})\,\rangle\Big|_{S^3} = e^{-i\pi/4}\,, \tag{18.68}$$

and

$$\langle\, W(C_1;\, \Psi[1/2])\,\rangle\Big|_{\mathcal{P}} = 1\,, \tag{18.69}$$

$$\langle\, W(C_2;\, \Psi[1/2])\,\rangle\Big|_{\mathcal{P}} = i\,. \tag{18.70}$$

When $k = 4$, one obtains

$$\langle\, \widetilde{W}(\mathcal{L})\,\rangle\Big|_{S^3} = -e^{-i3\pi/8}\,, \tag{18.71}$$

$$\langle\, W(C_1;\, \Psi[I\,])\,\rangle\Big|_{\mathcal{P}} = \begin{cases} i\sqrt{2} & \text{if } I = 1/2; \\ 1 & \text{if } I = 1; \end{cases} \tag{18.72}$$

$$\langle\, W(C_2;\, \Psi[I\,])\,\rangle\Big|_{\mathcal{P}} = \begin{cases} -\sqrt{2}e^{i3\pi/8} & \text{if } I = 1/2; \\ -1 & \text{if } I = 1. \end{cases} \tag{18.73}$$

When $k = 5$, one finds

$$\langle\, \widetilde{W}(\mathcal{L})\,\rangle\Big|_{S^3} = \sqrt{\frac{2}{5}}\,\sin(\pi/5)\,(1+i)\,d\,, \tag{18.74}$$

where

$$d = 1 + 2\cos(\pi/5) + 4e^{-i\pi/5}\cos^2(\pi/5)\,. \tag{18.75}$$

Moreover,

$$\begin{aligned}\langle\, W(C_1;\, \Psi[1/2])\,\rangle\Big|_{\mathcal{P}} &= \langle\, W(C_1;\, \Psi[1])\,\rangle\Big|_{\mathcal{P}} = \\ &= d^{-1} 4\cos(\pi/5)\left[\cos(\pi/5) + e^{-i\pi/10}\cos(\pi/10)\right],\end{aligned} \tag{18.76}$$

$$\langle\, W(C_1;\, \Psi[3/2])\,\rangle\Big|_{\mathcal{P}} = 1\,, \tag{18.77}$$

$$\langle W(C_2; \Psi[1/2]) \rangle \Big|_{\mathcal{P}} = -i \, \langle W(C_2; \Psi[1]) \rangle \Big|_{\mathcal{P}} = \\ = i\, d^{-1} \left[1 - 4e^{i2\pi/5} \cos(\pi/10) \, \sin(\pi/10) \right] , \tag{18.78}$$

$$\langle W(C_2; \Psi[3/2]) \rangle \Big|_{\mathcal{P}} = -i \, . \tag{18.79}$$

Finally, for $k = 6$, one gets

$$\langle \widetilde{W}(\mathcal{L}) \rangle \Big|_{S^3} = -i \, , \tag{18.80}$$

and

$$\langle W(C_1; \Psi[I]) \rangle \Big|_{\mathcal{P}} = \begin{cases} \sqrt{3} \, , & \text{if } I = 1/2 \, ; \\ 2e^{i\pi 2/3} \, , & \text{if } I = 1 \, ; \\ \sqrt{3} \, , & \text{if } I = 3/2 \, ; \\ 1 \, , & \text{if } I = 2 \, ; \end{cases} \tag{18.81}$$

$$\langle W(C_2; \Psi[I]) \rangle \Big|_{\mathcal{P}} = \begin{cases} 3e^{i\pi 3/4} \, , & \text{if } I = 1/2 \, ; \\ 2e^{i\pi 2/3} \, , & \text{if } I = 1 \, ; \\ 3e^{-i\pi/4} \, , & \text{if } I = 3/2 \, ; \\ 1 \, , & \text{if } I = 2 \, . \end{cases} \tag{18.82}$$

18.5 The manifold $T^2 \times S^1$

The manifold $T^2 \times S^1 \cong S^1 \times S^1 \times S^1$ has a surgery presentation described by the Borromean rings $\mathcal{B}$ with surgery coefficients $r_1 = r_2 = r_3 = 0$. Clearly, $\pi_1(T^2 \times S^1)$ is the free abelian group of rank 3; a particular generator of this group, which is related to one of the S^1 factors, is shown in Fig. 18.24.

When $k = 1$, the surgery operator is

$$\widetilde{W}(\mathcal{L}) = W(\mathcal{B}; \chi_0, \chi_0, \chi_0) \, , \tag{18.83}$$

and is defined for the components of $\mathcal{B}$ with preferred framings. One obtains

$$\langle \widetilde{W}(\mathcal{L}) \rangle \Big|_{S^3} = 2\sqrt{2} \, . \tag{18.84}$$

When the knot C, shown in Fig. 18.24, has preferred framing and is labelled by $\chi[I]$, one finds

$$\langle W(C; \chi[I]) \rangle \Big|_{T^2 \times S^1} = (-1)^{2I} \, (2I + 1) \, \tfrac{1}{2} \left[1 + (-1)^{2I} \right] , \tag{18.85}$$

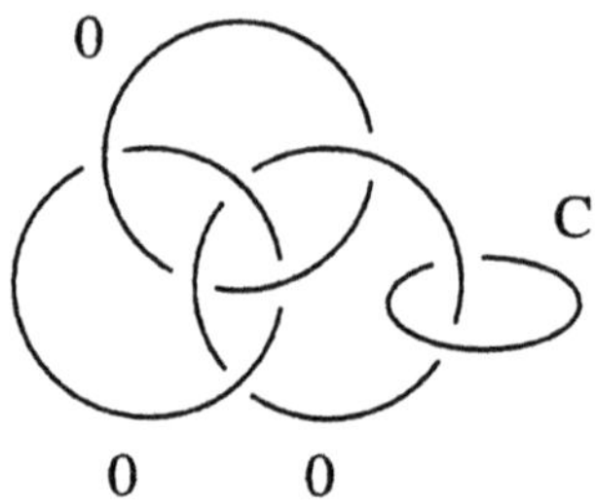

Figure 18.24.

For $k \geq 2$, the surgery operator is

$$\widetilde{W}(\mathcal{L}) = W(\mathcal{B}; \Psi_0, \Psi_0, \Psi_0), \tag{18.86}$$

where the components of $\mathcal{B}$ have preferred framings.
When $k = 3$, one has

$$\langle \widetilde{W}(\mathcal{L}) \rangle \Big|_{S^3} = 2\sqrt{2}, \tag{18.87}$$

and

$$\langle W(C; \Psi[1/2]) \rangle \Big|_{T^2 \times S^1} = 0. \tag{18.88}$$

When $k = 4$, one finds

$$\langle \widetilde{W}(\mathcal{L}) \rangle \Big|_{S^3} = 6, \tag{18.89}$$

and

$$\langle W(C; \Psi[1/2]) \rangle \Big|_{T^2 \times S^1} = 0, \tag{18.90}$$

$$\langle W(C; \Psi[1]) \rangle \Big|_{T^2 \times S^1} = \frac{1}{3}. \tag{18.91}$$

When $k = 5$, one gets

$$\langle \widetilde{W}(\mathcal{L}) \rangle \Big|_{S^3} = 4\sqrt{\frac{5}{2}} \frac{1}{\sin(\pi/5)}, \tag{18.92}$$

$$\langle W(C; \Psi[1/2]) \rangle \Big|_{T^2 \times S^1} = \langle W(C; \Psi[3/2]) \rangle \Big|_{T^2 \times S^1} = 0, \tag{18.93}$$

$$\langle W(C; \Psi[1]) \rangle \Big|_{T^2 \times S^1} = \frac{1}{2}. \tag{18.94}$$

Finally, when $k = 6$, one has

$$\langle \widetilde{W}(\mathcal{L}) \rangle \Big|_{S^3} = 10\sqrt{3}, \tag{18.95}$$

$$\langle W(C; \Psi[1/2]) \rangle \Big|_{T^2 \times S^1} = \langle W(C; \Psi[3/2]) \rangle \Big|_{T^2 \times S^1} = 0, \tag{18.96}$$

$$\langle W(C; \Psi[1]) \rangle \Big|_{T^2 \times S^1} = \frac{3}{5}, \tag{18.97}$$

$$\langle W(C; \Psi[2]) \rangle \Big|_{T^2 \times S^1} = \frac{1}{5}. \tag{18.98}$$

Chapter 19

Three – manifold invariant

The construction of three-manifold invariants is of particular interest in mathematics and, recently, new important results on this subject have been obtained. By using different methods, several authors have defined three-manifold invariants [7,8,67 - 71]. As a matter of fact, all these new invariants are strictly related and represent, in slightly different forms, essentially the same quantity: the Reshetikhin-Turaev invariant [7]. In view of the universality character of the CS theory, it is not surprising to find that the invariant defined by the CS theory is a particular version of the Reshetikhin-Turaev invariant.

In this chapter, we shall define a three-manifold invariant $\mathcal{I}(\mathcal{M})$ naturally associated with the Chern-Simons theory. We shall consider several examples of three-manifolds $\mathcal{M}$ and the corresponding values $\mathcal{I}(\mathcal{M})$ will be computed.

19.1 Improved partition function

In the topological field theory approach, the natural candidate for being an invariant of three-manifolds is the so-called "partition function"

$$Z(\mathcal{M}) = \int_{\mathcal{M}} \mathcal{D}A \; e^{iS} . \tag{19.1}$$

In equation (19.1), it is understood that one has to integrate over all the fields (gauge connection, auxiliary and ghost fields) which are needed to define the theory in a given manifold $\mathcal{M}$. The expression (19.1) represents the vacuum-to-vacuum amplitude for the field theory model in $\mathcal{M}$

$$Z(\mathcal{M}) = \langle 0 | 0 \rangle \Big|_{\mathcal{M}} . \tag{19.2}$$

Of course, the "partition function" (19.1) by itself cannot represent any meaningful invariant at all. Since the normalization of the path-integral is arbitrary, the expression (19.1) has no intrinsic meaning. This fact is in perfect agreement with the fundamental principles of quantum mechanics. Indeed, the states of a quantum system are described by the rays and not by the vectors of a linear vector space. Consequently, the value of the vacuum-to-vacuum amplitude (19.2) has no physical meaning. Unlike the expression (19.1), the expectation values of the observables,

$$\langle W(L) \rangle \Big|_{\mathcal{M}} = \frac{\langle 0 | W(L) | 0 \rangle \Big|_{\mathcal{M}}}{\langle 0 | 0 \rangle \Big|_{\mathcal{M}}} , \tag{19.3}$$

are well defined because they are properly normalized. Equation (19.3) suggests that, in order to define a meaningful invariant of three-manifolds, it would be better to consider the ratio

$$\mathcal{R}(\mathcal{M}, \mathcal{M}_0) = \frac{\langle 0 | 0 \rangle \Big|_{\mathcal{M}}}{\langle 0 | 0 \rangle \Big|_{\mathcal{M}_0}} , \tag{19.4}$$

where $\mathcal{M}_0$ is a given three-manifold which is chosen to fix the normalization. Clearly, we need to introduce a consistent procedure to give a meaning to this normalization; otherwise, the expression (19.4) is still meaningless. In the computation of the expectation values (19.3) of the observables, this procedure is represented by the Feynman rules derived from the action of the model. In the case of equation (19.4), we have no action, nor do we have a set of fields whose fluctuations connect ***different*** three-manifolds. In fact, the action of the CS theory tells us how to compute the observables in a given manifold only.

Suppose that, in order to define the partition function of the CS theory in different manifolds, one introduces some regularization and computes the vacuum-to-vacuum diagrams defined by the CS action. It should be clear that this method is just a waste of time. The reason is very simple. In the computation of the partition function, any given regularization represents a way of factorizing a certain (formally infinite) part. Consider now the CS theory defined in two different three-manifolds. In general, the same regularization prescription will factorize, in the two manifolds, different (formally infinite) parts. Consequently, the comparison of the obtained finite results is completely meaningless because these finite parts refer to two different choices of the "zero-energy point". The comparison of the finite results is meaningful if and only if the chosen zero-point for the vacuum energy is rigorously the same for all the different manifolds. So, we have reached the crucial issue: one needs to find some method for connecting different manifolds (or, if you like, different theories) in such a way that the zero energy point remains fixed.

It should be noted that, in spite of their apparent similarity, expressions (19.3) and (19.4) have completely different meanings. In equation (19.3), the amplitudes which appear in the numerator and in the denominator are defined on the same manifold. These two amplitudes differ because of the presence of source terms in the path-integral which represents the numerator. On the other hand, the two amplitudes which appear in equation (19.4) correspond to two path-integrals which are defined by the same Lagrangian density but refer to different manifolds. For this reason, the standard Feynman rules, which are perfectly consistent in the computation of the expression (19.3), are of no help in trying to compute the expression (19.4).

This kind of problem is not new in physics; the Casimir effect is a typical example in which, with a given fixed Lagrangian, two different vacuum energies are compared. In the Casimir effect, the vacuum energy is controlled by a parameter d which represents the distance between two conducting surfaces. In this case, the different theories of interest are connected by varying continuously this distance d. The observable physical quantity is not the absolute value of the vacuum energy, of course. The observable physical quantity is the derivative of the vacuum energy with respect to d or, equivalently, the difference between the vacuum energies for different values of d.

Let us now consider two different (topologically inequivalent) three-manifolds $\mathcal{M}_0$ and $\mathcal{M}$ which are closed, connected and orientable. Is there any parameter t such that, by varying t, $\mathcal{M}_0$ can be smoothly connected with $\mathcal{M}$?

In a purely three-dimensional context, such a parameter t does not exist. However, in a four-dimensional setting, $\mathcal{M}_0$ and $\mathcal{M}$ can easily be connected. Indeed, $\mathcal{M}_0$ and $\mathcal{M}$ are always associated with a cobordism of a four-dimensional manifold.

In a naive extension of the field theory formalism from three to four dimensions, however, one finds some problems. For any nontrivial cobordism, there are critical points for the (Morse) parameter t; thus, a canonical field theory approach cannot be used. Furthermore, one should reformulate the three-dimensional CS theory in a certain enlarged four-dimensional context; this does not seem to be a simple task.

In conclusion, we prefer to remain in three dimensions and give a prescription which is consistent with all the results that we have already obtained.

As we have already stated, any given three-manifold $\mathcal{M}$ is connected with S^3 by a four-dimensional cobordism. Let us consider the three-dimensional aspects of this fact. Firstly, $\mathcal{M}$ can always be obtained by means of an "honest" surgery on S^3 performed along a surgery link $\mathcal{L}$. Secondly, a surgery operator $\widetilde{W}(\mathcal{L})$, which represents surgery as shown in equation (17.78), exists. Thirdly, surgery can be understood in terms of ***symmetry transformations*** acting on the system. The crucial point now is that, under the action of well defined symmetry transformations, the zero energy point can easily be controlled. Thus, in order to fix the normalization of the partition function, we can use surgery transformations to connect different manifolds.

In quite general terms, the method we shall adopt to give a meaning to the partition function of the CS theory in different manifolds consists of

(1) choice of a reference point. For convenience, we will take S^3 as the reference manifold. Thus, some regularization will be used to give a meaning to the partition function in S^3.

(2) we shall introduce a set of well defined transformations, acting on the partition function in S^3, in such a way that, by means of these transformations, S^3 can be connected with an arbitrary three-manifold $\mathcal{M}$.
(3) we shall take the ratio between the modified partition function, obtained according to some given transformation defined in (2), and the partition function of S^3. At this stage, all the cutoffs introduced in the regularization will be removed and the result will be expressed in terms of the renormalized coupling constant k.

We shall consider in detail the case in which the transformations mentioned in point (2) correspond to surgery operations. As we have seen in Chapter 17, the new pairing (between the state vectors of the field theory) resulting from surgery is obtained by inserting the operator $\widetilde{W}(\mathcal{L})$ in the expectation values. Therefore, for any given surgery instruction described by the surgery link $\mathcal{L}$ with integer surgery coefficients, the ratio specified in point (3) is given by

$$V(\mathcal{L}) = \frac{\langle 0 \,|\, \widetilde{W}(\mathcal{L}) \,|\, 0 \rangle \Big|_{S^3}}{\langle 0 \,|\, 0 \rangle \Big|_{S^3}} = \langle\, \widetilde{W}(\mathcal{L}) \,\rangle \Big|_{S^3} , \tag{19.5}$$

where $\widetilde{W}(\mathcal{L})$ is the surgery operator of equation (17.77). By construction, $V(\mathcal{L})$ shows how the partition function gets modified under the surgery operation corresponding to $\mathcal{L}$. For this reason, $V(\mathcal{L})$ a priori depends on the transformation that we are considering and does not necessarily represent a three-manifold invariant. In fact, as we have seen in Chapter 17, the expression (19.5) is not invariant under Kirby moves and, therefore, does not describe an invariant of the manifold. All the problems are due to the presence of the non-trivial phase factor $e^{\pm i\theta}$,

$$e^{\pm i\theta} = \begin{cases} \exp\left[\left(\pm\frac{i\pi}{4}\right) \frac{3(k-2)}{k}\right] & \text{if } k \geq 2, \\ \exp\left[\mp\left(\frac{i\pi}{4}\right)\right] & \text{if } k = 1, \end{cases} \tag{19.6}$$

in equations (17.66) and (17.69). Let us consider the behaviour of $V(\mathcal{L})$ under a Kirby move. Suppose that one component of the surgery link $\mathcal{L}$ is the unknot; we shall denote this component by $\mathcal{L}_1$ and its framing by $\mathcal{L}_{1f}$. We shall be interested in the case in which the linking number of $\mathcal{L}_1$ and $\mathcal{L}_{1f}$ is given by $\chi(\mathcal{L}_1, \mathcal{L}_{1f}) = \pm 1$. The remaining components of $\mathcal{L}$ belong to the complement solid torus $\mathcal{V}_1$ of $\mathcal{L}_1$ in S^3. Consider now the new link $\mathcal{L}_{(\mp)}$ which is obtained from $\mathcal{L}$ by removing the component $\mathcal{L}_1$ and by transforming the remaining components according to a twist homeomorphism $\tau_\mp$ of $\mathcal{V}_1$. As we have seen in Chapter 16, the transformation

$$\mathcal{L} \Longrightarrow \mathcal{L}_{(\mp)} \tag{19.7}$$

corresponds to a Kirby move. From equations (17.55) and (17.62) it follows that

$$V(\mathcal{L}_{(\mp)}) = e^{\pm i\theta}\, V(\mathcal{L})\,. \tag{19.8}$$

As we have already mentioned in Chapter 17, it is not possible (by means of the Wilson line operators of the $SU(2)$ CS theory) to achieve a complete invariance under Kirby moves for the expression (19.5). The phase factors $e^{\pm i\theta}$ could be interpreted as the three-dimensional counterpart of the central charge in two dimensions. The origin of these phase factors and their specific values are related to the properties of the Gauss sums, see Chapter 17. In our approach, the presence of these phases is not naively related to the so-called "framing of the three-manifold" but is strictly connected with the twist homeomorphism transformations of solid tori. This fact presents strong analogies with what happens in two dimensions.

A genuine invariant of three-manifolds can be obtained from the expression (19.5) by multiplying $V(\mathcal{L})$ by an appropriate prefactor which, under a Kirby move, cancels out the undesired phase factors. This method has been used by several authors [7,67-71]. Here, we shall adopt the prescription introduced by Lickorish [67] and by Morton and Strickland [68]. For any given surgery link $\mathcal{L}$ in S^3, with components $\{\mathcal{L}_i\}$ and framings $\{\mathcal{L}_{if}\}$, let us introduce an orientation for the different link components. Consider then the linking matrix $\mathcal{L}_{ij}$ which is defined by

$$\mathcal{L}_{ij} = \begin{cases} \chi(\mathcal{L}_i, \mathcal{L}_j) & \text{if } i \neq j; \\ \chi(\mathcal{L}_i, \mathcal{L}_{if}) & \text{if } i = j; \end{cases} \tag{19.9}$$

where $\chi(\mathcal{L}_a, \mathcal{L}_b)$ is the linking number of $\mathcal{L}_a$ and $\mathcal{L}_b$. The ***signature*** $\sigma(\mathcal{L})$ of the linking matrix (19.9) is independent [67,68] of the choice of the orientation of the components of $\mathcal{L}$. Under a Kirby move (19.7), one has [68]

$$\sigma(\mathcal{L}_{(\mp)}) = \sigma(\mathcal{L}) \mp 1\,. \tag{19.10}$$

Therefore, the product

$$\mathcal{I}(\mathcal{M}) = \exp\left[\, i\,\theta\,\sigma(\mathcal{L})\,\right]\; \langle\, \widetilde{W}(\mathcal{L})\,\rangle\Big|_{S^3} \tag{19.11}$$

is invariant under Kirby moves and represents an ***invariant*** of the three-manifold $\mathcal{M}$ corresponding to the "honest" surgery instruction described by $\mathcal{L}$.

The expression (19.11) represents the value of the improved partition function of the CS theory in the three-manifold $\mathcal{M}$. With respect to the naive definition of the partition function, the improvement that we have introduced consists of using symmetry transformations (corresponding to surgery on S^3) to connect different manifolds. This procedure is the analogue of the modification of the distance d between the two conducting surfaces in the ordinary Casimir effect.

It should be noted that the orientation of each manifold, obtained by surgery on S^3, is uniquely determined by the orientation of S^3. As we have already mentioned, in our notations the orientation of S^3 has been fixed by adopting the usual right-handed convention to compute linking numbers. By construction, the CS invariant of three-manifolds $\mathcal{I}(\mathcal{M})$ is invariant under orientation-preserving self-homeomorphisms of $\mathcal{M}$.

Let us now consider the behaviour of $\mathcal{I}(\mathcal{M})$ under the action of a self-homeomorphism $h^* : \mathcal{M} \to \mathcal{M}$ which is not orientation-preserving. In the CS theory, defined by the action (4.3), a modification of the orientation of the manifold is equivalent to exchanging the coupling constant k with $-k$. Since the coupling constant k multiplies the whole action, the value of the partition function is transformed into its complex conjugate. Therefore, since $\mathcal{I}(\mathcal{M})$ represents the value of the CS partition function, $\mathcal{I}(\mathcal{M})$ should transform under the action of the homeomorphism h^* into its complex conjugate $\mathcal{I}^*(\mathcal{M})$. We shall now prove that this is indeed the case.

Theorem R9. *Let $\mathcal{L}$ be the surgery link corresponding to a given three-manifold $\mathcal{M}$. Then, $\mathcal{I}(\mathcal{M})$ defined in* equation (19.11) *is invariant under orientation-preserving self-homeomorphisms of $\mathcal{M}$. Moreover, under the action of a self-homeomorphism $h^* : \mathcal{M} \to \mathcal{M}$, which is not orientation-preserving, $\mathcal{I}(\mathcal{M})$ transforms into $\mathcal{I}^*(\mathcal{M})$.*

Proof. We have already seen that, by construction, $\mathcal{I}(\mathcal{M})$ is invariant under Kirby moves. Let $\widetilde{\mathcal{L}}$ be the mirror image of the surgery link $\mathcal{L}$. From the definition of the extended Jones polynomial it follows that, for fixed irreducible representations of $SU(2)$ assigned to the components of $\mathcal{L}$, one has

$$E(\widetilde{\mathcal{L}}) \;=\; E^*(\mathcal{L})\,. \tag{19.12}$$

Note that the elements Ψ_0 and χ_0 are real; i.e. they have real expansion coefficients in the standard basis of the reduced tensor algebra or of the tensor algebra. Thus, equation (19.12) implies

$$\langle\, \widetilde{W}(\widetilde{\mathcal{L}})\,\rangle\Big|_{S^3} \;=\; \Big[\langle\, \widetilde{W}(\mathcal{L})\,\rangle\Big|_{S^3}\Big]^*\,. \tag{19.13}$$

On the other hand,

$$\sigma(\widetilde{\mathcal{L}}) \;=\; -\,\sigma(\mathcal{L})\,. \tag{19.14}$$

Therefore, by combining equations (19.13) and (19.14) with the definition (19.11), it follows that $\mathcal{I}(\mathcal{M})$ transforms under h^* into $\mathcal{I}^*(\mathcal{M})$. □

19.2 Values of the invariant

In this section, we will show that the invariant $\mathcal{I}(\mathcal{M})$ is nontrivial. Several examples of three-manifolds will be analysed and the value of the associated invariant will be reported. Apart from two examples in which the computations can easily be done for arbitrary values of k, the integer values of the coupling constant that we shall consider are given by $1 \leq k \leq 6$. Since for $k = 2$ the surgery operator is the identity, we will ignore this trivial case.

Let us begin with the three-sphere S^3. Our normalization (19.5) of the invariant (19.11) implies that, for any value of k,

$$\mathcal{I}(S^3) = 1 . \tag{19.15}$$

For the manifold $S^2 \times S^1$, one has

$$\mathcal{I}(S^2 \times S^1) = \begin{cases} \sqrt{2} & \text{when } k = 1; \\ \sqrt{\frac{k}{2}} \sin^{-1}(\pi/k) & \text{when } k \geq 2. \end{cases} \tag{19.16}$$

For the manifold RP^3, one finds

$$\mathcal{I}(RP^3) = \begin{cases} \frac{1}{\sqrt{2}} \cos^{-1}(\pi/4n) & \text{if } k = 2n; \\ 0 & \text{if } k = 2n + 1. \end{cases} \tag{19.17}$$

Let us now consider the lens spaces $L(p, 1)$; clearly, we only need to analyse the cases in which $p \geq 2$. When $k = 1$, one obtains

$$\mathcal{I}(L(p, 1)) = \begin{cases} \sqrt{2}\, e^{-i\pi/4} & \text{if } p \equiv 0 \pmod 4; \\ 1 & \text{if } p \equiv 1 \pmod 4; \\ 0 & \text{if } p \equiv 2 \pmod 4; \\ -i & \text{if } p \equiv 3 \pmod 4. \end{cases} \tag{19.18}$$

When $k = 3$, one has

$$\mathcal{I}(L(p, 1)) = \frac{1}{\sqrt{2}} \left[1 + (-i)^p \right] e^{i\pi/4} . \tag{19.19}$$

For $k = 4$, one gets

$$\mathcal{I}(L(p, 1)) = \left[\tfrac{1}{2} \left(1 + (-i)^p \right) + e^{-i3\pi p/8} \right] e^{i3\pi/8} . \tag{19.20}$$

When $k = 5$, one finds

$$\mathcal{I}(L(p,1)) = \sqrt{\frac{2}{5}} \sin(\pi/5) \Big[1 + (i)^p + + 4e^{-i3\pi p/10} (1 + (-i)^p) \cos^2(\pi/5) \Big] e^{i9\pi/20} . \tag{19.21}$$

Finally, for $k = 6$, one has

$$\mathcal{I}(L(p,1)) = \frac{i}{2\sqrt{3}} \left[2 + 4e^{-i2\pi p/3} + 3(e^{-i\pi p/4} + e^{i3\pi p/4}) \right] . \tag{19.22}$$

The two components of the surgery link shown in Fig. 19.1 have surgery coefficients $r_1 = a$ and $r_2 = b$ with nonvanishing integers a and b. Let $a > 0$, $b > 0$ and $ab > 1$. The corresponding three-manifold is the lens space $L(ab-1, a)$; clearly, $L(ab-1, a) \simeq L(ab-1, b)$.

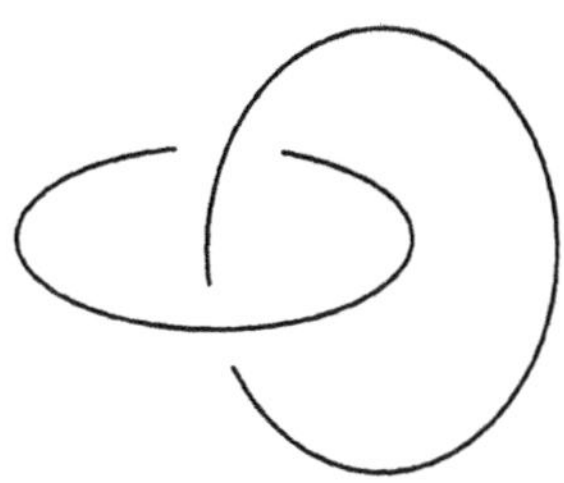

Figure 19.1.

The associated values of the invariant are in order:
when $k = 1$,

$$\mathcal{I}(L(ab-1,a)) = -\frac{i}{2} \left[2 - \left(e^{ia\pi/2} - 1\right) \left(e^{ib\pi/2} - 1\right) \right] ; \tag{19.23}$$

for $k = 3$,

$$\mathcal{I}(L(ab-1,a)) = \frac{i}{2} \left[2 - \left(e^{-ia\pi/2} - 1\right) \left(e^{-ib\pi/2} - 1\right) \right] ; \tag{19.24}$$

for $k = 4$,

$$\mathcal{I}(L(ab-1,a)) = e^{i3\pi/4}\,\tfrac{1}{4}\Big\{\Big[1+(-1)^a\Big]\Big[1+(-1)^b\Big]+ + 2e^{-i3a\pi/8}\Big[1-(-1)^b\Big] + 2e^{-i3b\pi/8}\Big[1-(-1)^a\Big]\Big\}; \tag{19.25}$$

for $k = 5$,

$$\mathcal{I}(L(ab-1,a)) = e^{i9\pi/10}\left(\frac{2}{5}\right)\sin^2(\pi/5)\Big\{1+i^a+i^b-i^{a+b}+ + 4\cos^2(\pi/5)\Big[e^{-i3a\pi/10}\Big(1-i^b+(-i)^a+i^b(-i)^a\Big)+ + e^{-i3b\pi/10}\Big(1-i^a+(-i)^b+i^a(-i)^b\Big)+ + e^{-i3(a+b)\pi/10}\Big(1-(-i)^a-(-i)^b-(-i)^{a+b}\Big)\Big]\Big\}; \tag{19.26}$$

and for $k = 6$,

$$\mathcal{I}(L(ab-1,a)) = -1 + \tfrac{2}{3}\left(e^{-i2a\pi/3}-1\right)\left(e^{-i2b\pi/3}-1\right)+ - \frac{\sqrt{3}}{4}\left(e^{-ia\pi/4}-e^{i3a\pi/4}\right)\left(e^{-ib\pi/4}-e^{i3b\pi/4}\right). \tag{19.27}$$

The Poincaré manifold $\mathcal{P}$ is a homology sphere but, as we have seen in Section 18.4, it is not a homotopy sphere. The values taken by the invariant (19.11) are

$$\mathcal{I}(\mathcal{P}) = 1, \tag{19.28}$$

for $k = 1, 3, 6$. When $k = 4$, we find

$$\mathcal{I}(\mathcal{P}) = -1; \tag{19.29}$$

and, for $k = 5$, we obtain

$$\mathcal{I}(\mathcal{P}) = \frac{2}{\sqrt{5}}e^{i3\pi/10}\sin(\pi/5)\left[1+2\cos(\pi/5)+4e^{-i\pi/5}\cos^2(\pi/5)\right]. \tag{19.30}$$

In the case of the manifold $T^2 \times S^1$, when $k = 1, 3$ we get

$$\mathcal{I}(T^2 \times S^1) = 2\sqrt{2}. \tag{19.31}$$

Moreover

$$\mathcal{I}(T^2 \times S^1) = \begin{cases} 6 & \text{if } k = 4; \\ 4\sqrt{\frac{5}{2}}\sin^{-1}(\pi/5) & \text{if } k = 5; \\ 10\sqrt{3} & \text{if } k = 6. \end{cases} \tag{19.32}$$

Let us now consider two new examples of cyclic coverings of S^3 branched over the trefoil. The manifold X_3, which is homeomorphic with the 3-fold cyclic covering of S^3 over the trefoil, is described [12] by the surgery instruction shown in Fig. 19.2. The fundamental group of X_3 is the finite group of order 8 which is isomorphic with the quaternionic group.

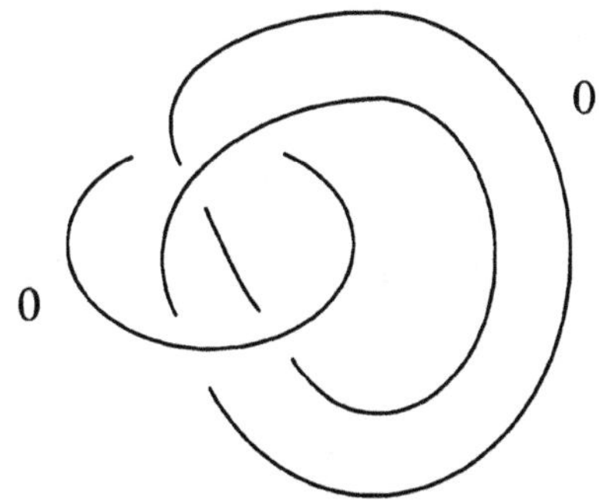

Figure 19.2.

For $k = 1$ and $k = 3$, we find

$$\mathcal{I}(X_3) = 2. \tag{19.33}$$

For $k = 4$, we obtain

$$\mathcal{I}(X_3) = 1 - i. \tag{19.34}$$

When $k = 5$, we have

$$\mathcal{I}(X_3) = 2\left(1 - e^{i\,2\pi/5} - e^{i\,4\pi/5}\right); \tag{19.35}$$

and, for $k = 6$, the value of the invariant is

$$\mathcal{I}(X_3) = 1 - i\,3\sqrt{3}. \tag{19.36}$$

The surgery instruction described by the right-handed trefoil with surgery coefficient $r = +3$ corresponds to the three-manifold X_4, which is homeomorphic [12]

with the 4-fold cyclic covering of S^3 over the trefoil. The fundamental group of X_4 is isomorphic with the binary tetrahedral group. In this case, the values of the invariant (19.11) are:
for $k = 1$,

$$\mathcal{I}(X_4) = -i\,; \tag{19.37}$$

for $k = 3$,

$$\mathcal{I}(X_4) = i\,; \tag{19.38}$$

for $k = 4$,

$$\mathcal{I}(X_4) = -\frac{1+i}{\sqrt{2}}\,; \tag{19.39}$$

for $k = 5$,

$$\begin{aligned}\mathcal{I}(X_4) = \frac{2}{\sqrt{5}} \sin(\pi/5) \Big[& -i + 2\cos(\pi/5)\, e^{-i\pi/10} + \\ & + 4\cos^2(\pi/5)\, e^{-i3\pi/10} \Big] e^{i7\pi/10}\,;\end{aligned} \tag{19.40}$$

and, for $k = 6$,

$$\mathcal{I}(X_4) = i\sqrt{3}\,. \tag{19.41}$$

Chapter 20

Abelian surgery invariant

Topological quantum field theories are characterized by a remarkable property: the solution of the theory in a generic (closed, connected and orientable) three-manifold $\mathcal{M}$ is determined by the solution of the theory in S^3. In fact, surgery on S^3 can be used to connect different manifolds and surgery can be understood in terms of symmetry transformations. The structure of the possible representations of this symmetry depends on the specific topological model under consideration. Each representation of surgery must satisfy the consistency conditions following from the invariance under Kirby moves.

In the previous chapters, we analysed in detail the non-Abelian $SU(2)$ CS model, which illustrates the main properties of the non-Abelian CS theory based on a generic compact simple Lie group.

As we have seen in Chapter 19, the prefactor multiplying $\langle\, \widetilde{W}(\mathcal{L})\,\rangle\big|_{S^3}$ in equation (19.11) has been introduced to obtain invariance under Kirby moves. We shall now consider a particular set of $U(1)$ models, defined by the Abelian CS action, in which the invariance under Kirby moves does not require the introduction of this prefactor.

20.1 Compact Abelian theory

Let us reconsider the action of the Abelian CS theory in $\mathbb{R}^3$,

$$S_0 = \frac{k'}{8\pi} \int d^3x \, \epsilon^{\mu\nu\rho} \, A_\mu \partial_\nu A_\rho \, . \tag{20.1}$$

The expression (20.1) and the gauge transformations (3.2) do not fix the gauge group uniquely. As in any gauge theory, equations (20.1) and (3.2) determine the algebra of the universal covering of the gauge group. In order to specify the gauge group of the model, some information concerning the observables of the system must be introduced. We shall be interested in the Abelian $U(1)$ CS theory. This means that a complete set of gauge invariant observables is represented by the Wilson line operators (3.18) with charges $\{\, e_i \,\}$ which are multiples of a unique elementary charge e_0. That is, any permitted value e_i of the charge is given by

$$e_i = n_i \, e_0 , \tag{20.2}$$

for some integer n_i. This being the case, we can introduce the vector fields B_μ defined by

$$B_\mu(x) = e_0 \, A_\mu(x) , \tag{20.3}$$

and rewrite the action (20.1) in terms of B_μ:

$$S_0 = \left(\frac{k'}{e_0^2}\right) \frac{1}{8\pi} \int d^3x \, \epsilon^{\mu\nu\rho} \, B_\mu(x) \, \partial_\nu B_\rho(x) . \tag{20.4}$$

Then, the Wilson line operators will be defined in terms of the fields B_μ and the charges associated with the different components of the links will be represented by integer numbers $\{ n_i \}$. In these new notations, the coupling constant is

$$k = \frac{k'}{e_0^2} . \tag{20.5}$$

We shall consider the particular set [72] of Abelian $U(1)$ CS models in which the value of the coupling constant k, appearing in equation (20.5), takes integer values and satisfies the constraints

$$k = \text{odd prime} , \qquad k \equiv 1 \; (\text{mod} \, 4) . \tag{20.6}$$

The first condition is just a technical requirement which simplifies our discussion. The second condition is imposed in order to have invariance under Kirby moves without needing to introduce any prefactor in the expression (19.11). The precise meaning of conditions (20.6) will be explained in detail in what follows.

The solution of the Abelian CS theory in $\mathbb{R}^3$ was derived in Chapter 3. The expectation value of the Wilson line operator, associated with the link L, is shown in equation (3.21). The expression (3.21) depends exclusively on the linking matrix L_{ij} of the link L.

Any oriented knot C in $\mathbb{R}^3$ represents the (oriented) boundary of an oriented surface $S(C)$, called the Seifert surface. The linking number $\chi(C_i, C_j)$ simply counts the number of intersection points of C_i with $S(C_j)$, or of C_j with $S(C_i)$. Note that each intersection point has a sign $+1$ or -1 depending on the way in which the oriented knot crosses the oriented surface.

In the Abelian CS theory, each Wilson line operator $W(C)$ can also be expressed in terms of the integral over the Seifert surface $S(C)$ of the field strength associated with B_μ. The curl of the propagator (3.4) represents an intersection form of a one-dimensional manifold (a given component of a link) and a two-dimensional manifold (the Seifert surface associated with a different component of the link) in $\mathbb{R}^3$. For this reason, the expression (3.19) contains the factors $\{\chi(C_i, C_j)\}$ at the exponent. The self-interaction parts $\{\chi(C_i, C_{if})\}$ originate from the definition of the composite Wilson line operator by means of the framing procedure. In the expression (3.19), the interaction and self-interaction contributions are combined in such a way that the satellite relations hold. It should be noted that the satellite relations are valid in any three-manifold because they depend exclusively on the structure of the Wilson line operators associated with links in a solid torus.

The fact that the curl of the expression (3.4) represents an intersection form is a consequence of the Schwinger-Dyson equations. Indeed, let us consider the expectation values of the fields computed with the action (20.4). The equations of motion of the model imply that an insertion (in the vacuum expectation values) of the field strength operator must be equivalent to the action of the operator which is conjugate to the insertion of a vector field [18,20]

$$\partial_\mu B_\nu(x) - \partial_\nu B_\mu(x) \Longrightarrow i\,\epsilon_{\mu\nu\rho}\left(\frac{4\pi}{k}\right)\frac{\delta}{\delta B_\rho(x)}\,.$$

Let us now consider the Abelian CS theory in S^3. Since S^3 is a homology sphere, any knot in S^3 is the boundary of a Seifert surface. Consequently, the linking number of two oriented knots is well defined in S^3. Since the CS model in S^3 is defined by the same Lagrangian density which appears in equation (3.1), the structure of the Schwinger-Dyson equations in S^3 is the same as in $\mathbb{R}^3$.

Note that any given link in $\mathbb{R}^3$ belongs to the interior of a three-ball and any link in S^3 also belongs to the interior of a three-ball. The topological properties of links in S^3 and of links in $\mathbb{R}^3$ are equivalent. Therefore, the expression (3.21) also represents the solution of the Abelian CS theory in S^3. Indeed, the Schwinger-Dyson equations imply that the expectation values must contain the factors $\{\chi(C_i, C_j)\}$ at the exponent. Moreover, because of the satellite relations, the self-interaction terms $\{\chi(C_i, C_{if})\}$ must also be present and must have the same coefficients shown in equation (3.21). Unlike the case in which the gauge group is $G = SU(N)$, for the $U(1)$ theory there are no constraints (coming from gauge invariance) to be imposed in S^3. Therefore, the solution of the Abelian $U(1)$ CS theory in S^3 coincides with the solution in $\mathbb{R}^3$.

For any oriented framed link L, whose components $\{C_i\}$ are labelled by the charges $\{n_i\}$, one has

$$\langle W(L)\rangle\Big|_{S^3} = \exp\left[-i\left(\frac{2\pi}{k}\right)\sum_{i,j} n_i\,L_{ij}\,n_j\right], \qquad (20.7)$$

where L_{ij} is the linking matrix defined by the link L. Notice that, according to the definition (3.20), the matrix L_{ij} is real, symmetric and has integer matrix elements.

Equation (20.7) represents the solution of the Abelian $U(1)$ CS theory in S^3. Our main purpose now is to produce the solution of the theory in a generic manifold $\mathcal{M}$ which is closed, connected and orientable.

In order to find the surgery rules for the Abelian $U(1)$ CS theory, we need to discuss some basic properties of the solution (20.7). We shall consider, first the basic satellite relations and secondly the reduced tensor algebra of the Abelian $U(1)$ CS theory in S^3.

Let L be a framed link in S^3 and let its components $\{C_1, C_2, \dots, C_t\}$ be labelled by the charges $\{n_1, n_2, \dots, n_t\}$. The associated Wilson line operator is denoted by $W(C_1, \dots, C_t; n_1, \dots, n_t)$. Suppose that the link $L' = \{C_1^{(2)}, C_2, \dots, C_t\}$ is a satellite of a given link $L = \{C_1, C_2, \dots, C_t\}$. This satellite L' has been constructed with the pattern link B shown in Fig. 6.6; i.e., $C_1^{(2)} = h_1^\circ(B)$ and h_1° is the homeomorphism introduced in Chapter 2 and associated with the link component C_1. The link components $\{C_2, \dots, C_t\}$ and their associated charges are kept fixed. If the two components of $C_1^{(2)}$ are labelled by the charges n and m respectively, one has

$$\begin{aligned} \langle\, W(C_1^{(2)}, C_2, \dots; n, m, n_2, \dots) \,\rangle\Big|_{S^3} &= \\ = \langle\, W(C_1, C_2, \dots; n_1 = n+m, n_2, \dots) \,\rangle\Big|_{S^3} . \end{aligned} \tag{20.8}$$

By taking into account the solution (20.7), it is easy to verify that equation (20.8) is satisfied.

As we have seen in Chapter 15, the general structure of the satellite relations is described by the tensor algebra $\mathcal{T}$. The inequivalent irreducible representations of $U(1)$ are characterized by an integer n. For each n, we shall introduce an element $\chi[n]$ in the algebra $\mathcal{T}$ which is defined over the complex numbers. The multiplication operation in $\mathcal{T}$ is defined according to the decomposition rule of the tensor product of the representations of the group

$$\chi[n]\, \chi[m] \;=\; \chi[n+m] . \tag{20.9}$$

As a linear space, $\mathcal{T}$ is generated by $\{\chi[n]\}$ and the structure constants (20.9) determine the form of the satellite relations.

Let C be a framed knot labelled by the element $\chi = \sum_n d(n)\chi[n]$ of $\mathcal{T}$. We shall define

$$W(C;\chi) \;=\; \sum_n d(n)\, W(C;n) , \tag{20.10}$$

and, according to equation (20.10), for a link L with m components the associated expectation value $\langle W(L) \rangle$ will be understood as a multilinear function on $\mathcal{T}^{\otimes m}$.

The tensor algebra $\mathcal{T}$ is of infinite order. For generic (not rational) values of k, $\mathcal{T}$ can be interpreted as the space of the physical (gauge invariant) states of the theory associated with the boundary of a solid torus.

Let us now consider the case in which the coupling constant k is an integer. For fixed integer k, if the value of one of the charges n_i, appearing in equation (20.7), is replaced by

$$n_i \Longrightarrow n_i + k , \tag{20.11}$$

the expression (20.7) is invariant. Therefore, the elements $\chi[n]$ and $\chi[n']$ of $\mathcal{T}$, where $n' - n = p\,k$ (with integer p), cannot be distinguished. Since $\{ \langle W(L) \rangle \}$ is a complete set of observables, the labels $\chi[n]$ and $\chi[n + p\,k]$ must be identified.

In abstract terms, the elements $\{ \chi[n] \}$ with the multiplication law (20.9) form an Abelian group X which is isomorphic with Z. The subgroup Y generated by $\chi[k]$ is an invariant subgroup and the elements of $X/Y \simeq Z_k$ are denoted by $\{ \Psi[n] \}$ with $n = 0, 1, \ldots, k - 1$. For the Abelian $U(1)$ CS theory with integer coupling constant k, the reduced tensor algebra $\mathcal{T}_{(k)}$ coincides with the group algebra of Z_k. The elements $\{ \Psi[n] \}$ form the standard basis of $\mathcal{T}_{(k)}$ and

$$\Psi[n]\,\Psi[m] \;=\; \Psi[p], \quad p \equiv m + n \;(\text{mod}\; k) . \tag{20.12}$$

$\mathcal{T}_{(k)}$ is a commutative algebra of order k over the complex numbers. Any element Ψ of $\mathcal{T}_{(k)}$ can be written as $\Psi = \sum_{n=0}^{k-1} d(n)\Psi[n]$ and each class $\Psi[n]$ can be represented by $\chi[n]$ with $n = 0, 1, \ldots, k - 1$. For any given link L, with m components, $\langle W(L) \rangle$ will be interpreted as a multilinear function on $\mathcal{T}_{(k)}^{\otimes m}$. For fixed integer k, $\mathcal{T}_{(k)}$ represents the space of the physical states of the theory associated with the boundary of a solid torus.

Let L be a link in a solid torus $V \subset S^3$; for fixed integer k, the analogue of equations (13.18) and (17.43) reads

$$W(L) \;=\; \sum_{n=0}^{k-1} \xi(n)\; W(C; \Psi[n]) \tag{20.13}$$

where C is the oriented core of V, with some fixed framing, and the (complex) coefficients $\{ \xi(n) \}$ depend on the structure of $L \subset V$.

Since a modification of the orientation of S^3 is equivalent to replacing k with $-k$, we may assume that k is positive. Furthermore, the value $k = 0$ must be excluded because, in this case, a meaningful CS theory does not exist. Hence, k is a natural number and we are interested in the residue classes of congruent integers mod k. A complete residue system contains k elements which are in one-to-one correspondence with the elements $\{ \Psi[n] \}$ of the standard basis of $\mathcal{T}_{(k)}$.

Let us recall that if one element of a given residue class is coprime with k, then [66] every element of the same class is coprime with k. Consider now a complete residue system; the subset of all its elements which are coprime with k is called a reduced residue system. For fixed $k > 1$, let m be an integer which is coprime with k; if $x^2 \equiv m \pmod{k}$ has a solution (with integer x), m is called a quadratic residue mod k. Otherwise, m is called a quadratic non-residue mod k.

The expression (20.7) shows that the case $k = 1$ is not interesting because it corresponds to a completely trivial situation. So, we shall assume that $k > 1$.

Let us consider an oriented framed component C of a given link L. The value n of the charge associated with C represents an internal quantum number which characterizes the link component C. A modification of the orientation of C is equivalent to replacing n with $-n$. If we modify the framing of C in such a way that the linking number of C and its framing changes by a factor ± 1, the expectation value of the associated Wilson line operator gets multiplied by

$$[\,\alpha(n)\,]^{\pm 1} = q^{\pm n^2} , \tag{20.14}$$

where

$$q = \exp\left(-i\,\frac{2\pi}{k}\right) . \tag{20.15}$$

Thus, for fixed integer k, the intrinsic properties of each link component are described by

$$\beta = n^2 \pmod{k} . \tag{20.16}$$

Clearly, β is invariant under the transformation $n \to -n$. Let us suppose that, apart from this ambiguity associated with the sign of the charge, β determines the absolute value of the charge uniquely. In this case, apart from the unavoidable symmetry $n \leftrightarrow -n$, there is no additional degeneracy in the parameter which characterizes the intrinsic properties of the link components. Furthermore, let us suppose that the sign of the charge of a given link component can be determined unambiguously by considering the interactions with the remaining components of the link. This is the simplest and the most fundamental situation which can be realized. Now we shall concentrate on this case. Let us summarize the conditions which must be imposed.

In order to eliminate the degeneracy related to the sign of the charge, we shall assume that $n \not\equiv -n \pmod{k}$ for any nontrivial value $1 \leq n \leq k-1$ of the charge; this implies that k must be odd. Moreover, the set of all possible values of β must contain $(k+1)/2$ different elements. For completeness, this set should coincide with the set of the quadratic residues mod k plus the trivial (vanishing) value. These conditions are satisfied when

$$k = \text{odd prime} . \tag{20.17}$$

Indeed, it is known [66] that, when equation (20.17) holds (with $k > 1$), there are $(k-1)/2$ quadratic residues, and $(k-1)/2$ quadratic non-residues in any reduced residue system. Moreover, $\{1^2, 2^2, \ldots, [(k-1)/2]^2\}$ form a complete set of quadratic residues. So, we shall assume that equation (20.17) is satisfied. This restriction simplifies our subsequent analysis of the model. Condition (20.17) can be understood as a convenient technical requirement which removes any kind of "accidental" degeneracy.

20.2 Abelian surgery rules

The surgery operator $\widetilde{W}(\mathcal{L})$ will be determined by using the method illustrated in Chapter 17. We shall concentrate on the Abelian $U(1)$ CS theory with coupling constant k satisfying condition (20.17).

Let us consider a surgery link $\mathcal{L}$ with m components $\{\mathcal{L}_i\}$ and integer surgery coefficients $\{r_i\}$. As we have seen in Chapter 17, the general structure of the surgery operator $\widetilde{W}(\mathcal{L})$ is given by

$$\widetilde{W}(\mathcal{L}) = \prod_{i=1}^{m} W(\mathcal{L}_i \,;\, \Psi_0) \,. \tag{20.18}$$

In this equation, $W(\mathcal{L}_i\,;\,\Psi_0)$ is the Wilson line operator associated with the framed component $\mathcal{L}_i$ which is labelled by the universal surgery element Ψ_0 of the reduced tensor algebra $\mathcal{T}_{(k)}$. The framing $\mathcal{L}_{if}$ of the component $\mathcal{L}_i$ satisfies equation (17.72); i.e. the linking number of $\mathcal{L}_i$ and $\mathcal{L}_{if}$ is equal to the value of the surgery coefficient r_i.

Since Wilson line operators are defined on oriented knots, an orientation for the components $\{\mathcal{L}_i\}$ will be introduced. Since the surgery instruction is characterized by the unoriented link $\mathcal{L}$, the particular choice of these orientations must be irrelevant. This implies that the expansion coefficients $\{\phi(n)\}$ of Ψ_0,

$$\Psi_0 = \sum_{n=0}^{k-1} \phi(n)\, \Psi[n] \,, \tag{20.19}$$

must satisfy

$$\phi(n) = \phi(k-n) \,, \qquad \forall n \neq 0 \,. \tag{20.20}$$

Let us now consider the elementary surgeries $S_{\pm}$ and S_- and the associated surgery operators which are denoted by $\widetilde{W}(U;+1)$ and $\widetilde{W}(U;-1)$ respectively. In this case, the surgery instruction is given by the unknot U with surgery coefficient $r=+1$ ($r=-1$). As we have seen in Chapter 16, the elementary surgery S_+ (S_-) is equivalent (by homeomorphism) to a left-handed (right-handed) twist homeomorphism τ_- (τ_+) of the complement solid torus of U in S^3. Therefore, the element $\Psi_0 \in \mathcal{T}_{(k)}$ will be determined by imposing that, for any link L,

$$\frac{\langle \widetilde{W}(U;\pm 1)\; W(L)\rangle\Big|_{S^3}}{\langle \widetilde{W}(U;\pm 1)\rangle\Big|_{S^3}} = \langle W(L^{(\mp)})\rangle\Big|_{S^3}, \tag{20.21}$$

where $L^{(\mp)}$ is the image of L under $\tau_\mp$. Equation (20.21) is a necessary condition which must be satisfied if the operator $\widetilde{W}(\mathcal{L})$ represents surgery. Moreover, as we have seen in Chapter 17, equation (20.21) implies that the results obtained according to equation (17.78) are invariant under Kirby moves. In equation (20.21), Ψ_0 is fixed up to an overall rescaling by a nonvanishing complex parameter.

For the non-Abelian $SU(2)$ CS theory, the numerator (and, separately, the denominator) of the ratio (17.78) is not invariant under Kirby moves. (Of course, the ratio (17.78) **is** invariant.) This invariance is broken by the nontrivial phase factor (19.6) which is associated to each twist homeomorphism of a solid torus. We shall now require this phase factor to be absent in the Abelian $U(1)$ CS theory. This means that a single insertion of $\widetilde{W}(U;\pm 1)$ in the expectation values should generate the twist homeomorphism $\tau_\mp$.

Let C be the oriented core of the complement solid torus of U in S^3 and let C have a preferred framing. The orientations of U and C will be fixed in such a way that the linking number of U and C is equal to unity, $\chi(U, C) = 1$. Under a twist homeomorphism $\tau_\mp$, the framing of C is modified and, according to equation (20.14), we find

$$\langle \widetilde{W}(U;\pm 1)\; W(C;\Psi[m])\rangle\Big|_{S^3} = q^{\mp m^2}\,\langle W(C;\Psi[m])\rangle\Big|_{S^3}. \tag{20.22}$$

These two equations, which must hold for any m with $0 \le m \le k-1$, will determine Ψ_0. By using the solution (20.7), equation (20.22) can be written as

$$\sum_{n=0}^{k-1} \phi(n)\, q^{\,n^2+2nm} = q^{-m^2}, \tag{20.23a}$$

and

$$\sum_{n=0}^{k-1} \phi(n)\, q^{-n^2+2nm} = q^{m^2}, \tag{20.23b}$$

where q is shown in equation (20.15). Let us denote by H the Hopf matrix entering the linear systems (20.23); H is a $(k \times k)$ symmetric matrix with matrix elements

$$H_{nm} = q^{2nm}. \tag{20.24}$$

Since k satisfies equation (20.17), we have

$$\left(H^{-1}\right)_{nm} = \frac{1}{k} H_{nm} . \tag{20.25}$$

Therefore, the solution of equation (20.23a) is given by

$$\phi(n) = \frac{1}{k} \sum_{m=0}^{k-1} q^{-(m-n)^2} , \tag{20.26a}$$

and the solution of equation (20.23b) is

$$\phi(n) = \frac{1}{k} \sum_{m=0}^{k-1} q^{(m+n)^2} . \tag{20.26b}$$

Let us recall that, for odd values of k, the Gauss sum $S(1, k)$ is given by [66]

$$S(1,k) = \sum_{n=0}^{k-1} \exp\left(\frac{2\pi i}{k} n^2\right) = \begin{cases} \sqrt{k} & \text{if } k \equiv 1 \pmod 4; \\ i\sqrt{k} & \text{if } k \equiv 3 \pmod 4). \end{cases} \tag{20.27}$$

Therefore, both equations (20.26a) and (20.26b) are satisfied if and only if

$$k \equiv 1 \pmod 4 . \tag{20.28}$$

When condition (20.28) holds, the surgery element $\Psi_0 \in \mathcal{T}_{(k)}$ is given by

$$\Psi_0 = \frac{1}{\sqrt{k}} \sum_{n=0}^{k-1} \Psi[n] . \tag{20.29}$$

The explicit expression (20.29) of Ψ_0 shows that condition (20.20) is indeed satisfied. To sum up, we have proved that

Theorem R10. *When conditions* (20.6) *are satisfied, the solution of the Abelian $U(1)$ CS theory in any fixed three-manifold $\mathcal{M}$, which is closed, connected and orientable, is given by the surgery rule* (17.78) *in which the surgery operator $\widetilde{W}(\mathcal{L})$ is defined in* equation (20.18) *and the element Ψ_0 is shown in* equation (20.29).

Therefore, the expectation value of the observable associated with a generic link L in $\mathcal{M}$ can easily be obtained by using equation (17.78) and the solution (20.7) of the CS theory in S^3.

Clearly, for the models that we are considering, the numerator and the denominator of the ratio (17.78) are separately invariant under Kirby moves. Indeed, suppose that one component of the surgery link is the unknot with surgery coefficient $r = \pm 1$. Let us denote this component by $\mathcal{L}_1$. All the remaining components of $\mathcal{L}$ and the "true" link L belong to the complement solid torus $\mathcal{V}$ of $\mathcal{L}_1$ in S^3. From equation (20.22) it follows that an insertion of the operator $\widetilde{W}(\mathcal{L}_1; \pm 1)$ in the expectation values generates a twist homeomorphism $\tau_{\mp}$ of $\mathcal{V}$. Therefore, $\mathcal{L}_1$ can be eliminated provided $\mathcal{V}$ is twisted by $\tau_{\mp}$. This means that the invariance under Kirby moves is satisfied.

20.3 Abelian surgery invariant

From the results of the previous section, it follows that

Theorem R11. *Let the three-manifold $\mathcal{M}$ have a Dehn surgery presentation described by the surgery link $\mathcal{L}$ in S^3 with integer surgery coefficients $\{r_i\}$. The expectation value $\mathcal{I}(\mathcal{M})$, defined by*

$$\mathcal{I}(\mathcal{M}) = \langle \widetilde{W}(\mathcal{L}) \rangle \Big|_{S^3} , \tag{20.30}$$

where $\widetilde{W}(\mathcal{L})$ is shown in equation (20.18) *and Ψ_0 is given by* equation (20.29), *represents a* ***topological invariant*** *of $\mathcal{M}$.*

Proof. We have seen in Section 20.2 that the expression (20.30) is invariant under Kirby moves. Therefore, the invariance under Kirby moves combined with the ambient isotopy invariance of the CS expectation values in S^3 imply that $\mathcal{I}(\mathcal{M})$ is invariant under orientation-preserving self-homeomorphisms of $\mathcal{M}$. □

Let us now consider a self-homeomorphism $h^* : \mathcal{M} \to \mathcal{M}$ which is not orientation-preserving. Let L_{ij} be the linking matrix associated with a given link L in S^3. Under a mirror reflection transformation, the sign of each matrix element L_{ij} is modified. Consequently, under a mirror reflection, the expectation value (20.7), which is defined for fixed values $\{n_i\}$ of the charges, is transformed into its complex conjugate. The surgery element Ψ_0 has real expansion coefficients with respect to the standard basis of $\mathcal{T}_{(k)}$. Therefore, under h^*, $\mathcal{I}(\mathcal{M})$ is transformed into its complex conjugate $\mathcal{I}^*(\mathcal{M})$.

We shall now prove that $\mathcal{I}(\mathcal{M})$ is also invariant under the action of self-homeomorphisms of $\mathcal{M}$ which are not orientation-preserving.

Theorem R12. *The invariant $\mathcal{I}(\mathcal{M})$, defined in* equation (20.30), *is real.*

Proof. Suppose that the three-manifold $\mathcal{M}$ admits a surgery presentation corresponding to the surgery instruction described by the surgery link $\mathcal{L}$ in S^3. The

m components (with $m \geq 1$) of $\mathcal{L}$ are denoted by $\{ \mathcal{L}_i \}$ and the integer surgery coefficients are denoted by $\{ r_i \}$. According to the definition (20.30), one has

$$\mathcal{I}(\mathcal{M}) = (k)^{-m/2} \sum_{n_1=0}^{k-1} \cdots \sum_{n_m=0}^{k-1} \mathcal{A} , \tag{20.31}$$

where the amplitude $\mathcal{A}$ is given by

$$\mathcal{A} = \exp \left\{ -i \left(\frac{2\pi}{k} \right) \left[\sum_i n_i^2 \, r_i \; + 2 \sum_{i<j} n_i n_j \, \chi(\mathcal{L}_i , \mathcal{L}_j) \right] \right\} . \tag{20.32}$$

and the coupling constant k satisfies the constraints (20.6).

When $m = 1$, one has

$$\begin{aligned} \mathcal{I}(\mathcal{M}) &= (k)^{-1/2} \sum_{n=0}^{k-1} \exp \left[-i \frac{2\pi}{k} r \, n^2 \right] \\ &= (k)^{-1/2} \left\{ 1 + \sum_{n=1}^{k-1} \exp \left[-i \frac{2\pi}{k} r \, n^2 \right] \right\} . \end{aligned} \tag{20.33}$$

As we have already mentioned, $\{ 1^2, 2^2, \ldots , [(k-1)/2]^2 \}$ form a complete set of quadratic residues mod k. Because of conditions (20.6), if x is a quadratic residue mod k then [66] $-x$ also is a quadratic residue mod k. Consequently, the sum

$$\sum_{n=1}^{k-1} \exp \left[-i \frac{2\pi}{k} r \, n^2 \right]$$

is real because it is equal to twice the sum over a complete set of quadratic residues mod k. Therefore, when $m = 1$, $\mathcal{I}(\mathcal{M})$ is real.

Let us now consider the case in which $m > 1$. The expression (20.31) can be written as

$$\mathcal{I}(\mathcal{M}) = (k)^{-m/2} \sum_{n_2=0}^{k-1} \cdots \sum_{n_m=0}^{k-1} \mathcal{B} , \tag{20.34}$$

where

$$\mathcal{B} = \sum_{n_1=0}^{k-1} \mathcal{A} . \tag{20.35}$$

One has

$$\mathcal{B} = F \sum_{n_1=0}^{k-1} \exp \left\{ -i \frac{2\pi}{k} \left[r_1 \, n_1^2 + 2 n_1 \, c \right] \right\} , \tag{20.36}$$

where

$$c = \sum_i{}' n_i \; \chi(\mathcal{L}_1, \mathcal{L}_i) \,, \tag{20.37}$$

and

$$F = \exp\left\{ -i \left(\frac{2\pi}{k}\right) \left[\sum_i{}' n_i^2 \, r_i \; + 2 \sum_{i<j}{}' n_i n_j \; \chi(\mathcal{L}_i, \mathcal{L}_j) \right] \right\} . \tag{20.38}$$

The symbol $\sum'$ denotes the sum over the values of the index i (or of the indices i and j) from 2 to m.

If $r_1 \not\equiv 0 \pmod k$, let us denote by $[r_1]^{-1}$ the integer which is the solution of the congruence

$$r_1 \, [r_1]^{-1} \equiv 1 \pmod{k} \,, \tag{20.39}$$

with $1 \leq [r_1]^{-1} \leq k - 1$. In this case,

$$\sum_{n_1=0}^{k-1} \exp\left\{ -i \frac{2\pi}{k} \left[r_1 \, n_1^2 + 2 n_1 \, c \right] \right\} = K(r_1) \, \exp\left\{ i \, \frac{2\pi}{k} [r_1]^{-1} c^2 \right\} , \tag{20.40}$$

where

$$K(r_1) = \sum_{n_1=0}^{k-1} \exp\left\{ -i \frac{2\pi}{k} \left[r_1 \, n_1^2 \right] \right\} . \tag{20.41}$$

As we have seen, $K(r_1)$ is real. From equations (20.36) and (20.40), it follows that

$$\mathcal{B} = K(r_1) \, \exp\left\{ -i \left(\frac{2\pi}{k}\right) \left[\sum_i{}' n_i^2 \, r_i' \; + 2 \sum_{i<j}{}' n_i n_j \; \phi_{ij} \right] \right\} , \tag{20.42}$$

where

$$r_i' = r_i - [r_1]^{-1} \left[\chi(\mathcal{L}_1, \mathcal{L}_i) \right]^2 , \tag{20.43}$$

and

$$\phi_{ij} = \chi(\mathcal{L}_i, \mathcal{L}_j) - [r_1]^{-1} \, \chi(\mathcal{L}_1, \mathcal{L}_i) \, \chi(\mathcal{L}_1, \mathcal{L}_j) \, . \tag{20.44}$$

Note that r_i' and ϕ_{ij} are integers. Therefore, apart from the real multiplicative factor $K(r_1)$, $\mathcal{B}$ has the same structure as the amplitude $\mathcal{A}$ which is shown in equation (20.32).

If $r_1 \equiv 0 \pmod k$, we need to analyse the two possible cases

(1) $\chi(\mathcal{L}_1, \mathcal{L}_i) \equiv 0 \pmod{k} \qquad \forall i = 2, \ldots, m;$

(2) $\chi(\mathcal{L}_1, \mathcal{L}_i) \not\equiv 0 \pmod k$ for some i, with $i = 2, \ldots, m$.

In case (1), we have

$$\mathcal{B} = k\, F$$

and, apart from the real multiplicative factor k, $\mathcal{B}$ has again the same structure as $\mathcal{A}$.

In case (2), we obtain

$$\sum_{n_1=0}^{k-1} \exp\left\{ -i\,\frac{2\pi}{k} 2\, n_1\, c \right\} = \begin{cases} k & \text{if } c \equiv 0 \bmod k; \\ 0 & \text{if } c \not\equiv 0 \bmod k. \end{cases} \tag{20.45}$$

Suppose now that $\chi(\mathcal{L}_1, \mathcal{L}_i)$ is not equivalent to zero mod k when, for example, $i = 2$. In this case, let us denote by d the solution of the congruence

$$\chi(\mathcal{L}_1, \mathcal{L}_2)\, d \equiv 1 \pmod k, \tag{20.46}$$

with $1 \leq d \leq k - 1$. From equations (20.34), (20.36), (20.38) and (20.45), it follows that

$$\mathcal{I}(\mathcal{M}) = (k)^{1-m/2} \sum_{n_3=0}^{k-1} \cdots \sum_{n_m=0}^{k-1} F', \tag{20.47}$$

where

$$F' = \exp\left\{ -i \left(\frac{2\pi}{k} \right) \left[\sum_i{}'' n_i^2\, r_i'' + 2 \sum_{i<j}{}'' n_i n_j\, \phi'_{ij} \right] \right\}, \tag{20.48}$$

$$r_i'' = r_i + r_2\, d^2\, [\chi(\mathcal{L}_1, \mathcal{L}_i)]^2 - 2d\, \chi(\mathcal{L}_1, \mathcal{L}_i)\, \chi(\mathcal{L}_2, \mathcal{L}_i), \tag{20.49}$$

$$\begin{aligned} \phi'_{ij} = {} & \chi(\mathcal{L}_i, \mathcal{L}_j) + r_2\, d^2\, \chi(\mathcal{L}_1, \mathcal{L}_i)\, \chi(\mathcal{L}_1, \mathcal{L}_j) \\ & - 2d \left(\chi(\mathcal{L}_1, \mathcal{L}_i)\, \chi(\mathcal{L}_2, \mathcal{L}_j) + \chi(\mathcal{L}_1, \mathcal{L}_j)\, \chi(\mathcal{L}_2, \mathcal{L}_i) \right). \end{aligned} \tag{20.50}$$

The symbol $\sum''$ denotes the sum over the values of the index i (or of the indices i and j) from 3 to m. Note that r_i'' and ϕ'_{ij} are integers. Therefore, F' has the same structure as $\mathcal{A}$.

Now, the argument we have just presented can be used recursively to analyse the multiple sum appearing in equation (20.31). Consequently, the three-manifold invariant $\mathcal{I}(\mathcal{M})$ is real. □

To sum up, $\mathcal{I}(\mathcal{M})$ is real; therefore, $\mathcal{I}(\mathcal{M})$ is invariant under a generic self-homeomorphism of the three-manifold $\mathcal{M}$.

By definition, the value of the three-manifold invariant (20.30) for the three-sphere S^3 is

$$\mathcal{I}(S^3) = 1. \tag{20.51}$$

Let us consider a three-manifold $\mathcal{M}$ which corresponds to a surgery instruction described by a knot C in S^3 with surgery coefficient r. In this case, the value of the invariant is given by

$$\mathcal{I}(\mathcal{M}) = \frac{1}{\sqrt{k}} \sum_{n=0}^{k-1} q^{r n^2}. \tag{20.52}$$

Clearly, this expression depends on the residue class of r mod k. When $r \equiv 0$ (mod k), one finds

$$\mathcal{I}(\mathcal{M}) = \sqrt{k}. \tag{20.53}$$

Let us now consider the case in which $1 \leq r \leq k-1$ and, therefore, r and k are relatively prime. Let us introduce Legendre's symbol $\left[\frac{r}{k}\right]$ defined by [66]

$$\left[\frac{r}{k}\right] = \begin{cases} +1 & \text{if } r \text{ is a quadratic residue mod } k; \\ -1 & \text{if } r \text{ is a quadratic non-residue mod } k. \end{cases} \tag{20.54}$$

When r and k are coprime integers and k is an odd prime, the Gauss sum $S(r,k)$ satisfies [66]

$$S(r,k) = \sum_{n=0}^{k-1} \exp\left(\frac{2\pi i}{k} r n^2\right) = \left[\frac{r}{k}\right] \sqrt{k}. \tag{20.55}$$

Therefore, for $1 \leq r \leq k-1$, one obtains

$$\mathcal{I}(\mathcal{M}) = \left[\frac{r}{k}\right]. \tag{20.56}$$

Thus, for the lens space $L(p,1)$ with integer p, for instance, which corresponds to the surgery instruction given by the unknot with surgery coefficient $r = p$, the value of the invariant (20.30) is

$$\mathcal{I}(L(p,1)) = \begin{cases} \sqrt{k} & \text{if } p \equiv 0 \text{ (mod } k); \\ +1 & \text{if } p \text{ is a quadratic residue mod } k; \\ -1 & \text{if } p \text{ is a quadratic non-residue mod } k. \end{cases} \tag{20.57}$$

The Poincaré manifold $\mathcal{P}$ corresponds to the surgery instruction described by the right-handed trefoil with surgery coefficient $r = +1$. Therefore

$$\mathcal{I}(\mathcal{P}) = 1. \tag{20.58}$$

Even if the value of the invariant $\mathcal{I}(\mathcal{P})$ is trivial, the expectation values of the Wilson line operators associated with links in $\mathcal{P}$ are not equal to the expectation values in S^3. Let us consider, for example, the knots C_1 and C_2 shown in Fig. 18.22 and Fig. 18.23. When C_1 and C_2 have preferred framings and are labelled by $\Psi[m]$, the surgery rule (17.78) gives

$$\langle W(C_1; \Psi[m]) \rangle \Big|_{\mathcal{P}} = q^{-4m^2} , \tag{20.59}$$

and

$$\langle W(C_2; \Psi[m]) \rangle \Big|_{\mathcal{P}} = q^{-m^2} . \tag{20.60}$$

The surgery instruction described by the right-handed trefoil with surgery coefficient $r = +3$ corresponds to the three-manifold X_4, which is homeomorphic with the 4-fold cyclic covering of S^3 over the trefoil. In this case, one has

$$\mathcal{I}(X_4) = \left[\frac{3}{k}\right] . \tag{20.61}$$

The manifold X_3, which is homeomorphic with the 3-fold cyclic covering of S^3 over the trefoil, is described by the surgery instruction shown in Fig. 19.2. A direct computation shows that

$$\mathcal{I}(X_3) = 1 . \tag{20.62}$$

Let us now consider the surgery link shown in Fig. 19.1; the surgery coefficients of the two components of the Hopf link are $r_1 = a$ and $r_2 = b$ with nonvanishing integers a and b. The corresponding three-manifold is the lens space $L(ab-1, a) \simeq L(ab-1, b)$. In this case, the three-manifold invariant (20.30) is given by

$$\mathcal{I}(L(ab-1, a)) = \frac{1}{k} \sum_{m=0}^{k-1} q^{bm^2} \sum_{n=0}^{k-1} q^{an^2} q^{2nm} . \tag{20.63}$$

When $a \equiv 0 \pmod{k}$, or $b \equiv 0 \pmod{k}$, or both a and b are congruent with 0 mod k, one has

$$\mathcal{I}(L(ab-1, a)) = 1 . \tag{20.64}$$

Let us now consider the case in which $1 \leq a \leq k-1$ and $1 \leq b \leq k-1$. Since a and k are coprime integers, the congruence

$$a\,[a]^{-1} \equiv 1 \pmod{k} \tag{20.65}$$

has a solution for an integer $[a]^{-1}$ with $1 \leq [a]^{-1} \leq k-1$. Then, one has

$$\sum_{n=0}^{k-1} q^{a n^2} q^{2nm} = \sqrt{k} \left[\frac{a}{k}\right] q^{-m^2 [a]^{-1}} . \tag{20.66}$$

Consequently, expression (20.63) can be written as

$$\mathcal{I}(L(ab-1,a)) = \left[\frac{a}{k}\right] \frac{1}{\sqrt{k}} \sum_{m=0}^{k-1} q^{m^2 (b-[a]^{-1})} . \tag{20.67}$$

At this point, we can use the results concerning the sum (20.52). If $(b-[a]^{-1}) \equiv 0 \pmod{k}$, then

$$\mathcal{I}(L(ab-1,a)) = \left[\frac{a}{k}\right] \sqrt{k} . \tag{20.68}$$

In the remaining cases, we obtain

$$\mathcal{I}(L(ab-1,a)) = \left[\frac{a}{k}\right] \left[\frac{b-[a]^{-1}}{k}\right] . \tag{20.69}$$

When $a=3$ and $b=2$, for instance, we get the lens space $L(5,2)$. For $k=5$, we have

$$\mathcal{I}(L(5,2)) = -\sqrt{5} , \tag{20.70}$$

whereas, for $k=13$, we find

$$\mathcal{I}(L(5,2)) = -1 . \tag{20.71}$$

The manifold $S^2 \times S^1$ corresponds to the surgery instruction described by the unknot with surgery coefficient $r=0$; therefore

$$\mathcal{I}(S^2 \times S^1) = \sqrt{k} . \tag{20.72}$$

The Borromean Rings with surgery coefficients $r_1 = r_2 = r_3 = 0$ describe the manifold $T^2 \times S^1$. Since the linking number of two generic components of this link is vanishing, we find

$$\mathcal{I}(T^2 \times S^1) = k^{3/2} . \tag{20.73}$$

Equations (20.72) and (20.73) can easily be generalized. Let us consider the three-manifold $\Sigma_g \times S^1$ where Σ_g is the Riemann surface of genus g. The manifold $\Sigma_g \times S^1$ admits a surgery presentation described by a surgery link which has $2g+1$ components and in which all the surgery coefficients are equal to 0. As R. Benedetti

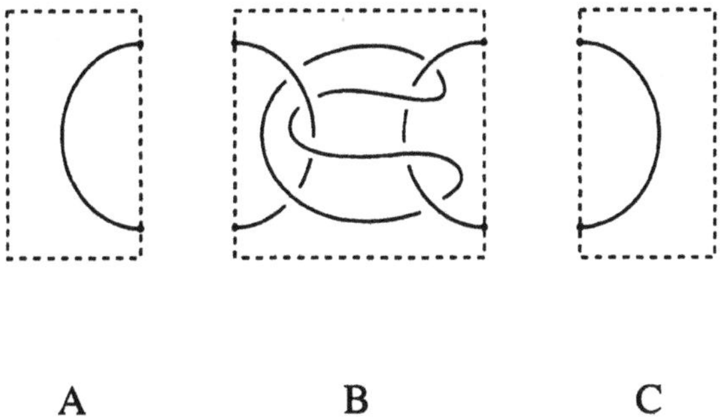

Figure 20.1.

suggested to me, this surgery link can be obtained by combining an appropriate number of the three different tangles shown in Fig. 20.1.

Consider g consecutive copies of picture B; sew them together and add picture A on the left and picture C on the right. The final picture that one obtains represents a link diagram D_g which can symbolically be written as

$$D_g = A + g\,B + C\,. \tag{20.74}$$

According to the procedure (20.74), one obtains for $g = 0$ the known surgery presentation of $S^2 \times S^1$ given by the unknot with surgery coefficient $r = 0$. For $g = 1$, one finds the Borromean Rings with all the surgery coefficients equal to 0 ; this is precisely the known surgery presentation of $T^2 \times S^1$. Notice that for the link described by D_g the linking number of two generic components is vanishing. Thus, for the manifold $\Sigma_g \times S^1$, the value of the invariant (20.30) is

$$\mathcal{I}(\,\Sigma_g \times S^1\,) = k^{(2g+1)/2}\,. \tag{20.75}$$

According to equation (17.78), the invariant (20.30) can be interpreted as the value of the improved partition function of the Abelian $U(1)$ CS theory defined in $\mathcal{M}$ with the coupling constant k satisfying constraints (20.6). To be more precise, the invariant (20.30) represents the ratio of the partition function of the CS theory in $\mathcal{M}$ and the partition function in S^3.

Since the action of the Abelian CS theory is quadratic in the fields, the partition function can be expected to be written as a ratio of determinants. Because of gauge invariance and the presence of zero modes in $\mathcal{M}$, however, an explicit interpretation of the invariant $\mathcal{I}(\mathcal{M})$ in terms of determinants is not obvious.

Let us recall that, in the Landau gauge, the gauge-fixed action of the Abelian CS theory in $\mathbb{R}^3$ is given by

$$S = \frac{k}{4\pi} \int d^3x \left\{ \tfrac{1}{2} \epsilon^{\mu\nu\rho} A_\mu \partial_\nu A_\rho - B \partial^\mu A_\mu + \overline{c} \partial^\mu \partial_\mu c \right\} , \qquad (20.76)$$

where B is the auxiliary field and c and $\overline{c}$ are the ghost fields. In the gauge fixing and ghost parts of the action, the ordinary flat metric of $\mathbb{R}^3$ has been used to contract the vector indices. The equation of motion related to the auxiliary field gives the gauge-fixing constraint

$$\partial^\mu A_\mu = 0 . \qquad (20.77)$$

The gauge-fixing introduced in (20.76) is a consistent and complete gauge-fixing in $\mathbb{R}^3$. In fact, the result (3.19) has been obtained from the action (20.76) by means of a direct resummation of the perturbative expansion associated with the observables.

The field variables $\{ A_\mu \}$ and B can be understood as the components of the bosonic field $\Phi = (A_\mu , B)$. The operator $\mathcal{D}$, which enters the quadratic action (20.76) and acts on the field Φ, is given by

$$\mathcal{D} = \begin{pmatrix} \epsilon^{\mu\nu\rho} \partial_\nu & \partial^\mu \\ -\partial^\rho & 0 \end{pmatrix} . \qquad (20.78)$$

Therefore, the vacuum-to-vacuum amplitude in $\mathbb{R}^3$, obtained with the action (20.76), is proportional to

$$G = \mathrm{Det} (\partial^\mu \partial_\mu) \, \mathrm{Det}^{-1/2} (\mathcal{D}) . \qquad (20.79)$$

The expression (20.79) can be extended formally [1] from $\mathbb{R}^3$ to a generic (connected, closed and orientable) three-manifold $\mathcal{M}$. In this case, it has been argued [1] that, by introducing a suitable procedure to give a meaning to the expression (20.79), G is proportional to the Ray-Singer torsion. This formal extension of G has very interesting properties. Unfortunately, this definition of G cannot represent the partition function of the Abelian $U(1)$ CS theory defined in $\mathcal{M}$. The reason is that the gauge-fixing constraint (20.77), which is perfectly consistent in $\mathbb{R}^3$, is not correct in a generic three-manifold $\mathcal{M}$. Because of the zero modes of the Laplacian, constraint (20.77) is not a complete gauge fixing in $\mathcal{M}$. Moreover, the global structure of the $U(1)$ gauge group is not taken into account by the usual gauge-fixing (20.77). This structure is important in $\mathcal{M}$. Indeed, as we have seen in the previous sections, the global structure of the gauge group plays a crucial role in the construction of a linear representation of standard cobordism.

In conclusion, the real meaning of the three-manifold invariants (19.11) and (20.30) has to be found in the deep connection between twist homeomorphism transformations and the structure of the gauge group. From the algebraic point of view, the three-manifold invariants defined by the CS theory can easily be understood. However, the topological meaning of these invariants is not completely known.

It would be important to find the precise relations [3] between the invariants in two, three and four dimensions. As far as the two dimensions are concerned, several suggestions [4] have been made. From the physical point of view, the four dimensions appear to be particularly interesting; realistic applications of the CS theory in quantum gravity are worth studying.

References

[1] A.S. Schwarz, *The Partition Function of Degenerate Quadratic Functional and Ray-Singer Invariants*, Lett. Math. Phys. 2 (1978) 247-252.
———, *The Partition Function of a Degenerate Functional*, Commun. Math. Phys. 67 (1979) 1-16.

[2] C.R. Hagen, *A New Gauge Theory without an Elementary Photon*, Ann. Phys. (N.Y.) 157 (1984) 342-359.
A.M. Polyakov, *Fermi-Bose Transmutations Induced by Gauge Fields*, Mod. Phys. Lett. A3 (1988) 325-328.

[3] M.F. Atiyah, *New Invariants of Three- and Four-Dimensional Manifolds*, in *The Mathematical Heritage of Hermann Weyl*, Proc. Symp. Pure Math. 48, ed. R. Wells (Amer. Math. Soc., 1988).
———, *The geometry and physics of knots*, Cambridge University Press (Cambridge, 1990).

[4] E. Witten, *Quantum Field Theory and the Jones Polynomial*, Commun. Math. Phys. 121 (1989) 351-399.

[5] V.G. Drinfeld, *Quasi-Hopf Algebras and Knizhnik-Zamolodchikov Equations*, in Problems of Modern Quantum Field Theory, ed. A.A. Belavin, A.V. Klinyk and A.B. Zamolodchikov (Springer-Verlag, 1990).

[6] V.F.R. Jones, *A Polynomial Invariant for Knots via von Neumann Algebras*, Bull. Amer. Math. Soc. 12 (1985) 103-111.
———, *Hecke Algebra Representations of Braid Groups and Link Polynomials*, Ann. Math. 126 (1987) 335-388.

[7] N.Y. Reshetikhin and V.G. Turaev, *Ribbon Graphs and Their Invariants Derived from Quantum Groups*, Commun. Math. Phys. 127 (1990) 1-26.
———, *Invariants of 3-manifolds via Link Polynomials and Quantum Groups*, Invent. Math. 103 (1991) 547-597.

[8] V.G. Turaev, *Quantum invariants of 3-manifolds*, to be published by Walter de Gruyter, Berlin.

[9] E. Guadagnini, *The Universal Link Polynomial*, Int. Journ. Mod. Phys. A7 (1992) 877-945.

[10] E. Guadagnini, *Surgery Rules in Quantum Chern-Simons Field Theory*, Nucl. Phys. B375 (1992) 381-398.

[11] E. Guadagnini and S. Panicucci, *Topological Field Theory and Surgery on Three-Manifolds*, Nucl. Phys. B388 (1992) 159-226.

[12] D. Rolfsen, *Knots and Links*, (Publish or Perish, 1976).

[13] L.H. Kauffman, *On Knots*, (Princeton Univ. Press, 1987).

[14] J.H. Conway, in *Computational Problems in Abstract Algebra*, ed. J. Leech, (Pergamon Press, 1969).

[15] P. Freyd, D. Yetter, J. Hoste, W.B.R. Lickorish, K. Millett and A. Ocneanu, *A New Polynomial Invariant of Knots and Links*, Bull. Amer. Math. Soc. 12 (1985) 239-246.
J.H. Przytycki and P. Traczyk, *Invariants of Links of Conway Type*, Kobe J. Math. 4 (1987) 115-139.

[16] A.S. Lipson, *A Note on Some Link Polynomials*, Bull. London Math. Soc. 20 (1988) 532-534.

[17] W.B.R. Lickorish and K.C. Millett, *A Polynomial Invariant of Oriented Links*, Topology 26 (1987) 107-141.

[18] P. Cotta-Ramusino, E. Guadagnini, M. Martellini and M. Mintchev, *Quantum Field Theory and Link Invariants*, Nucl. Phys. B330 (1990) 557-374.

[19] C. Becchi, A. Rouet and R. Stora, *Renormalization of the Abelian Higgs-Kibble Model*, Commun. Math. Phys. 42 (1975) 127-162.
——————, *Renormalization of Gauge Theories*, Ann. Phys. (NY) 98 (1976) 287-328.

[20] E. Guadagnini, M. Martellini and M. Mintchev, *Perturbative Aspects of the Chern-Simons Field Theory*, Phys. Lett. B227 (1989) 111-117.

[21] G. Calugareanu, *Sur les Classes d'Isotopie des Noeuds Tridimensionnels et leurs Invariants*, Czech. Math. J. 11 (1961) 588-624.

[22] E. Guadagnini, M. Martellini and M. Mintchev, *Wilson Lines in Chern-Simons Theory and Link Invariants*, Nucl. Phys. B330 (1990) 575-607.
——————, *Recent Developments in Chern-Simons Theory and Link Invariants*, Nucl. Phys. B (Proc. Suppl.) 16B (1990) 588-590.

[23] G. Calugareanu, *L'Intégrale de Gauss et l'Analyse des Noeuds Tridimensionnels*, Rev. Math. Pures Appl. (Bucarest) 4 (1959) 5-20.
W.F. Pohl, *The Self-Linking Number of a Closed Space Curve*, J. Math. Mech. 17 (1968) 975-985.

[24] O.I. Zav'yalov and A.M. Malokostov, *Universal Regularizations. I. Prescriptions and single-loop Approximation*, Teoreticheskaya i Matematicheskaya Fizika 84, No.1 (1990) 46-63.
——————, *Universal Regularizations. II. Gauge Theories*, Teoreticheskaya i Matematicheskaya Fizika 84, No.2 (1990) 195-204.

[25] L. Alvarez-Gaumé, J.M.F. Labastida and A.V. Ramallo, *A Note on Perturbative Chern-Simons Theory*, Nucl. Phys. B334 (1990) 103-124.

[26] F. Delduc, O. Piguet, C. Lucchesi and S.P. Sorella, *Exact Scale Invariance of the Chern-Simons Theory in the Landau Gauge*, Nucl. Phys. B346 (1990) 313-328.

[27] A. Blasi and S. Collina, *Finiteness of the Chern-Simons Model in Perturbation Theory*, Nucl. Phys. B345 (1990) 472-492.

[28] F. Delduc, F. Gieres and S.P. Sorella, *Supersymmetry of the d=3 Chern-Simons Action in the Landau Gauge*, Phys. Lett. B225 (1989) 367-371.

[29] E. Guadagnini, N. Maggiore and S.P. Sorella, *Supersymmetry of the Three-Dimensional Einstein-Hilbert Gravity in the Landau Gauge*, Phys. Lett. B247 (1990) 543-548.
———— , *Supersymmetric Structure of Four-Dimensional Antisymmetric Tensor Fields*, Phys. Lett. B255 (1991) 65-73.

[30] E. Guadagnini, M. Martellini and M. Mintchev, *Braids and Quantum Group Symmetry in Chern-Simons Theory*, Nucl. Phys. B336 (1990) 581-609.
———— , *Link Invariants from Chern-Simons Theory*, Nucl. Phys. B (Proc. Suppl.) 18B (1990) 121-134.

[31] J.L. Gervais and A. Neveu, *The Slope of the Leading Regge Trajectory in Quantum Chromodynamics*, Nucl. Phys. B163 (1980) 189-216.

[32] R.F. Dashen, B. Hasslacher and A. Neveu, *Semiclassical Bound States in an Asymptotically Free Theory*, Phys. Rev. D12 (1975) 2443-2458.

[33] I.S. Gradshteyn and I.M. Ryzhik, *Table of Integrals, Series, and Products*, Edited by A. Jeffrey, (Academic Press, 1980).

[34] E. Artin, *Theory of Braids*, Ann. of Math. 48 (1947) 101-126.

[35] J. Birman, *Braids, Links and Mapping Class Groups*, Ann. Math. Stud. 82 (1974).
———— , *On the Stable Equivalence of Plat Representations of Knots and Links*, Can. J. Math. 28 (1976) 264-290.

[36] P. De La Harpe, M. Kerwaire and C. Weber, *On the Jones Polynomial*, L'Enseignement Mathématique 32 (1986) 271-335.

[37] V.G. Turaev, *The Yang-Baxter Equation and Invariants of Links*, Invent. Math. 92 (1988) 527-553;
———— , *The Yang-Baxter Equation and Invariants of Links*, LOMI preprint E-3-87, Steklov Institute, Leningrad 1988.

[38] W.B.R. Lickorish, *Some Link Polynomial Relations*, Math. Proc. Camb. Phil. Soc. 105 (1989) 103-107.

[39] V.G. Drinfeld, *Quantum Groups*, Proc. Int. Congr. Math, Berkeley, 1 (1986) 789-820.
M. Jimbo, *A q-difference Analog of U(g) and the Yang-Baxter Equation*, Lett. Math. Phys. 10 (1985) 63-69.
A.N. Kirillov and N.Y. Reshetikhin, *q-Weyl Group and a Multiplicative Formula for Universal R-Matrices*, preprint 1990.

[40] P.P. Kulish, N. Yu. Reshetikhin and E.K. Sklyanin, *Yang-Baxter Equation and Representation Theory: I*, Lett. Math. Phys. 5 (1981) 393-403.
A.A. Belavin and V.G. Drinfeld, *Solutions of the Classical Yang-Baxter Equation for Simple Lie Algebras*, Funct. Anal. Appl. 16 (1982) 159-180.

[41] V.G. Drinfeld, *Hamiltonian Structures on Lie Groups, Lie Bialgebras and the Geometric Meaning of the Classical Yang-Baxter Equations*, Sov. Math. Dokl. 27 (1983) 68-71.

N. Yu. Reshetikhin and L.D. Faddeev, *Hamiltonian Structures for Integrable Models of Field Theory*, Theor. Math. Phys. 56 (1984) 847-862.
M. Jimbo, *Quantum R Matrix for the Generalized Toda Systems*, Commun. Math. Phys. 102 (1986) 537-547.

[42] N.Y. Reshetikhin, *Quantized Universal Enveloping Algebras, the Yang-Baxter Equations and Invariants of Links, I and II*, LOMI preprints E-4-87 and E-17-87, Steklov Institute, Leningrad 1987.

[43] A.N. Kirillov and N.Y. Reshetikhin, *Representations of the Algebra $U_q(sl(2))$, q-orthogonal polynomials and invariants of links*, LOMI preprint E-9-88, Steklov Institute, Leningrad 1988.

[44] M. Rosso, *Finite Dimensional Representations of the Quantum Analog of the Enveloping Algebra of a Complex Simple Lie Algebra*, Commun. Math. Phys. 117 (1988) 581-593.
————, *An Analogue of P.B.W. Theorem and the Universal R-Matrix for $U_h sl(N+1)$*, Commun. Math. Phys. 124 (1989) 307-318.
G. Lusztig, *Quantum Deformations of Certain Simple Modules Over Envelopig Algebras*, Adv. in Math. 70 (1988) 237-249.
N. Burroughs, *The Universal R-Matrix for $U_q sl(3)$ and Beyond!*, Commun. Math. Phys. 127 (1990) 109-128.
H. Wenzel, *Quantum Groups and Subfactors of Type B, C, and D*, Commun. Math. Phys. 133 (1990) 383-432.

[45] T. Kohno, *Monodromy Representations of Braids Groups and Yang-Baxter Equation*, Ann. Inst. Fourier, Grenoble, 37 (1987) 139-160.
————, *Hecke Algebra Representations of Braids Groups and Classical Yang-Baxter Equation*, Adv. Studies in Pure Math. 16 (1988) 255-269.
————, *Quantized Universal Enveloping Algebras and Monodromy of Braids Groups*, Nagoya University preprint No 3, 1989.

[46] V.G. Knizhnik and A.B. Zamolodchikov, *Current Algebra and Wess-Zumino Model in Two Dimensions*, Nucl. Phys. B247 (1984) 83-103.

[47] Y. Akutsu and M. Wadati, *Exactly Solvable Models and New Link Polynomials. I. N-State Vertex Models*, J. Phys. Soc. Jpn. 56 (1987) 3039-3051.
Y. Akutsu, T. Deguchi and M. Wadati, *Exactly Solvable Models and New Link Polynomials. II. Link Polynomials for Closed 3-Braids*, J. Phys. Soc. Jpn. 56 (1987) 3464-3479.
————, *Exactly Solvable Models and New Link Polynomials. III. Two-Variable Topological Invariants*, J. Phys. Soc. Jpn. 57 (1988) 757-776.

[48] A. Liguori and M. Mintchev, *Spectral Parameters of the Quantum Yang-Baxter Equation*, Phys. Lett. B275 (1992) 371-374.

[49] E. Guadagnini, M. Martellini and M. Mintchev, *Chern-Simons Holonomies and the Appearance of Quantum Groups*, Phys. Lett. B235 (1990) 275-281.

[50] L.A. Takhtadzhjan and L.D. Faddeev, *The Quantum Method of the Inverse Problem and the Heisenberg XYZ Model*, Usp. Mat. Nauk. 34:5 (1979) 13-63.

P.P. Kulish and E.K. Sklyanin, *Solutions of the Yang-Baxter Equations*, J. Sov. Math. 19 (1982) 1596-1620.

[51] M. Wadati, T. Deguchi and Y. Akutsu, *Exactly Solvable Models and Knot Theory*, Phys. Rep. 180 (1989) 247-332, and references therein.

[52] B. Schroer, *New Methods and Results in Conformal QFT_2 and the "String idea"*, Cargèse Lectures 1987 (Plenum Press, 1988).

[53] J. Fröhlich, *Statistics of Fields, the Yang-Baxter Equation and the Theory of Knots and Links*, Cargèse Lectures 1987 (Plenum Press, 1988).
J. Fröhlich and C. King, *The Chern-Simons Theory and Knot Polynomials*, Commun. Math. Phys. 126 (1989) 167-199.
———— , *Two-Dimensional Conformal Field Theory and the Three-Dimensional Topology*, Int. J. Mod. Phys. A4 (1989) 5321-5399.

[54] A. Tsuchiya and Y. Kanie, *Vertex Operators in Conformal Field Theory on P^1 and Monodromy Representations of Braid Group*, Adv. Stud. Pure Math. 16 (1988) 297-372.
G. Moore and N. Seiberg, *Polynomial Equations for Rational Conformal Field Theories*, Phys. Lett. B212 (1988) 451-460.
G. Moore and N. Reshetikhin, *A Comment on Quantum Groups Symmetry in Conformal Field Theory*, Nucl. Phys. B328 (1989) 557-574.

[55] L. Alvarez-Gaumé, C. Gomez and G. Sierra, *Hidden Quantum Symmetries in Rational Conformal Field Theories*, Nucl. Phys. B319 (1989) 155-186.
M. Wadati, Y. Yamada and T. Deguchi, *Knot Theory and Conformal Field Theory: Reduction Relations for Braid Generators*, J. Phys. Soc. Jpn. 58 (1989) 1153-1161.
A.C. Ganchev and V.B. Petkova, *$U_q(sl(2))$ Invariant Operators and Minimal Theories Fusion Matrices*, Phys. Lett. B233 (1989) 374-382.

[56] G. Moore and N. Seiberg, *Taming the Conformal Zoo*, Phys. Lett. B220 (1989) 422-430.
S. Elitzur, G. Moore, A. Schwimmer and N. Seiberg, *Remarks on the Canonical Quantization of the Chern-Simons-Witten Theory*, Nucl. Phys. B326 (1989) 108-134.
E. Witten, *Gauge Theories and Integrable Lattice Models*, Nucl. Phys B322 (1989) 629-697.
M. Bos and V.P. Nair, *Coherent State Quantization of Chern-Simons Theory*, Int. Journ. Mod. Phys. A5 (1990) 959-988.

[57] E. Guadagnini, *The Link Polynomials of the Chern-Simons Theory*, Phys. Lett. B251 (1990) 115-120.

[58] L.H. Kauffman, *State Models and the Jones Polynomial*, Topology 26 (1987) 395-407.

[59] E. Guadagnini, *Chern-Simons Theory and Fusion Algebras*, Phys. Lett. B260 (1991) 353-358.

[60] D. Gepner and E. Witten, *String Theory on Group Manifolds*, Nucl. Phys. B278 (1986) 493-549.

[61] A.A. Belavin, A.M. Polyakov and A.B. Zamolodchikov, *Infinite Conformal Symmetry in Two-Dimensional Quantum Field Theory*, Nucl. Phys. B241 (1984) 333-380.
D. Gepner, Nucl. Phys. B287 (1987) 111.
A. Tsuchiya and Y. Kanie, *Vertex Operators in Conformal Field Theory on P^1 and Monodromy Representations of Braid Group*, Adv. Stud. Pure Math. 16 (1988) 297-372.
G. Moore and N. Seiberg, *Classical and Quantum Conformal Field Theory*, Commun. Math. Phys. 123 (1989) 177-254.
A. Tsuchiya, K. Veno and Y. Yamada, Adv. Studies in Pure Math. 19 (1989) 459.

[62] A.N. Kirillov and N.Y. Reshetikhin, Adv. Ser. Math. Phys. 7 (1988) 285.
V. Pasquier and H. Saleur, *Common Structures Between Finite Systems and Conformal Field Theories Through Quantum Groups*, Nucl Phys. B330 (1990) 523-556.
P. Furlan, A.C. Gancev and V.B. Petkova, *Quantum Groups and Fusion Rules Multiplicities*, Nucl. Phys. B343 (1990) 205-227.

[63] C. De Concini and V.G. Kac, *Representations of Quantum Groups at Roots of 1*, Progress in Math. 92, Birkhäuser, (1990) 471-506.
G. Lusztig, *Quantum Groups at Roots of 1*, Geom. Ded. 35 (1990) 89-114.
C. De Concini, V.G. Kac and C. Procesi, *Quantum Coadjoint Action*, Journ. Am. Math. Soc. 5 (1992) 151-190.
C. De Concini and V.G. Kac, *Representations of quantum groups at roots of* 1, in Infinite Analysis (World Scientific, 1992), 141-150.

[64] W.B.R. Lickorish, *A Representation of Orientable Combinatorial 3-Manifolds*, Ann. Math. 76 (1962) 531-540.

[65] R. Kirby, *A Calculus for Framed Links in S^3*, Invent. Math. 45 (1978) 35-56.
R. Fenn and C. Rourke, *On Kirby's Calculus of Links*, Topology 18 (1979) 1-15.
D. Rolfsen, *Rational Surgery Calculus: Extension of Kirby's Theorem*, Pacific Journ. Math. 110 (1984) 377-386.

[66] Hua Loo-Keng, *Introduction to Number Theory*, (Springer-Verlag, New York 1982).

[67] W.B.R. Lickorish, *Three-manifold Invariants from the Combinatorics of the Jones Polynomial*, Pacific J. Math. 149 (1991) 337-347.
———, *Three-Manifolds and the Temperley-Lieb Algebra*, Math. Ann. (to appear).
———, *Variants on 3-manifold Invariants*, in Knots 90, edit. A. Kawauchi, (Walter de Gruyter, Berlin 1992).

[68] H.R. Morton and P.M. Strickland, *Satellites and Surgery Invariants*, in Knots 90, edit. A. Kawauchi, (Walter de Gruyter, Berlin 1992) .

[69] V.G. Turaev and O.Y. Viro, *State Sum Invariants of 3-Manifolds and Quantum 6J-Symbols*, Topology 31 (1992) 865-902.

[70] T. Kohno, *Topological Invariants for 3-Manifolds Using Representations of Mapping Class Groups I*, Topology 31 (1992) 203-230.

[71] R. Kirby and P. Melvin, *Evaluations of the 3-Manifold Invariants of Witten and Reshetikhin-Turaev for sl(2)*, Invent. Math. 105 (1991) 473-545.

[72] E. Guadagnini, *Abelian Surgery Invariant of Three-Manifolds*, Pisa University preprint IFUP-TH. 39/92.

Subject Index

Bei Fragen zur Produktsicherheit wenden Sie sich bitte an:
If you have any questions regarding product safety,
please contact:

Walter de Gruyter GmbH
Genthiner Straße 13
10785 Berlin
productsafety@degruyterbrill.com